To: Kim

Always write. You are the only one who can tell your stories

Ned P. Grickenhocker
AKA
Jeffry D. Van Bm

MR. MIZITHRA'S
OLYMPIC ODYSSEY

by
Ned P. Guickenhacker

valuable tidbits by
Calliope Anne Piffl

© Ned P. Guickenhacker. All rights reserved. No part of this publication may be reproduced, distributed, or transmitted in any form or by any means, including photocopying, recording, or other electronic or mechanical methods, without the prior written permission of the author, except in the case of brief quotations embodied in critical reviews and certain other noncommercial uses permitted by copyright law.

ISBN (Print): 978-1-09836-876-0
ISBN (eBook): 978-1-09836-877-7

DEDICATION

This book is dedicated to those
who see the beauty of nature and smile

PREFACE

This book is humbly written to enlighten the reader on a few themes: nature, a respect for it, and to get you out to enjoy the woods, wherever you may live. With determined aim, animals live life as they normally would in the wild. The exception is talking and a few added human traits. Reading sparks imagination. The goal, then, is to have you want to visit the Olympics, any mountains really, and view animals in the wild, not cages. It is to be thankful for the wilderness we have and to work for it's survival.

Characters in this story are fictitious, their world is not. Many years hiking, camping, tracking, and surviving the wild in umpteen places around the globe provided the author with a true appreciation for what Mother Nature is and to work as a human being to care for it. Animals inhabit the earth's forests. They do not have the ability as we to maintain them. It should be we who want fish, insects, plants, trees, birds, and animals to flourish for all to profit. This is simple common sense.

Please, enjoy *Mr. Mizithra's Olympic Odyssey*. It is a tale of love, friendship, joy, and adventure. Know it is written with you in mind, to take you away from home, to mountains, to heights, home of an interesting Olympic marmot, indeed. Here is an honest story built to teach you to love what is just outside your door, down the street, and in the woods. Oh yes, please enjoy this humble offering, my friend.

BOOK 1: 100 DAY TREK
Journey To Mount Olympus

Chapters

1. Hurricane Ridge...4
2. Mr. Mizithra Prepares..12
3. Murray..16
4. The Pond..27
5. What The Wind Blew In..35
6. The Madcap Marmot..44
7. Chance Encounter..50
8. A Wild Night..58
9. Cruel, Cruel World...67
10. Retribution...71
11. The Journey Continues..78
12. Choices..84
13. Above The Rapids..90
14. Dramatic News..95
15. On To White Mountain..101
16. Not So Fast..104
17. So Very Close..109
18. Climb, Climb, Climb..113
19. Back To The Rapids...118
20. A Murder Of Crows..126
21. Journey to a New Home...132
22. The Valley of the Stones...141

Chapter 1 Hurricane Ridge

Mr. Mizithra's Olympic marmot family sprang from ancient, distinguished lines. Spread out in a cordial, early summer sun on Hurricane Ridge in the Olympic Mountains, he lay on his belly, arms and legs completely splayed for maximum sunshine absorption. Little in life felt better than nabbing a few sweet, tasty, golden rays. He was in rodent Paradise on the wondrous peaks. Marmots are like an immense squirrel.

Their culture is based on the ground. They are not tree climbers like chipmunks or squirrels. Do not think them the cute little nut gatherers fed in the park. They are simply in the same animal family. Olympic marmots are the largest example of this critter. There are several other species of marmots around the world, Olympics the largest. It is Mr. Mizithra we follow. He is a character to be reckoned with, a rodent of great resolve and love. He lived a rather full marmot life.

My, oh my, what a fantastic day it was. He lay gnawing on a darling clump of juicy clover. A buoyant breeze softly swirled around the long, dusky brown hairs, busying his outer fur. He gazed at a large red tail hawk circling below, seeking a repast. What an interesting thing, he imagined, to watch the hunter at work. This taught an extra awareness of their adversary. Any raptor meant death to those below. They strike without warning. Woe to any marmot who did not pay careful attention to assassination from hawks.

A handful of shiny cirrus clouds soared above, wispy shapes ever changing, chasing across the dulcet, azure Pacific Northwest sky. His home high above the clouds suited him to perfection. His species demanded remote, lofty elevations.

The colony's relaxed collection of rodents lay strewn about radiant, sun baked rocks. Snow vanished from their little meadow save a few spots permanently hidden in shadows. The thirty odd members of this loving clan employed an excellent preemptive system: four distinct whistles warning of impending doom: raptors above, coyotes or bears on the ground, a fire, or, they dreaded the thought, man.

In the world of wild animals, no police exist to intervene in the bad deeds of brutes. You defend hearth and home. No coming to the rescue.

History spells this in exhaustive detail. Those who fail to heed danger end up supper for coyotes, bobcats, cougars, and bears.

His mother's family established this colony many generations long forgotten. They were a proud, simple family of marmots, doing the best to thrive in a very unforgiving environment. His father's people found the clan when their home in a stone riddled thicket faced ambush by black bears. Marmots fear those dark, ebony beasts most.

They are so big, so strong, so destructive. Rodents have little defense against them. If one were to breech the tunnel system by intense digging with powerful claws all would be lost. Thankful to find safety, his father's emigrating survivors reinforced those in residence, assimilating nicely. There is strength in numbers.

Mr. Mizithra's wife and two daughters were devoured by a vicious carnivore. The terror, the pain haunted him. What he endured afterward was most difficult. There was self-doubt, guilt, and shame. Often humans go through grief. Rather than talk about it with others, they internalize emotions with dire results. Death demands we accept it and express the pain to those able to care and assist to alleviate the inner hurt.

These past, tragic events made him what he became: an adventurer, a yarn spinner. Telling vivid tales with vibrant descriptions kept the tender ones alert to ever present dangers in the world. One learns more from a lively teacher. It was a role cherished with all his heart. He was a colorful rodent. He painted, armed with a palette full of adverbs and adjectives.

The marmot cubs loved him. They would gather, begging for some story, narrative, saga, anything to fascinate young, eager imaginations. Only he could tell them. An honest to goodness yarn spinner and raconteur, he expertly enhanced with graphic descriptions, heightening the story's impact while delivering truth. The colony's elders gave him this charge once he began to tell tales.

They found the words and lessons essential for pups. The parents of the adolescents saw him as their Aesop. He was able to reach children as no other. He was happy to oblige. He found telling stories therapy. Helping others assisted with the deep pain life dished.

Placing a fresh sprig of Flett's violets in the side of his fuzzy mouth, he would pause, looking afar, deeply searching for a whopper to bless them. All the while, he knew the adventure to entertain and teach: a life lesson essential to survival. As he got going, the audience was enraptured by his phraseology. He stacked excitement to a fever pitch. Adults were known to listen. They, too, were moved to dread by his telling of tales.

As the colony's resident sage, Mr. Mizithra certainly looked the part. A sizable chunk of his left ear was missing. An absent patch of fur permanently marred a deep gash along his left side. A toe was missing on his front right paw, another on his rear left foot. A scar dotted over his left eye, shaded to provide a sinister look in the proper light.

Each scar had it's own, well-earned story. The marks bore testament to survival skills. He was the ultimate example of what it took to live as an Olympic Marmot, to thrive in mountain country. The Olympics are no place for the timid or weak. The weather and animals are unforgiving. Eking out a life here meant finding food and keeping the devil at bay.

His life was one of endurance. Nature is relentless. Any species has only themselves and compadres to accept the harshness life is happy to serve. He took nothing for granted. The important sagas were true. They were designed for the colony to learn and apply for survival.

Mr. Mizithra did not teach to punish, but enlighten. The point was to give the youth of the tribe a good look at the cold, cruel world. He told these narratives with a sense of humor, a gleam in his eye, a sprig in his gob. He told any who would listen with a wide grin, "You don't get to look like me without having a survival story or two to tell."

An essential element to understanding marmots is they happen to be nature's best wrestlers. This is not the silly, phony stuff with gaudily garbed, muscle bound clowns bouncing in a ring, encircled by screeching morons driven mad by the outlandish show. The colony held multiple matches daily. It was a communal activity enjoyed constantly, all day, every day. Snag whomever you want to grab, then tumble, rumble, and jumble on the mountain. Sometimes, it would be three or four, spilling over a hillside meadow, in and out of holes, happily wrestling the day away. This activity's additional benefit produced living weapons.

At the end of each match, combatants hug and kiss. It was all in play. The training was serious. The result was not to win, but build muscles. Male or female, mattered not at all, it was their own unique way of family unity. Marmots are a very loving, affectionate species. Kinfolk bonded, the colony grew closer and stronger by tussling and hugging. Their single endeavor was to be one big, happy rodent clan. The sight of grappling, for other animals, was one of hilarity and laughter. The actions gave others joy, a wonderful by-product. Other species often paused their busy day to watch a match or two.

On this particularly bright, sunshiny day, a group of tender ones jostled, cajoled, and jockeyed for position to garner the best hold on eager opponents. Deeply engaged in their actions, they failed to keep warning signals ready. They were distracted by the immediate events. Wrestling is an excellent form of exercise. Be that as it may, forgetting to maintain guard all too often spelled death.

The youngsters reveled in the activity. Those standing and watching were busy shoving handfuls of greenery in their gobs, chewing avidly away. It is like popcorn with humans. They had to eat to endure. Winter loomed. Still, they were not paying attention to immediate dangers surrounding them. This gave him pause. Mr. Mizithra knew it was a perfect time to snag this pack of pups and teach them about life.

He moved to attract attention. It worked. Spying Mr. Mizithra out for a good tanning, they ceased play, waddling for their sagacious friend or Uncle. Nothing stopped play like an opportunity to listen to an anecdote from him. No one could tell a perspicacious tale better than the king of the yarn spinners. A story from him was better than any old wrestling match. His pointed, lazy plan had it's desired result. They arrived enmasse to see the favorite old colony Olympic marmot.

"Oh, please, Mr. Mizithra, please tell the ancient tale of marmots and man."

"Yes, yes, mother and father told us to have you share the story."

"Yes, Uncle Mr. Mizithra, my parents said the same thing. Why is this lesson so important?"

"Now, now, young ones", proffered the sagacious friend, grasping a handy, sweet Flett's violet, "to tell this legend, you must follow me. It is a long one. We must not be caught above for the safety of our colony. This tale must be first. It is the most important if you want to understand why man is who he is today."

Without arousing fear, he accomplished the goal of getting them underground. The delighted instructor led in a passageway to a central room beneath the surface. They gathered about to heed this vital lesson. Safety is of paramount importance. They do not have the same ways of protection available to human beings. This lack of refuge is one crucial thing separating us from beasts.

Animals have no standing army to repel invaders or a police force paid through taxes to watch out for you and nail bad folks. We take these for granted. Creatures have to find workable, common sense ways. The colony is the marmot way, their culture.

"You see, my innocent charges, many years ago, before any marmots dwelt on the Olympic Mountains, animals and humans were friends. I mean, coyotes still ate marmots, as did hawks and anything else bigger than us. In fact, smaller animals had a go at us, too. We marmots are very tasty", uttered with a wry, educating smile.

"To understand the Why, we must first establish the What. For you see, in the olden days man was much the same as today, eating everything like pigs, monkey, raccoons, and crows do. Called omnivores: anything they find tasty goes down their gullet."

"We only consume vegetables, grasses, flowers, and the odd insect, so our meat is sweet. Everything wants to eat us. This makes us an extremely cautious species. Only being constantly on guard ensures the safety of the whole. Why fear man? Humans have many recipes for baked, stewed, or roasted marmot. Don't ever be fooled by man. If it can be cooked and eaten, they will do so. They will make a tasty sauce and serve you up with a walnut in your teeth!"

At this horrific revelation, the cubs gasped, looking at each other in shock and amazement. The notion of being a meal was fright filling. To immature, developing minds, everything bigger wanted to eat them; smaller critters wanted to gobble them up, too. Marmots are wee, but cubs are little things, barely a nibble. Many clutched each other in fear.

The way to get to young, malleable, mush minds is shock treatment. The goal was to firm awareness. They had to know in no uncertain terms, life is not fair. They better learn before it is too late. Death is truth and permanent. They could be killed at any time by a myriad of foes. His task was preparation for the world they would face.

Mr. Mizithra desired this precise reaction from the innocents. The story's opening certainly grabbed attention; goal accomplished. They would go home and tell their family the story he shared full of vim, vigor, and vellication. This would give parents comfort, knowing he did the job.

Offspring profited from the lessons, in spite of delivery. Explaining reality to those we are honored to raise is common sense. In this way, he was an asset to the others. Mr. Mizithra was greatly appreciated.

He knew from what his parents taught him and life lessons due his own survival these lessons needed to be impressed in their minds. Were he and the rest of the colony to shirk their responsibility here, the future of their family would be in the gravest peril. It behooves human parents to teach their own children well to ensure their safety as well.

Mr. Mizithra gathered the young closer to share his yarn with them. For their part, the kiddies were eager to hear the story. It sounded like a really good one. A few chomps on his violets, then he began.

"Well, in olden times there was a big difference between the people today versus the past. You see kids, in ancient days people did not have thumbs. Oh, to be sure they were there, but their hands were more like ours, the thumb not really of much use."

"They would eat a marmot when really hungry and could catch it. For the most part they ate fruit, vegetables, meat or fish close at hand. The problem was they wanted more variety. The lack of a working thumb was the restriction in need of fixing. If they found a way to be better hunters, life would be easier. They could have free time to invent things and advance their species."

"One day though, my young students, a horrible thing happened. It shook the animal world from aardvarks to zebras. The world was permanently shattered when a man was born with four separate fingers and an unattached thumb. It was separate from the four fingers. There it was, swinging loose on it's own. With the use of this final digit, man did things no animal could. And I don't mean hitchhiking."

"Dogs, cows, chickens, pigs, sheep, goats, horses, donkeys, ducks, geese, and even cats, sort of, decided it wisest to be of man's world, eaten by them over cougars, bears, and hawks and other nasty beasts of nature. They surrendered freedom for "domestication", a terrible and degrading fate. To these animal species the way they were killed in the wild was a far worse fate than a farmer or rancher doing the dirty deed."

"Of course, dear ones, you know from the bobcats, lynx, and cougar in our world how unpredictable, silent, and cruel they are. Felines toy with their prey before killing. To them, it is not a quick kill, it is shear delight in taking time. They torture. Beware the cat!"

"One important point, kids: everything wants to eat us. So always be on guard and learn your four whistles.", he said, vigorously plucking the violet from his mouth, stern voice, focused, glowering eyes, and firm jaw. Here the scar over the eye worked to dramatize fright.

Every young cub knew this fact from the very earliest days of life. They had to master the four whistles to ensure safety for the colony's survival. Each was a soldier of the army. No one had the luxury of non-participation. There can be no conscientious objectors in nature. They instantly blew the four in unison. This species relied on a concerted effort. It took the cooperation of all to ensure the safety of the whole.

Individuality was dangerous. Those who did so soon found a home in the belly of a wily coyote. Doing the same as everyone else may seem boring because you are a human being. For animals to do so meant not just peril, but potential elimination of a whole family of marmots.

Mr. Mizithra nodded his head in approval, placing a fresh violet in his orifice. The best thing he could accomplish was to drill this fact in their immature, impressionable minds. If they learned while young, they might live to be old. He taught simple lessons based on common sense and hard earned knowledge. He continued the tale.

"With a thumb, man could build and manipulate tools. He made weapons to kill more efficiently. Soon they trapped beaver and the like. They had spears to slaughter any animal within eyesight. They invented the bow and arrow to lengthen the killing field. Finally, they created evil, noisy all-powerful guns. Pistol, rifle, or shotgun bring us down, ending lives in droves. Fear man, young ones, fear them and live. It was the final separation between man and wild creatures. From the day man first got thumbs, our fragile relationship dissolved. The bridge was burned."

"Nowadays, well, you see man's intrusion in our world. The pathetic way they try to "help" us. Usually, it ends in their good intention and bad results for us. They may come and take a picture with a camera or try to feed you. They have dogs. They leave garbage all over. They make trails through our lands and trample over our tunnels."

"Years ago, man planting more trees took out meadows, forcing our families to flee. We had to build new colonies in smaller areas, many soon too distant to visit. Our ancient forms of staying in touch were abolished, intruding in our world. This caused families to lose contact with each other. It was too far to travel. Called, "reforestation", it is where we previously had a home. The great designs of man too often end as disaster for those forced to live with their decisions. They seem incapable of leaving things well enough alone."

"The humans come here to "hike" and "camp", which usually mean devastation for us. People drive "cars". The noise is terrible. They lumber about, heavy, trodding boots shaking our homes. They build tracks and paths, making ours disappear. Humans leave garbage behind in piles and in huge bins. They erect centers to gather, buy trinkets, expensive food, and coffee. Their viewpoints ruin ours. People feel no guilt or shame for destruction and intrusion. If they have the T-shirt or hat, that will suffice. They have no qualms at what they do. It is an arrogance highly offensive."

"Always with them, always *dogs*. Canines hit with vicious hate. Dogs are evil creatures. They come in sizes not much bigger than us to massive, absolute monsters. Some can even dig in a tunnel and kill those found within. They are swift and strong. It is an animal you can't outrun, either. They are too fast."

"Dogs kill and eat us. Always run for cover from any breed. They are not friends. They used to speak animal languages, but surrendered long ago for the comfort of an old bone, a pat on the head, and warm fire. Now, they make guttural noises and bark. If you spy one get out of sight and scent quick as possible. Whistle a warning and run."

Mr. Mizithra came to the conclusion of the lesson. Every young face turned to him, completely focused on every syllable. The story would remain their whole lives. Any contact with man brought the same end for any slip up. The impact of man's destructive behavior was enough for the youth to understand. So, it was here he lowered the boom. He did it with love. What is called a velvet hammer. Sometimes teaching harsh truth, rough methods are required. This was one such time.

He turned slowly, reaching up, pulling the remnants of the sprig from his mouth. Then he leaned towards them, speaking in most somber tones. Turning his head side to side, bearing no particular sign of emotion on his face or eyes. Mr. Mizithra glowered and slowly spoke.

"And what, my young charges, was the name of this first man with thumbs, this source of evil and separation for our kind? Adam."

Chapter 2 Mr. Mizithra Prepares

The obligatory, fundamental tales were from personal experience. Mr. Mizithra was no spring chicken. His life was rough and tumble from the beginning. Experience transformed into tales. The hope: the next generation to have a better life. When a marmot reaches three years of age, they are ready to mate. It is then, armed with new found, mature feelings, the urge and need comes to begin a family. This chapter was some time long ago. For him, a time of both joy and pain.

Such was the case for Mr. Mizithra. His parents, Mr. Gingersnap and Mrs. Cupcake, were proud. He was the only one of his litter to live to maturity. His sister, Miss Maple Syrup, drowned as a cub in a flood. Her portion of the cave collapsed in a terrible storm. Weather high in the Olympic Mountain Range is not for the meek or timid. Conditions can change in a matter of moments, last for days, and kill. If snowfall is not deep enough, things like this were bound to happen. There was really nothing they could do. The mountains were rugged, but home. The thought of moving did not occur. The Olympics were Paradise.

One curious fact concerning Olympic marmots you simply must understand, to let you in on a little secret, so please keep it to yourself, they always named cubs after food wrappers, magazines, and the like left by the Thumbs.

Contact with humans produced this quirk from way back. While having little time for man, they nonetheless were intrigued by words on paper. This oddity began in olden times with the first Olympic marmot named: Mr. Copenhagen.

Loggers were the first to traipse up and litter the land with empty chewing tobacco cans, sardine tins, beer bottles, candy wrappers, and magazines. They chopped down trees and built roads without regard to the animal inhabitants. Man had eminent domain over the world. He said with assurances from arrogant pride animals were his to exploit. He could eat them, turn them into soap, belts, hand purses, makeup, or nail hides to his walls to display the kill. Ain't that right, John Jacob Astor?

After loggers ripped out as many trees as they wanted, someone got smart for once and turned the entire mountain range into a national park. The rules changed. Soon paved roads replaced dirt ones. Parking areas arose with a massive one complete with a park center up top of Hurricane Ridge, stocked with coffee, souvenirs, and garbage. People came in droves, hiking anywhere they pleased. Animals of every species were soon scattering away from the area as quickly as possible.

This breed's odd quirk is how his name came. While pregnant, his mother had a label from a syrup bottle and a small container for Greek cheese to help pad her nest. His sister got the sweet. Mother liked the sound of the cheese as it rolled off her tongue. His father agreed, thinking it noble and unique. No other marmot had ever been given this title. With this name, it was hoped he would be a marmot of outstanding character to make his parents proud. They did their best to prepare him for the harsh world to increase his survivability.

Marmots made do with what they found. Scavenging is a natural course for wild animals. They have no stores to go buy things they need. There are no midnight markets or drive throughs. A rabbit cannot pop down to the marketplace and pick up a bunch of carrots for supper. They must forage for anything they can find. He cannot order a parsnip burger and a side of butter lettuce at a diner.

This is the major reason they are vulnerable. When they go looking for something to eat, something is looking to eat them. Animals must live on resources found close by. If man disrupts this fragile world, tragedy ensues. If they venture too far from safety, life is often forfeit. Those who do not heed the rules in the wild pay with their lives. No safety net exists in the forest.

Some species, like squirrels, store up nuts and such, but they are not the same as marmots, deer, ducks, or coyotes. Each animal has it's own way to survive. This teaches while we arrive at the same goal, each must take their own path to a common destination. It means the way we must live is not necessarily the same for another group. To truly believe in diversity, one must accept different cultures for what they are, not try and change them. A marmot cannot live as does a giraffe.

Why expect a person from a culture in a warm climate from the Sahara in Africa to have the identical society as one from a cold climate in Mongolia in Asia? If we want to be loving and accepting of the human world, we must embrace difference not condemn or demand altering to fit ours. Are we superior? This type of practice did not do the Native American tribes or Europeans who tried to employ their culture much good in the long run. Acceptance is difficult. What works for animals ought to be good enough for us.

Storms are frequent, often violently vicious in these alpine peaks. The Olympic Peninsula is the only true, moderate rainforest on the North American Continent. It rains or snows many, many days. In fact, it usually sheds some form of precipitation about two thirds of the time. This is not where one goes for sun and fun. Hurricane Ridge received it's name when an early white explorer said the winds were strong as a hurricane. It's vistas are breathtaking.

The cloudy skies part once in a while; the sun rains down its golden rays. One sights Canada north across the Strait of Juan De Fuca, it's entrance to Puget Sound. From an elevation at sea level, a drive up to Hurricane Ridge rises to 1,600 meters, nearly a mile. A true wonder of nature gladly exposes itself to you. Naked nature, good stuff.

As for marmots, it was heaven, home, theirs. These native creatures, birds, other mammals, fish, and insects have no where else to live. Man really has little business there, venturing in from time to time to gum up the works. They stick a nose in where it is neither wanted or needed.

The animals of the wild are very different than you. Perhaps they do not make much sense from your culture. It is, however, not your country, it is theirs. Go in nature. Observe the lives of creatures who inhabit these places innocently. This way, you learn what it takes for these humble beings to survive and thrive. We honestly ought to be doing our best to better their world. Often this means staying out of their country and in our own. We have plenty of land, let's give them all they need to live happy lives.

For Mr. Mizithra, the meadow where he had always lived was home. His family was there. His parents had other litters. Brothers and sisters inhabited burrows all around. These family members looked at him as a novelty. He was the family character. It was true, in his heart secretly relishing this position.

Traveling led him over Olympic peaks and pinnacles, covering huge tracts of land. Each venture into the unknown was as unique as it was similar. This one was special. He was busy getting ready for the journey. The imagination began to wander. He became excited at the prospect of any foray in the wild and crazy world of the Olympics. What lay down the lane, around the next corner, over the high hill, these thoughts drove him onward and upward.

Time, time was his enemy. You see, marmots hibernate about nine months of a year. He only had 100 days to accomplish any journey or perish in cold and snow. He was bound by the necessity of hibernation. He had to prepare while on the way around the mountains. In front of his mind was the need to gain weight. Were he not to, winter would kill him by starvation. There is no forgiveness in the high Olympics. Nature is just and honest, but harsh.

Marmots have two distinct layers of fur: one underneath to provide warmth and protection from cold, the outer to shed wet and wind. This makes them the perfect species to endure the cold climes of this lofty wonderland. During those 100 days above ground, these feisty rodents eat, eat, eat. Wrestle, eat, drink, wrestle, eat, eat, eat. They balloon in size and must live off fat stored during dormancy.

If you chance to spot one in their natural environment, they will probably be chomping away at some form of greenery. By the time they are ready to disappear for the long sleep, one practically has to roll down the tunnel due to the immense size. They are bloated butterballs at the time of the lengthy slumber.

While on sojourns hither and thither, Mr. Mizithra was usually feeding his face with violets, clover, grasses, and flower blossoms. Yummy. Lupines were the favorite. If a dandy bug passed by, gulp, down it went. Whenever they paused for a break, something tasty was shoved down his gullet. Eating was what he enjoyed doing, except exploring new places. However, for 100 days, mostly he fed his face. He loved to eat. He lived to tell stories. He existed to find new adventures. Saying good-bye to family and friends, he loaded up a few items, heading to find a friend.

The next adventure was about to begin.

Chapter 3 Murray

Mr. Mizithra's constant companion was a friendly woodrat named, Murray. His family lived under the wood shed next to the ranger shack. It was a lovely spot up on a small knoll, perfectly placed with an excellent view all around. Pleasure and safety combined to make it an ideal abode for a small household of humble rodents. The ranger came up four times a year to check the area.

The ranger only stayed a few nights each time. It was not too bad for them, usually leaving a few tidbits to swipe. A scarf or sock was donated thanks to light paws. These were used to make the inhabitants warm in the cold, snowy winter in the hills. If the ranger came, they hid. If caught: instant death.

Years before, they lived down the way in an old logging cabin. The lumber company came, knocking it down for a new, cement building. They burned the debris. The rodents were displaced and forced to seek other shelter. Since it is so harsh up in alpine peaks, were they unable to find a safe place to live rapidly death would arrive. Murray and his family were forced to find a new home. Under the wood shed did the trick.

They dug a good entrance on the side, where the cement was not totally firm. Home had escape tunnels, one opposite side the entrance, the other out back under a small sisal bush. Past this, lay a safety chamber were the home demolished. It had a pair of safety tunnels. Otherwise safe underneath, it was a splendid home as woodrats saw things.

The journey was not long. He chewed away at what was handy to grab, waddling to his buddy's place. He was lost in mind meanderings, serving to pass the time. He thought long and hard about the trek before them and new adventures to follow. This anticipation really got the juices flowing. He doddered faster, all the way up to a meander. Along the path to Murray's place Mr. Mizithra came across Ernie, an old friend.

"Ernie, here, here, here, old friend. My, my, my, it has been quite a while since our paths crossed. How are things with you?"

"Oh, Mr. Mizithra," slobbered Ernie. He salivated perpetually. Ernie was an adult Pacific Northwest banana slug, twenty-three centimeters long, a lovely shade of yellow. Every word spoken had a dose of drool. A banana slug is naturally covered in a thin layer of slime for protection. They have no bones.

"Things are sliding along nicely. Sniffed out a lovely patch of leaves rotting on the south side a while back. Kept me busy munching for days. Yummy stuff, don't you know. Why, I recall a time when it would take a whole week to plow through a patch that large. How are things with you, my fine fellow?."

"Well, sir, I'm going to see Murray. We are off on another adventure. Ernie, we are heading to the White Mountain to see what we can see. Stay in touch and have fun chewing away.", sauntering off towards Murray. Of course, Mr. Mizithra snagged a handful of succulent, blossoming Flett's violets on the way. He crammed the whole wad in the right side of his mouth, gnawing joyfully away. He strode towards his bosom buddy.

Ernie muttered something incoherent in return about saying hello to Murray for him. Ernie had a nasty habit of speaking so low few could hear unless rather close. It was due to being alone so much, spending days consuming rotting leaves. He had little interest in anything else. The life of a fellow like this, a slug, really had a small purpose: eat, reproduce, eat, die. People plan careers and long range projections, slugs have a very simple philosophy: chew well, mate better, and live on.

Mr. Mizithra waddled up a small hill towards the ranger's shack. He stopped to eat several times, pausing to munch away in total glee. Food was fun. He found Murray resting, perched on a Sitka spruce pine cone outside the entrance to his home under the end of the wood shed farthest from the shack. He liked these particular cones because they were long enough to stretch his bad left leg along it. Cones comforted his ache. This helped ease old wounds earned avoiding a fox.

Murray wore a kippah, a Jewish hat always worn when a male was outside. His family were descended from a long, long line of woodrats who found their way to this high elevation a hundred generations ago. His family stowed away on a Russian fur trading expedition during the eighteenth century. They hopped off when the boat docked along the coast to trade with the native Lower Elwha Klallam tribe (the Strong People), site of present day Port Angeles. The family came to high country out of necessity for safety. These mountain peaks were a wonderful home for his small family, like the steppes back in the old country. Murray grinned, nodding joyful acknowledgment of his friend's arrival.

"Mizzy, my dearest friend, I see you already gained a gram or two. Time for you to fatten up. You are doing so nicely. It must be ze sign of a cold vinter, ja? If so, ve had best get schtarted on our journey over ze river und through ze woods. Heh, heh. I cannot vait to begin", Murray and Mr. Mizithra's friendship was based on picking on each other in a friendly way.

"First, though, you must come in und have some of mama's voodrat soup. I know you love it. Besides, if you don't she vould come after you mit a switch und pull vat's left of your ear.", Murray turned, laughing. Both knew mama's love for the big galoot.

He hopped off the pine cone, waving his chum forward. The entrance to the humble home was rather dark. The happy compadres descended quickly down the dirt tunnel lined with bits of wood, stone, and glass for support. It was arranged as a mural of refuse. The woodrats found it artistic. Murray lived here most of his life. The home was built by his grandparents. Only grandmother survived. Grandpa was a goshawk meal years ago. Clinging to the bark, he hunkered down. The raptor wrapped tightly around the trunk, snagging grandpa. Such is life.

The entryway wound around down a half a meter to a small room with three different passageways. Murray led down the hall to a large, cozy room lined with feathers and straw. It was aesthetic in an animal sort of way. It made the chamber warm, homey, and happy. Mr. Mizithra had been here on many occasions and enjoyed Murray's family. They had a comfy home. He always felt welcome. He was an adopted family member, albeit the largest in history, an ersatz woodrat.

An older female woodrat waddled in from a small room off the side of this larger chamber. A huge vat of fragrant, steaming liquid proceeded a pair of slender, furry arms. The inviting odor wafted to the ceiling, enveloping the entire abode. It induced hunger, simple considering the chubby goal. Two hands held the vessel, ring finger missing from the left hand. She came in view, an immediate smile creasing Mr. Mizithra's mug. He loved this little old lady. She was a gem, an absolute gem.

A gray-haired head hid her face behind the pleasing, piping pot. She had gray fur around the mouth as well, vivid proof of advanced age. An ample, plump rump waddled in, edging the steaming entity ever closer. The hot tureen landed on a pad on the table in the center of the room. A huge ladle lay to one side. Bowls were stacked to the other. Steam arose towards the ceiling some half a meter high.

It was supper time. Woe betide the marmot or woodrat who turned down her sumptuous repast. It was a labor of love, dammit. She would chide any who refused, telling how hard she slaved. They better eat before some horrible human came, disrupting their lives.

"Oh, Mr. Mizithra, my boy. Ven Murray told me you vas comink, oh, I tell you, my old heart vas so happy, you big galoot. Und you remember mama? Come bobeshi", begged Nediva. Her name meant, noble and kind-hearted, defining her so well.

"Come see Mr. Mizithra, mama Frema. He is Murray's friend. Ze boys are off on another mishegas escapade. You vill be ze death of me, you know zat, boys? Vy you insist going on zese horrible adventures is beyond me. Zere is no need for zis. You boys need to schtay home und find vives. All I vill do ven you are gone is sit in ze corner, chew old, moldy potato schkins und vorry.", her weepy voice quivering with guilt only a mother can summon.

The guilt trip Nediva laid on was exceedingly thick, like the skin on four day old pea soup. Murray remained unmoved. He was keen on another foray in the wilderness. For him, a confirmed bachelor, these summer trips made the rest of the year worthwhile. New panoramas, having adventures, meeting new creatures, and hanging out with his best marmot buddy were incentive enough to get away from home. He would have the rest of the year to lounge about under the ranger's shack.

In mountains 1800 meters above the nearby ocean, they saw waves crash on shore from rock cliffs on a clear day in the Olympic Mountains. You had to climb high enough, find a lovely perch, and the world spilled out before you. This was motive enough for the little guy.

"Ach, Murray, no no, you mustn't. Remember, remember how zose evil Cossacks chased us from our home und into zis, zis, zis HOVEL!"

One thing about Murray's grandmother, she could sure turn on the thick, heavy drama. She should be on stage. Her memory needed serious consideration. Not only did she exaggerate, but confusion ruled an aging mind. A rough life rattled her core and memory.

"Gram, it vas not ze Cossacks. Mizzy, ze Thumbs who tore down our old place ve lived in ven I vas a baby vas a Russian lumber company. It vas, Rummelov Logging. Zey vere not Cossacks. Gram loves theater. It is just zat our family came from Russia. Jews vere not vell treated by zem centuries ago."

Murray called humans, "Thumbs", a derogatory term of disdain, disrespect, and disgust. Human beings had a habit of destroying animal habitats, justifying it by building a "green zone". To animals, this is man's way of saying, prison camp or zoo. It was only logical they avoid man. The consequences were almost always lethal. Why, in harsh times human beings were known to eat woodrats.

They had no love of man, no reason to trust them. Human beings liked to tempt rodents with treats of nuts, cheese, or grain, bash them in the head, then kill them. This caused his family to repulse any urge to succumb to man offered temptations. Many followed them to their peril.

Soon the boys were on the way. Murray had to pull away from his mother's clinging hug. They waved good-bye. Mother and grandmother held each other and some moldy potato skins, clinging and weeping in unison. Guilt failed to work. Murray and Mr. Mizithra were on the way to new escapades.

The meadow was brimming with toasty morning summer sunshine. The Douglas firs, cedars, hemlocks, and pines swayed with a soft updraft lifted by winds below. The smell of life filled the air. Bugs of varying sorts flew about, pitching with the wafting breeze. Slowly, the world came alive. They were ready to go!

Mr. Mizithra carried a journal with exactly 100 pages. He had only 100 days to complete goals: gain enough weight to get through the winter, and get home in time to waddle down the tunnel with his chubby relatives. While they did not hurry, a clock ticked in his head. He had a goal to fulfill, the White Mountain (Mt. Olympus).

Accomplishment had to be in a short time. This was not merely an escape for our dynamic duo. Another goal was for the colony's marmots. Mr. Mizithra was the sole individual in the settlement with the talent and ability to accomplish their aim. It would benefit the entire body, holding the future in his marmot paws.

The task charged came from the head of the colony, Mrs. Lip Gloss. She was the longest-lived, most sagacious, fiercest, yet kindest of all the marmots. It was she who made any final decision. A counsel of the oldest and wisest amongst them, four in all, worked for the survival and benefit of the whole.

She was a serious marmot. It was her responsibility to oversee the family. She did not have the luxury of wanting only what she wanted or even what some faction of the group thought best. She had to think of every member. To this end, the counsel made a very important decision regarding their combined futures. He was given these instructions prior to leaving with Murray.

"Well, Mr. Mizithra, again you venture forth to see what you can see. Since the days", she paused, wiping at her aging eyes, "when my daughter, your wife, Mrs. Christmas Special, and darling granddaughters, your children, Miss Cake and Miss Cookie were killed and eaten by the menace from Bear Town, we fear intrusion. It will come either by bears, Thumbs, or weather. This place is not safe for the future of our colony."

"On your 100 day sojourn, you are hereby charged by the counsel to seek out a new, permanent settlement. It must be somewhere specifically suitable for our needs. I would imagine lots of rocks and exposed stones hidden against a bluff is ideal. Hopefully, there would be trees above and sufficient escape routes to guard our needs. There must be adequate water. Most of all, we need to get away from the infernal humans."

"It is clear with cameras and intrusions the aim is not a friendly one. It is only a matter of time, most feel, before they take us elsewhere or kill us outright. They are moving our good friends the mountain goats. We may be next. Too many humans spoils the colony, goes the old saying."

Mr. Mizithra knew well the needs of his rodent family. He assured her of his ongoing quest for another, suitable place to live. Thumbs were eying the meadow for what humans called, "Environmental Protection". All they knew was moving probably lay in the future for the "good" of wildlife, from a human standpoint.

He gave Mrs. Lip Gloss assurance to do his level best to secure their needs. He hugged his former mother-in-law good-bye, assuring he would be back in less than 100 days if possible. The priority was the mountain. A new home was added to his burden. He was proud of the confidence. Off he went, aware of a tremendous obligation, happy to be a vital part of the village. The marmots thought of the whole, not the individual. Theirs was a family unit. It's strength made him whom he became. Repayment came with a positive rather than a negative attitude. Duty to the colony: cool.

It's how you view responsibility, not of self, but others, that defines character. It is always easier to be hateful and angry versus thinking of and accomplishing the positive. The bulk of their rodent lives was spent underground. They had to have a place to inhabit which not only fit their needs, but was safe from forces constantly railed against them. This meant working together was a necessity. Life is lived differently if you live with those with whom you work.

By the time a marmot is ready to hibernate, they are many times bigger than when emerging come late spring. It is a habit, identical of other animals: bears, bats, skunks, snails, groundhogs, bees, and ladybugs. This ancient way was the method these specific creatures knew preserved life. Were they to thrive, a new, distant home must be found. They could not anticipate moves the Thumbs would make. Common sense did not seem common in human decisions as rulers of the world.

Once in burrows, marmots live off of fat reserves. When considering the life style of Olympic marmots, it works for them. It would not work for human beings. Think about how many different sized suits of clothes you would have to wear. It would make more sense to wear a one size fits all sack. We are not built for hibernation, though a lot of people do pack around a constant, healthy winter reserve.

For marmots, clothes are not used, so it is simply a matter of eating to the point of bursting, then deflating the balloon over the next nine months. Of course, we could not make it with such a long a nap, either. The world we inhabit would never survive if we slept that long. So, to know what animals lives are like is only afforded by observation. We cannot possibly understand it all, nor should we. Let them live their lives with a dash of mystery. That is only fair.

The verdant meadow spread out east before them, onto the woods, stream, laurel jungle, and land beyond up to the White Mountain. This incredible rising mass was a mystery, often seen when the high fog banks lifted. No rodent known ascended the fabled, distant, alabaster obelisk. The field ahead was the most familiar. They set a quick pace. The last thing needed was to lollygag. Were they to do so, Murray's mother would urge them to stay. The need to set out on this madcap undertaking was fervent in the mind. These two were off on another delirious escapade in the wild wilderness of the Olympics. They were off to have the time of their lives.

He and Murray were determined to be the first to ascend the peak. Resolve is what they possessed, a powerful emotion. Resolve meant they were not going to be deterred; nothing would stop them. This quest, the ascension of the White Mountain, was planned since last year's discovery of a small passage through the range. They saw this distant berg when making a ridge on the far east end of known lands. The difference being the last case they were further south. This time they headed due east. The path they chose was completely unknown past the edge of the pond.

Marmots ventured this way years ago. Some were never seen again, others told tales of animals and sights too wild to be imagined by small, simple rodent minds. In any case, this was new territory. Caution was paramount. A sense of adventure made them salivate with anticipation. They sallied forth with a smile on their faces and a song in their hearts. What waited down this path was unknown. That was half of it. Fun was down this road calling them to enjoy it's offering. They followed hook, line, and sinker.

Murray hugged his family. Mr. Mizithra waved at them, but his mind turned to the joyous task at hand. The thought of what potentially lay ahead filled his mind. His imagination engaged and expanded with glimpses of the unknown. He and Murray started downhill, turned the corner down the service road, then spied the familiar first step.

They initiated the new journey by crossing a large meadow past the ranger's shack from Murray's home. The little route led them past it as it went off towards the main road. Ernie had since munched his way down the path. He was nowhere in sight. The present concern, immediately to the right. This danger was the land south: Bear Town.

This thick, aphotic green screen of sisal berry bushes and rough shrubbery held real fear. Occasionally bears burst forth from this dense wall of greenery, giving no warning of impending doom. Big, ugly, smelly savages, thrashing forests as they go; eating anything, simply anything. Disgusting and filthy, they have no regard for others save filling their endless gullets with fruit, vegetable, honey, or meat. If they can shove it down their toothy holes, they would and did.

They fed on garbage of human beings and stole their pets to eat as well. It was observed they were the crudest of nature's creatures in these mountains and valleys. No other decent animal would ever associate with one. They would find themselves supper for the effort. Bears know only eating. Friendship is not an option.

Nothing is safe: ants, honey from bees, berries, birds, fish, and any animal it can get it's paws on is a meal. They have been known to kill and eat man. Their language is deep, dark, primitive. Most animals only know their words by hiding and evading detection. A bear has little to hold them at bay. Marmots make a fine meal. Some had been consumed this way to the pain of survivors. The image of a beast rending a loved one asunder in front of you would never leave the memory of the survivors.

Evidence of a recent invasion was all around. Feathers from some poor, lazy dove, a smattering of fresh blood, and ravaged berry bushes were telltale signs. It is amazing a monster of this size could sneak up on anything, much less a bird who is perhaps the most sensitive of all species. A dove will fly off at the slightest hint of trouble. Yet here the remnants of success lay strewn about for all to see. The smell of death was faint, but the large pile of coiled, black scat was proof the local bully passed by not long ago. This way those who came by would know what happened and who did it. These beasts were proud of achieving a successful kill.

Wisely, they pushed north through stands of tall grass and small bushes. There was no scent by air nor sound of movement in the green wall beside them. They felt it best to move along as quickly as possible. A marmot in full dodder is a sight to behold. The bad part: exposure from above. Raptors of various sorts from kestrels to bald eagles perused the ground below from the air, ever ready to shoot down for a meal.

In the case of Murray, he was small enough and his coat blended in the surroundings nicely. Mr. Mizithra, on the other hand, was a waddling mass of brown. He cut a wide swath moseying through the grassland. He was meals on wheels for any passing, wide-eyed bird of prey. A bald eagle could hoist him aloft in one fell swoop. Speed when in the open was paramount. They pushed along swiftly.

In spite of the knowledge he was wide open from above, Mr. Mizithra had the habit of eating along the way. Packing on the pounds, cramming on the grams was essential to survival. He had to keep it up. He could not to get caught too thin to endure the cold Olympic Mountain winter. Always in the front of his mind was the fact he had to gain more and more weight. The only way to do this was to eat on an almost constant basis. No one in his colony, though they loved him, could do anything. Each had to exist on their own fat reserves.

This ought to teach us sometimes we can rely only on ourselves. It should help us think strongly about preparedness in times of trouble. If something bad happens what if your parents are not available? If you are separated from those who guard you, it is imperative you are ready. Learn the world around you: forests, rivers, desert, in your immediate sphere. It is important. Staying home all day playing video games alienates you from the most wonderful world around you. It can become a friend instead of a mystery.

They skirted rapidly past black elderberry bushes for protection, buds soon to produce nourishing, sweet mountain fruit. They pushed onward. Both companions conceived thoughts of gorging on the juicy bombs of intense flavor and delightful sweetness. It was easy for Mr. Mizithra to wander off in a fantasy of large handfuls of berries, nectar dripping from the sides of his furry mouth. He loved to eat them and envisioned what culinary pleasures they would offer later in summer. Once ripened, he would devour them until he could feast no more. Ah, berry dreams.

Shade provided cover from above. Taking time, ever keeping vigil, not just looking up, but all around was the rule. Safety for animals is never guaranteed. They must always lookout for an unseen enemy. His colony had four individual whistles for warning for good reason.

They were not able to do much to defend themselves. About the only thing a woodrat could do is climb up to higher ground. That was their sole method of defense. Not too bright as any animal or person could see them trying to appear inconspicuous. It is not very easy to look nonchalant peering down from a tree limb.

The bright sun burned down. The natural heat was wonderful. The duo reveled in the thrill of the trek and good fortune of a lovely warm day in the Olympics. The rays brightened the path forward through the far northern side of the meadow. This was the way they usually took. It was the farthest from Bear Town. No one wanted to get too close to potential consumption.

This was same path they had taken in journeys before, always a different way than the one planned. It was this new route which gave so much excitement. The anticipation of seeing different things and finding new animals gave them pleasure. They wanted to go see the world round about. They were not content simply to live in the same place, never experiencing what else the world had to offer.

They made it to the end of the verdure, wisely pausing, scanning off behind them. Safety first was drilled in their minds since birth. It was a short lived critter who did not take superior precaution in all they did. While it may seem tedious, those who exercised the best safety methods tended to live the longest. This would ensure they were quite alone and give one long, last look at home. It would be many days until they would see it again, hopefully. They could just make out the top of the crest of the hill. Smiles adorned their mugs as they turned in unison towards the future. Ahead, awesome adventure awaited.

This being a dandy spot, they agreed to set up for the night. They began by digging a small burrow in a nearby stand of trees. A few stones provided the desired location. They dug for a while, pausing often to keep a vigil for enemies from above or about. Soon the temporary abode had a fine entrance, a large room with a sleeping area, and two escape tunnels. Marmots are tunneling fools let me tell you. The fellas were pleased with the combined efforts.

Over time and many travels they became economical with their methods. They could dig a night's lodging and be eating a meal before most other travelers struggled to figure things out. Experience bred wisdom. Applied logic was secured by the process of elimination. What did not work was not repeated. They became efficient by necessity.

An evening's scrumptious repast: Flett's violets, dandelions, lupines, and clover with a handful of pine nuts was enough to fill them. A flying bug or two nabbed by our woodrat friend garnered a bit of pure protein for the journey. Burning this many calories meant they needed a whole bunch more to make up the difference. Of course, the marmot had to eat about five times as much as his much smaller counterpart because of size and hibernation.

Murray brewed some cattail tea from a pool of water under a nearby cedar. It had been a tremendous first day. They got past the fear riddled Bear Town zone and were in the relative safety afforded by their little hovel. Mr. Mizithra wrote in his journal. Both lay down for a well earned night of rest.

Chapter 4 The Pond

Day Two found our fellas rising early, stretching, and yawning. The sun rose from it's usual position in the east. A sweet shard of light shone through the back of the tunnel. Warm air wafted merrily about, filling the air with solar heat, rather pleasing, indeed. The transitory lodging proved safe. They learned by trial and few errors to be quite adept at designing safeguards. Picking out a good, defensible spot was first. Next, it needed at least two good, quite separate escape tunnels. One thing most vital was a good way out of danger.

They feasted on a meal of bugs, lupines, grass, and clover, then were off quickly at a good clip. The direction was straight towards the glare due east. The sun provided an advantageous, shadowed vantage point. They headed down a narrow pathway, clearly used by other critters. Brightness made it much more difficult for any bird descending from above. The haze the sun made was a good cover for our amigos. It was best to start early. It gave additional shade from evil forces lurking higher up.

They could see the early birds catch worms. They appeared to be robin red-breasts. The ancient aviary delights were very friendly. Robins worked early morning and late of day, just before twilight. In between, they slept and cared for their young. Robins were good at telling stories. If you could get one to hang around, a wonderful tale would be told.

Deer and wapiti tracks, raccoon, skunk, squirrels, chipmunks, mice, and various birds came through, individual stamps on the soft soil of the path. It wound around, down along a natural way worn by the nightly trampling of they who find darkness the best defense. This was a rather long paw path to say the least.

The wanderers made good time. The pace set was steady, Murray in front. They paused often to survey the surroundings, to be safe. Night travelers retired, morning shift went on guard. Along the path, a mouse or two, Douglas squirrels, and a doe with two young ones crossed the path, no one willing to stop for a conversation.

Grabbing handfuls of roughage as they went, Mr. Mizithra spied a shimmering light stab in his eyes. He squinted and made out a pond ahead. This was a welcomed site. This was a goal and a place of friends, food, fun, and safety. Traveling for some hours, found them parched and hungry. Excitement gripped. At the same time Mr. Mizithra crammed more clover in his salivating mug. He managed to spit out a word or two.

"Let's head that way," he said between chews, "the pond is down on the left. Do you see it, Murray?"

Nodding in hearty agreement, Murray's tail vanished, dust swirling about as speed increased. Murray could flat out fly when called upon. It was his best tool when fleeing danger. He had seen up close and personal how climbing a tree did not work when his uncle did. A bobcat nabbed him anyway. Speed was his secret weapon. He wanted to get to water as soon as possible. Woodrats love to get in the drink and swim. Food in abundance awaited his eager mouth beneath the surface. He could see proof bobbing up and down in the sweet mountain water.

He dove in immediately, kippah flying off his head, finding a lovely, fat water lily bulb. He bit hard, shook his head to free it from it's roots, and made for the surface. Spraying water out of his nostrils, shaking his head and ears dry, Murray swam swiftly to the shore. A water bulb dangled from his marshy mug.

You know, the little guy had a pretty good stroke. He made for the shore swimming in the warm water. The shoreline was soggy. He climbed on a handy flat rock to munch on the bulb. He shook off the access water from his fur, savoring the meal. As his bulbous friend finally approached, gasping for air, Murray was finishing the delightful repast. There is nothing like a fresh, fat, luscious water lily bulb. Delicious!

"Oh, yummy. Zat vas ze best. All vet und juicy. I am full. For you, old Mizzy, you have to feast on all zis." waving his arm in a circular motion to show the abundance spring in nature can only provide.

"Eat, zen ve vill find a schpot to rest. I feel ve have done enough for today. Zis vas a shorter valk, but ve covered a fair amount of land und needn't tire ourselves out for no reason. Ve can build a shelter quickly enough. Eat, eat, my friend, eat."

No need to be told twice, or thrice for that matter, Mr. Mizithra's nodding noggin, dove headlong in a patch of lovely lavender hued lupines growing along the shoreline. Murray could operate as guard for a few minutes while he got filled. His chum, finishing the bulb, burped, was happy to oblige. It was going to take a lot of food to sate this appetite. They walked for hours, building up a natural need to gain weight. It was an excellent spot to drink their fill. Cool, clear mountain water was simply the best and sweetest tasting. Mmmm, delicious. After this Murray made tea from the handy flora.

A grove of cattails lined the sunniest portion of the sizable body of water. Murray utilized these to make his famous brew. They enjoyed the warmth it provided the insides. The taste was appealing. While sipping the concoction, activity swarmed alive all about. Red-winged blackbirds and tree swallows, shiny, shimmering blue bodies hiding the beautiful white underbelly, flitted about the long stems catching abundant bugs. Turtles sat upon thick tree branches jutting from the water or rocks providing sunshine, warmth, and joy.

Trees and bushes hugged the far east end of the tarn, showing the way up a steep grade in the distance. The ground sloped swiftly upward at a severe angle past the pond. Rocks dotted various spots in between thinning trees. A few minor stone cliffs appeared on the north side of the stream. Further up in elevation, a much sparser forest. They saw new and exciting things.

It was the river. Rapids were evident above the hill, swirling wicked, white water churned into softness in the fat pool. These lined the rocky waterway as it disappeared, rising over the vast mound a great distance away. Water cascaded down the rapids, alabaster spray glistening in a thousand directions serving up a rainbow of colors in the summer sun.

Ahead, they spotted a new creature at the water's edge. Immediately, safety was a concern. Lollygagging ceased. You cannot be casual for long in the open. Murray shinnied up a tree, peering at the beast. The unidentified animal was oblivious to them and seemed quite busy unto itself. It had dense burnt umber colored fur, and was bigger than Mr. Mizithra. Any animal bigger was a threat. Caution was the first thought in mind. Murray called down a description to his wise friend. The creature had a long, thick, brown tail, extended, shiny whiskers, and webbed feet. The fur on the beast was thick, thick, thick.

Bursting through a sisal bush, Mr. Mizithra strode from the greens smiling and waving at the confused dark animal ahead. The larger critter squinted, raising a paw above his eyes to shade the sunlight. His marmot's waddle increased, making a beeline for the bewildered critter. Murray was stunned. His normally super cautious traveling companion acted in a bold fashion. He quickly followed behind unable to see around the marmot's wide load.

Twitching thick, long, black, shiny whiskers side to side, the stunned object of Mr. Mizithra's interest squinted harder at the large, grinning mass tottering rapidly towards him. He recognized his old chum, waving a webbed appendage to invite his furry friend to the pond. Feeble eyesight did not allow much of a field of vision for anything more than seven or eight meters in front of his face. He shook his old otter head, laughing heartily. He hugged a dear old friend. He was very pleased to see this marvelous Olympic marmot once again. They had a long and happy past.

"Well, well, well, so, tis yourself, tis it not, Mr. Mizithra. My, my, my, oh, it tis ye. Ha-ha-ha, I cannot believe my wee, bitty, dim otter eyes. Well, well, well, now ye shall have to go and see the Missus. If ye doesn't, t'will be hell to pay, oh to be sure, yes, yes, yes t'will", spouted Graedy McGrady O'Grady, an old otter pal.

"Ye know it, I know it, and little Murray over there knows it, too. How is that bum leg doing? Better, good. My bride is a lover, unless ye cross 'er. Then, look out. Oh, how loverly it is to see ye agin, laddie. Ah, yes, yes, and double yes, tis a good thing, indeed.", the otter's brogue thick and warm as his fur.

Murray arrived and figured out what was going on with his madcap marmot friend. He was panting, still frightened by Mr. Mizithra throwing caution to the wind. It was completely out of his normal, cagey character. When he arrived immediately behind and saw their old friend, he eased the fear, stepping forward to greet this true friend.

A slap on the back from Graedy and he motioned towards a small hole in a clump of bushes on the water's edge. The three made for safety and good conversation with old, dear friends. They left the shoreline behind, entering safety the otter's home provided. A long trek for the day brought them to familiar friends.

This pond was a new home for this couple. In olden days, they lived along the stream which fed to this loch from up north, not east like the one at the rapids. It lay uphill over a kilometer away. A pack of coyotes found their home and ravaged it. They had to move. Safe haven was found down on this large watering hole.

Once inside, the furry trio followed single file through a long, round, mud packed corridor. This was lined on the ground and sides with soft straw, leaves, and grass. It dipped sharply at one point. They had to cross some pond water. The tunnel rose again sharply to a large, warm chamber. To the right was a small room from which steam emerged. To the left lay another tunnel, which went dark after a few steps.

They could make out another smaller passageway which led to the otter's bed chamber. Lapping water was heard inside, lower and far off sounding. Escape plans always included water. An otter in the drink is a very powerful opponent. Most other animals cannot or will not go under water to retrieve one. The whole place smelled a bit fishy, which would make sense with otters

Straight ahead on the other side of this large room was yet another passageway, which rose slightly. A willowy beam of light emanated from the tunnel's entrance. Shadows wavered, interrupting the beacon in the dark. Voices followed shadows. They were muffled, quite indiscernible. Graedy stepped to the tunnel's entrance, shouting up towards the figure blotting out the light by weaving back and forth. His bounding voice was uniquely filled with a combination of love and anger.

"Mother! Mother! MOTHER!!! MOLLY O'SHEA O'GRADY!!! For the love of Mike!!! My, my, my I swear she must be going deaf."

From the shadows a form turned, blotting the light. She came ever closer, a look of irritated concern on such a sweet face. A dull, coal black nose and thick, black whiskers came first, then a gentle countenance and a smile of pure love. A thick pair of rimless glasses covered her soft, brown eyes. A doctor's light and white hat covered her head.

This was the forest's favorite saw bones. When their home was demolished by coyotes, others found this spot to build a new home. She earned respect. She provided care for the smallest mouse up to big wapiti. Her fame was known throughout the hills by any in need of looking after.

"Oh my, dear Mr. Graedy McGrady O'Grady, we have a wondrous visit today. Mr. Mizithra and the woodrat, Murray. Are either of you injured? Oh, splendid. A fine day. Oh my, I really must go. Be right back, difficult case. Oh my, my, my, terrible. Put on a pot of tea, there's a dear."

Molly disappeared in the vapor from whence she came. She slid quickly back up the tunnel and current concern. Ever vigilant, she was completely absorbed with the task at hand. Entertainment would have to wait until later.

At present, she had a patient who needed attention. They would have to wait and enjoy time with Graedy. This suited them fine. Graedy continued to look towards his wife after fully leaving the room.

"Aye," said Graedy, "tis a sad case, indeed. Please, first, my friends, sit and rest ye. Have some lovely, fresh clover tea. The best. I have fish cakes, but we do have some lovely honey scones for ye."

"Ye see, Molly heard of a tremendous accident up on the path ye came down to our pond. Tis a long, winding, treacherous thingie. Well, it seems a damnable bull wapiti trampled all o'er a wee centipede. Poor fellow, sad, sad, sad. Broke 17 of his legs and sprained 6 more. He even lost two towards the back. Yes, yes, yes, the wapiti simply amputated them wi' a single stomp o' his mighty hoof. She's been at it for hours trying to set them in place. The logistics alone are bloody hair raising!"

"And I, I, I, laddies, just today had to adjust the back o' a garter snake. My, my, my, do not ever do that, me buckaroos. Ye have to start at the top and pop yer way down. There is no such thing as a short snake, I'll tell ye that for sure. Each time ye pop, the bugger tries to bite ye wi' those fangs. Let me tell ye just betwixt ye and me, when yer done wi' a snake, yer done fer the day. Whew."

They laughed and chuckled at the predicament. The image of doing work on a serpent was hilarious. The talk continued with great ease. They were able to catch up on what was going on in their lives. Graedy related the attack by the pack of coyotes and move.

In time, the doctor arrived, kissing her chiropractor/acupuncturist husband with a rub of noses, whiskers and liquid flying around. She gave a hug to each visitor and plopped down, grabbing a fish cake and cup of tea simultaneously. She was clearly relieved finally being able to sit and rest after such an arduous day's labor. An otter doctor's life was not an easy one.

They sat for a time, chatting about the sadness of moving. She spoke with joy of the new home. Molly was born not far from this lake. She was happy a pair of muskrats found the site. An entire crew helped build the place. They did not have to dig so much as a handful of mud. The travelers filled them in on the gossip from home and reason for the current sojourn.

"So, boys, yer heading east, up the barrier hill along the stream, then on to the White Mountain? Now, now, now, have ye considered the evil perils which follow such a journey? Why, many a body has gone up that way ne'er to return."

"To the south, an open field is fraught with dangers from above. Up here, laddies, hawks, eagles, and owls are plentiful. Tsk, tsk, tsk. To the north, lies a road made by the Thumbs. The noisy monster thingies on wheels travel up and down killing far too many a forest creature. They fly by so fast. They stir up nothing but dust and noise. Evil thingies."

"Once up on the crest, ye must decide what to do, which way to go. To the east is a thick forest known to have armies of voles and eastern gray squirrels. They have a vile leader known as Ralston, a raccoon of ill repute. Ye know his name? Curious."

"Humans built a wall to hold up their foolish road off to the north. True, true, true. Twas covered wi' heaps of food containers and chop sticks left by their filthy habits. So, the animals living nearby named it, the "Great Wall of China". If ye choose this quicker path, it is arduous and affords the most exposure from enemies on all sides, human, raptor, or predator. It is, however, the fastest way east."

"We hear of a way underneath, unexplored, an old tube for water drainage left behind when the Thumbs built the road. Yes, yes, yes. Tis said an old marmot couple lives there, but no one is sure. If ye make it there, beware, for the pipe is crumpled and old."

"There exists along the southern route, an odd place, my, my, my. It is here ye will find rocks strewn about from where a glacier stopped it's flow. Stones abound. For a marmot and a certain woodrat, t'will be shear heaven, I wager, indeed. We know these truths because our patients come from all over these hills and vales. They tell us about where their homes are and the places they know. It is a good way for us to share information with you fellows."

"The scar along yer side Mr. Mizithra, ah, there. It looks healed from the last time ye and Murray wi' a limp leg came through bloody, battered, and beaten. Ye two buckaroos want to go tempt fate again? Tis none of my business what ye do, but I am sure yer ma and grandma do not approve, Murray. Mr. Mizithra, ye are becoming nothing but a vagrant. Ye must settle down and become a respectable marmot."

Mr. Mizithra nodded, wincing at the thought of the terror they met along the last journey barely avoiding a coyote and risk of death. Luck was with them this past time. They escaped, but not before Mr. Mizithra was sliced open along the side with the beast's claws. It was only by wits escape proved successful. The hunter vexed them nearly to death.

He was scarred a good deal and had a rancid, juicy tale for each of them. You could not get out in the wild world and not chance getting pursued by someone hungry. Knowing how much reminded them about their injuries, painful to both. Molly quickly changed the subject.

"Now, please, please, please, tell us the news of yer family, Murray, and of yer colony, Mr. Mizithra, for gossip is always welcomed here.", said Molly with a special salacious smile.

Otters relish hearing fresh, juicy, gossip more than any woodland critter, except, as everybody knows, crows. Crows, forget about it, they are tittle-tattle machines. Molly wanted to know every detail of every birth, death, injury, and illness Mr. Mizithra's family endured. She knew them ever since the visit she was called to perform years prior.

Of course, nothing could beat the maladies Murray's mother and grandmother suffered. Oh, the pain, the pain was worse than any other rat suffered before or could in the future. Murray came at it with full vim and vigor. He vented the trials and tribulations of his little tribe. He tossed his arms in the air, moving and bobbing his head about. He would emote for all his little body was worth to get the point across to his avid audience. One almost felt like offering applause for the performance.

After some time and a lot of talking, food, and smiles, they retired for the night. The end of day two found the happy travelers in very good company, belly full, heart swelled. They rested in the main chamber with a small bed of straw made for each. They were asleep in moments, spent by the day's activity.

While not going far, they nonetheless got to a place of ultimate safety. The pair reconnected with old friends, spending a grand evening retelling old and new tales. It was an excellent beginning to a long, mysterious journey.

Chapter 5 What The Wind Blew In

Having a fine breakfast with their hosts, Mr. Mizithra and Murray set off as the sun rose. It's loving warmth given to another fine day to sally forth, explore, and find new stuff. They turned, waving good-bye. Their friend's minds were already turned towards two new patients: a hobbled, young black-tailed deer sporting a sprained front fetlock from falling in a hole and a feisty pocket gopher, carried in on a stretcher by two Douglas squirrels, with a head wound from a deer hoof.

Seriously, what is the fun in what the travelers were doing were it not to find new, interesting vistas? They provided fodder for exciting tales to tell the cubs in the colony. From his earliest times, Mr. Mizithra had a colorful little twist to his thoughts. While most were content with life in the colony, he experienced an incurable wanderlust. His mind was constantly filled with a curiosity; he wanted to know what was around the next corner.

Their medical hosts: he had a deer fetlock, she a gopher noggin; Heaven.

The path led along the tarn's edge. It provided maximum cover as well as a constant food and water supply. The day was perfect for the task. They made quick time, heading due east towards the objective. Ever present in mind was this solitary goal: the White Mountain. The water shone brightly, it's reflection making it impossible for a bird to spot them. Anything attacking could only approach from three sides. The water gave minor assurance of safety.

A slight breeze wafted reeds and fuzzy cattails, inviting them into a casual swaying waltz. The plentiful lily pads sported lilliputian bugs buzzing about in a busy bossa nova.

Both sojourners were caught smiling.

Fish occasionally popped out of the water, snagging an errant insect midair, manna from above. Telltale splashes remained after submerging, lapping gently against the banks. Frogs leapt over wet, gleaming lily pads, pausing long enough to spy anything else to eat. The odd tongue shot out to snag passing gnats. Their daily lives were not interrupted, merely observed by our two. They forged ahead, cognizant of lively existence blossoming about them. Spending so much time in hibernation, Mr. Mizithra missed a lot more of life than his bosom buddy.

Dragonflies, a most ancient, highly revered member among denizens of the forest world: blue, red, orange, gray, black, brown, purple, yellow, or green, buzz about doing the great work to contribute to society. You see, now this is a real secret so please tell no one else, dragonflies are the couriers of the animal world.

Yes, when one wants word to get from one place to another, the easiest, least costly, and certainly swiftest method to get to your necessary party was the trusty dragonfly. Their company, DHL (Dragonfly Hasty Ltd.), was very, very, very old, indeed. They rarely lost a message or, for that matter, were they ever late.

Speed, speed was friend and ally. They were gone with the wind, darting here and there with currents, in between trees, bushes, and grass. The paths often crossed other critters or swarms of flying little bugs. Nothing could or would deter the mission to deliver a customer's needs.

These droning, living arrows would zip past any slow, unsuspecting animal so quickly, it would cause wee ones to whirl in their wake. It was marvelous. They had messages to deliver. Lore demanded carrying out tasks assigned. To the rest of the animal world, dragonflies were highly admired. Out of simple courtesy, they were left to their own devices. The message carried may well be one of great impact.

Fame came from prompt assurances saving many beasts from certain death. No one dared injure or harm a dragonfly. If they did, it would bring the wrath of fellow dragonflies. No one wants to face the anger of a widow/widower of a fallen messenger. Others who rely on the amazing flying wonders to make the world go round would despise and shun the killer. Often we hear the term, "shoot the messenger" because of the news they bring. This reaction is why many are unwilling to share bad tidings; not wishing to face the wrath of the receiver.

Red dragonflies were the swiftest, while blue could carry the most weight or longest telegram. Each had it's own talent and purpose. Yellow served as an alert message. Green meant joy for the recipient. Black brought a sad message, purple one of mystery. Brown was most common. Gray was usually from a relative. Orange was reserved for secrets.

Why things were so could only be told by dragonflies. They would never spill. It was a species secret. All anyone knew was they were prized for helping society as a whole. A dragonfly only lives for 24 hours. Most only had one good message in them. The service to the animal world was truly invaluable.

The fellas were caught getting a little lazy in the sunshine. An odd sight ahead was quite bewildering. Personal safety instantly came back for self-preservation. They turned full attention to the vision in front. A white bird with really long black legs was standing in the water near shore. His eyes were sharp, very much so. They were icy, piercing, and deadly cold. He was covered with a swarm of dragonflies, coming and going at an incredible pace even for aerial emissaries. He was sending messages at a rapid rate, totally self-absorbed in a flurry of excitement.

His beak was long, yellow, rapier like. It could slice a marmot from one side of the belly to the other with one slick slash. This large creature was intimidating merely by appearance. The duo moved forward with great caution. Intrigued by his activities, they wanted to learn more.

What in the world was he doing and why?

Immediately behind him was a flock of darker, bluish birds of the same relative size. They had equally sharp eyes, donning elaborate blue and gray head feathers. They milled about, laughing and playing around. A couple of smaller, greenish birds of similar description stood with them. Occasionally they would look towards the white bird, shake their heads, and squawk.

There was a certain nervousness in their movements. Something interesting was afoot. The fellas got curious and had to find what was going on with the big birds. No dragonfly was near them, just swarming the big, white bird. They slowly crept forward, terrified. Safety mixed with a need to know what was causing the tension. Murray had genuine fear. He was just a little bite. He would be a meal for these monsters regardless what color the feathers. He moved forward, quivering the whole time, edging ahead, compelled by an urge to solve the mystery.

Mr. Mizithra knew himself to be too fat to make a meal for the fellows. Sometimes girth is worth a lot more than meets the eye. Besides, he noted to Murray, their sole interest seemed to be in catching little fish. The rodents inched ever closer, trying to hear what was what and why this little scene was what it was.

Drawing slowly towards the ongoing ruckus, the large, white bird speedily issued commands to each messenger who buzzed up. No matter how many approached, he had a hot news flash to issue regardless of color. The furor had our vagabonds gobsmacked. The white bird was a whirl of flapping wings and snapping beak. He bent forward occasionally for a sip of water. He froze. His eyes did not move.

The dagger-like orbs were locked onto a spot below the waterline in the pond. The dragonflies hovered in dead silence, unwilling to break concentration. Suddenly, he made a thrust in the water with his swift, amber saber. Its speed was unbelievable. In a flash, he stabbed the blue liquid. Hoisting a baneful, saffron barb in the luminous sunshine, he was found clutching a small, wiggler.

He tossed the tender trout in the air, catching it in mighty yellow jaws. The fish's form desperately wriggled within his throat, disappearing down an eager gullet. Lunch was served. The white bird shuddered with delight. He flapped the alabaster wings, then resumed issuing notations to the eager dragonflies. His repast over, it was back to business. This guy was a pure moneymaker for DHL. They were paid in bugs.

The albescent bird turned towards the azure group. The causation of vital commands over, the couriers darted off on assigned missions each in a different direction throughout the forest. His messages were on their merry way. It was now time to address the rest of the gang still killing time off to the side. The laid back blueish gray and dusky green feathered frolickers jostled, hemmed and hawed, guffawed, chortled, and ambled over to the white chap. They purposely took their own sweet time. It was a curious scene for the fellas as it appeared an odd drama. The blue and green birds did not seem to be overly friendly towards the white one.

Dribbling along took considerable time. They seemed in no hurry to drift over. They even paused to look at something off across the pond. The alabaster one, for his part, stood patiently awaiting their eventual arrival. It seems he had seen this scene play out before and was clearly not impressed with the antics.

Once there, heads hung in unison, awaiting what he had to share. They had done their part to stretch out the news and show disapproval. It seemed to be something done on a regular basis. It did not alter his behavior. He stood in front of them and began to speak, turning his head from side to side, the beak's movement becoming hypnotic.

The other birds did not seem to buy what the white one was selling. Body language is the same no matter what species. It was obvious to Mr. Mizithra and Murray there was tension between them. The countenance was clearly standoffish. They shuffled from one long leg to the next in a nervous movement. Both adventurers were spellbound by the scene and wanted to know precisely what was going on. These were new birds to them. Part of the travels were geared toward gaining knowledge. This helped the colony is how they saw it. Plus, it was fun.

After a few moments, the dark flock meandered off to another part of the pond muttering, shaking feathery tetes. The ivory feathered chap squawked. Tension was graphic to our observers. An instant later, several various colored dragonflies appeared out of thin air. He began to foment commands, sending evangelists forth through the forest and beyond. All he did was convey information to the others whether they wanted to hear it or not.

On closer inspection, the white bird with the golden beak had fairly tattered plumage. A feather was missing here or there. A couple hung loose, biding time before free falling to the ground. His alabaster feathers were faded hither and thither, more a chalky hue than true white. Some feathers had smudges of dirt or grease.

The black legs were wrinkled, bearing a nick or two. The yellow schnozzola was sliced in places. Age showed on the face. He looked haggard. He had bags under eyes whose luster was duller than at first glance. His best days were in the rear view mirror, yet still a force to be dealt with as far as he was concerned.

Our duo now carried past him, keeping vigil both around and above ensuring safety. They had to know what was going on with the frustrated feathery flock. The blue and green fellows were tightly cloistered in close conversation. With the utmost stealth, they sidled along the shoreline in a determined effort to satisfy their piqued interest. It was important to find what the hubbub was, bub. Tossing their guard aside, they just had to know. This was unwise. They could not stop. Curiosity compelled.

"Look man, we jus' about had it with this dude. Now he done got us so lost, we gots no idea where we is. Not cool, Bobby, not cool.", voiced one to the clear leader of the blue gang.

He stood a bit taller than the rest, more muscled, feathers and mane plumage of superior quality. With these attributes and aura of command, the blue and smaller green waders turned to him for an answer to the sad dilemma. This was apparently a leader due to his tone of voice, size, and respect the others bestowed without question. As the first to speak, he set the tone for the rest. It was clear he was not easily appeased. The fact they were lost did not sit well. He did not hesitate voicing displeasure with the situation.

"Hey, y'all know that nasty wind blew us all over. We lucky jus' to be together. We even kept old Pee Wee with us. He's the oldest and a green heron. We blue birds really got NOTHING to complain about, Chuck. You know, ever time we get a little off course, miss a meal, or a game, you gots to be whinin.'"

"What we need now is to work with Dmitri, not fight him. The days when he was the best imported center in the Heron Leagues are long gone. Now he's doin' his best to be a coach. If it weren't for that dude, we wouldn't have become so famous. I mean, seriously, Charles, who hasn't heard of the Blue Heron Globetrotters?"

"Okay, okay, Reggie, take a deep breath, brother. You may be right on, but I can make a sure bet those two little, furry critters starin' at us over there, hidin' in da reeds ain't never heard of no Heron Globetrotters, have you boys?"

The voice emanating from out back directed attention of the entire team towards a cowering chubby ball of fur and vibrating, little friend. Poor Murray, even his ears quivered at the sight of a dozen pointed beaks. He gripped his kippah with all his strength. Add twenty-four sharper, glaring silver eyes aimed at his tantalizing fat, fuzzy body. Terror, Murray's only thought, was of being an appetizer for the feathery foes. Mr. Mizithra stepped up, his traveling companion in no shape to speak.

"I do believe we may be of some assistance, gentlemen. That is, provided you keep the beaks to yourselves. My name is Mr. Mizithra. My timid associate is Murray. This is our forest, our home. We chanced to pass by and detected your quandary. We did not mean to eavesdrop, but overheard your plight. Where, pray tell, are you seeking?".

"Yo, Mr. Mizithra? Man, you birds catch that killer, funky name? Yeah, listen, my name is Reggie. The small, green guy with the big beak is Pee Wee. This fellow here is so sweet, so bad, so fine, we just calls him, Air. The grumpy one is Charles. One more of these crazy ball players you got to meet. Larry, where is dat wacky bird?"

At this, a large, lighter hued Great blue heron crashed through his team mates, raising a feathered wing, one head feather askew and bent. From simple observation, he was clearly a country bird. He looked like he fell out of a hay loft and landed in front of them.

He was the same species, just a tad bit off in comparison to the rest. He stumbled forward, a goober looking specimen if there ever was, staring directly at Mr. Mizithra with great concern. It was evident by his expression the fuzzy mass presented a real challenge to his bird brain.

"Hey, you, the big dude, what the heck are you? You are like the biggest, baddest, coolest looking squirrel ever. Dang, man, you are huge."

"Larry, is it? Yes, Mr. Bird, I am not a squirrel, but a marmot. In fact, sir, to be very specific, an Olympic Marmot. These hills, valleys, the big meadow, this pond are our native homeland. These high peaks and valleys are our home. We are about during a tiny window of time above ground to explore. We are raconteurs, madcaps, vagabonds, travelers."

"If you large fellows want to find somewhere or place, simply ask us. Believe me, the presence of beasts of your tremendous size intimidates every living creature both above and below the surface. Those beaks of yours are amazing and terrifying weapons."

"Hey, Dmitri, Dmitri, Dmitri, you dippy, dumb dodo. Yo bro, these fuzzy wuzzy dudes say they can help find where you is and how to get to the game. If we miss tonight, we gonna forfeit not only the game, but your sorry egret butt."

Charles' voice was the loudest of the friendly teammates. At his voluminous voice, Dmitri fanned off the dragonflies, wading to the scene at the edge of the pond. The rodents' came into view. He immediately identified the large form of Mr. Mizithra, then the shaking teeny mass of Murray, poor fellow. Curious, he plowed through the water towards our traveling companions. He was a sight to behold. Murray was no less frightened. Dmitri approached, churning water as he went.

"Yo, Dmitri, these cool cats say they know what's what, you nut. So, get off yo' butt and give 'em an audition."

"Oy veh, vat a schtorm ve vas ink. It vas too much for us, so ve set down in zis verkakta vatering hole. Nice, tasty little fishies, though, I must say. My team here, ze Great Blue Heron Globetrotters, have a game tonight und ve must make it or forfeit ze contest. If ve do, oy veh."

Dmitri began stomping out a pace, exploiting the water to roil. This caused several small water critters to flee their hiding places. Both green herons immediately made stabs in the water, each retrieving a prize. Fish was instantly on the carte du jour.

A couple of the bigger, blue oafs were successful, but poor Charles and Larry came up empty with their attempts. Larry reeled back from a failed go with a face full of fresh water seaweed. The rest of the team fell off, laughing, slapping each other on the back, making fun of his predicament. He was quite the hayseed.

"Where are you heading?"

"You, zere, you vearing a kippah, are you of ze Hebrew faith?"

"Ja, und you, sir?"

"It is so. Vat finds a fine Jewish lad up here in zis high country?"

"My people, ve are woodrats or pack rats. Ve have been here for generations, living zat way by ze lumber camp. Our people live entirely off of nature. Ve do not rely on ze Thumbs for anything unless ve must. Zen ve scavenge, taking only vat ve need. Ve are not Hanger-ons!"

This he spat with rather strong, negative emphasis. Hanger-ons were animals neither tame, domesticated, or fully wild. Instead, they lived in limbo between the world of man and animal. Dogs, cats, surrendered to the ways of the Thumbs long ago.

Hanger-ons, in contrast, lived in nature, but very close to the homes of humans. Extra food and shelter generously provided by caring homo sapiens was their edge in life. However, there was a price to pay. If the Thumbs split the scene, Hanger-ons had to again fend for themselves fully. Many perished as a result having lost the art of survival in nature.

If you willingly surrender freedom of the wild for the comfort man shares, the penalty is a loss for the world of animals. Out there is a true world of survival. It is why many like the notion of being a little bit a part of the civilization of the upright man.

However, what freedoms you decide to give away can never be retrieved. This is the reason people are vigilant about their beliefs. If you consider a thing good, essential for success, you will not be shaken from such conviction. The same is true for those living in the forests, deserts, mountains, rivers, and oceans.

"Ve are lookink for, vat is zis schtupid verkakta town? Oh ja, here it is, Sequim? Ve are playink againscht the Sequim Schtormchaser Seagulls. Ve vill slaughter zose poor schlobs. Face it, a seagull is fantastic at sea, but zey can't play ball vorth a tinker's darn."

"Sure, you need to turn around und head due north. You see the schpot up zere vere zose two big pines split? If you head up zere, zen veer to the right, you vill see a city. Ze ocean und Sequim is beyond."

"Ze city has a nice park at ze far western edge, Dungeness Park. Zat is vere ve is playink. Too bad you can't fly. No vay you could schlep back up here. Ha-ha."

Dmitri signaled a handy red dragonfly, who, receiving instructions, buzzed off north, between the pines, veering to the right, disappearing in a blaze of speed, as Murray instructed. The solution to the situation was close at hand. Murray knew just the thing to pass the time.

It would be to his benefit. The players could sit back and relax. His dear chum would get a prized chance to do what he did best: tell a tale. Murray signaled to Mr. Mizithra. The audience was ready. It was time for a sweet story to an eager audience.

"It will be a while friends. Please, relax. I am a yarn spinner, a story teller. Do you fellas like a good tall tale? Yes, yes, okay, well, here is an old marmot legend to help you pass the time. It will help you to learn more about me and my kind. To share it with you is my personal delight."

Chapter 6 The Madcap Marmot

The siege of herons huddled around Mr. Mizithra, eagerly expecting this curious, chubby rodent's heroic tale. They had nothing else to do and he was an interesting, rather odd fellow. They were not easily impressed. His story better be something unique or they would give him a hard time. He was flocked on all sides by a sea of feathers. Murray rested under the brush of a nearby rock, off the water's edge. Here he could monitor the situation, chew on water bulbs, and listen to a magical tale spew forth with relish and delight.

As an honest to goodness yarn spinner, Mr. Mizithra had to get the assembly prepared for what was to come. He built anticipation. The best preamble was a gathering stirring with excitement. The opposite meant a listless mob virtually impossible to impress. Arm movements were coordinated for maximum effect. This was no sideshow, it was main stage material. He held a captive audience, moving slowly to build a little head of steam. Finally settling in place, he slowly looked about, seeking the exact concluding touch to this performance. Mr. Mizithra grabbed a handy clover sprig, placed it ever so gingerly in the corner of his mouth, and began.

"Years ago, a mighty, outstanding, brave marmot of distinction, Mr. Pilsner, roamed these hills. He lived a very high life in the mountains, full of adventure, as do I and my good friend, Murray."

"You would not think it to see us, but marmots are the madcaps of the rodent world. Oh yes, I see the doubt in your avid avian eyes. Yet, tis true. Why, we have been known to have parties for days on end in winter. With no where to go and nine months to kill, it can get a little wacky down there. We don't just snooze the wintertime away, you know."

"Well, my feathery friends, it seems Mr. Pilsner, and best friends, Mr. Beer Nuts and Mr. Poo Poo Platter, set out for walk about. They were intent on going around the pond to map it. Being a daredevil, Mr. Pilsner took the lead and was soon out of sight. His fellow travelers paused to eat an entire lovely patch of summer flowers."

"Without warning, the sky above him began to darken. As marmots, we have a sixth sense about being watched. The hairs on both layers of fur went up. Trouble lurked close at hand. The shadow was not from clouds over the sun, rather an animal ready to pounce!"

The birds leaned in, intense eyes glaring in fear. No longer were they concerned if it was going to be a good story, they had to know about the nasty beastie looming above Mr. Pilsner. Feathers fluttered in vapid anticipation. What could it be? What would happen to Mr. Pilsner? Where were his friends?

Already, the wily Mr. Mizithra held them in the palm of his paw. It was simply a matter of drawing out the story in such a fashion to hold attention to the bitter end. Then he would reel them in like a pro.

Now one thing you simply must understand about herons, they are wading birds. They move very slowly on the ground and do not like to be disturbed. All animals: flying, walking, crawling, or swimming fear this very thing: some unseen force, a dark, hovering entity ready eat them at a moment's notice. Mr. Mizithra spelled out a worst case scenario. It takes a good bit of flapping before a blue heron can get airborne. If something were lingering directly over the top of one of them: curtains.

The other thing is, since herons are so slow moving, they developed the ability to multitask. A heron can stand for a long, long time, patiently waiting for a meal to move in perfect pouncing position. They still listen intently to anything said. In this method, they are able to strike at elusive little fish with fair success. This is unique to the species. They flourish in the Pacific Northwest. As for reactions, well wait to the end, you will see.

With a tale instantly engaging, his wide eyed audience was caught in rapt silence, urging more, yet dreading what was to come. Spellbound, he paused to chomp down on some greens. He plucked a bit more and crammed a hunk in the side of his gob. He seemed to fumble recalling his place. He muttered a bit, then nodded, finally sorting out where he was.

"Oh, come on, Miz, come on, man. What is it?"

"No, no, what is it?"

"You gots to tell us, Miz. Come on, Miz, baby. Spill, baby, spill."

"Okay, the shadow, yes, the dark, slow moving specter crept across the air above him, freezing Mr. Pilsner in his tracks. Turning his head, ever so slowly mind you, yes, ever, ever so very slowly. Fear engulfed him. You could see the hairs of his fur move. Terror was soon on it's heels."

Mr. Mizithra again paused, a genuine look of terror ridden angst on his rosy rodent mug. He had them hanging by a heron claw, and knew it. He chomped down on the clover, swallowed hard, and grabbed a handy sprig. He took a gander at the new clump, plucking an unwanted twig from the midst of the wad. He shoved the mess into an awaiting mug, chomping a bit before continuing.

"Sorry to keep eating fellas, but I have hibernation ahead, so… Just as he was able to see a bit of deep, reddish fur out of the corner of his brown, quivering eyes, the beast spoke down to him."

"You there, my sweet, succulent, supper. My hunger is extreme. You will make a fine meal. I know you have two others with you. I want them, as well," said the lady fox.

Mr. Mizithra's voice eerily mimicked the vixen.

"I am a mother and need to feed my young. You are going to help my quest, fat little rodent. You will make a fine appetizer for the others to complete the meal. Now, if you will just hold still a moment, it will be over. You will not worry any more once you fill my belly."

"With that, the hovering fox pounced upon her entrapped quarry.", Mr. Mizithra's voice rising to a crescendo, stopping short for effect.

They collectively gasped. Beaks leaned in ever closer, fear causing nervous clacking. He saw terror in their orbs. Hardened, urban feathery fellows fell apart in fear. Sharp eyes, focusing even more, the birds shifted to and fro in expectation, feathers shuddering. They awaited each and every word wanting to learn what happened next.

"Now everyone knows the trouble with a vixen is they are not very good at grabbing large prey easily. They have a small mouth. Mr. Pilsner did the only thing to do to keep from being eaten."

At this moment, Mr. Mizithra took intermission, pulling the sprig from his mouth. He took a bite from it, grabbing another chewing away. He gazed lazily at the remnants, sampling the unique flavor of the meal. He paused to burp, apologizing for rudeness. The flight of herons took a collective deep breath, terror drenching the air. The normally stone cold, still, feather laden statues were fidgeting about. The eyes held real fear.

"WHAT, WHAT, WHAT?", they squawked in unison.

"He kicked up his hind legs. With all the power a marmot could muster, he thrust with exact precision. From the thigh bone to the knee bone, to ankle bone to foot bones, Mr. Pilsner put every bit of energy into it. Marmots are tremendous diggers. Our rear limbs are very powerful. And, might I add, no animal can out wrestle Olympic marmots. One truly good reason is the power of our limbs."

"The stunned fox dove down. The impact of the rear paws pushing backwards struck. The collision was such the impact met the hurdling vixen at her chest, throwing her back, allowing the marmot to scramble in a hole. Not wasting a second, the fox made for the opening, shoving her snout in, exclaiming, 'You are mine! You are mine! You are mine!'"

"Mr. Pilsner, had no where to go. Trapped, it was only a matter of time before the hungry witch would reach him. Fear held on hard. Escape was not possible. Were he to begin digging, the huntress would catch him soon enough."

Adding a choice dramatic pause to swallow a mouthful of clover and long sip of water, the herons stirred, moving violently. Again he begged forgiveness. Eating made him thirsty. He had to pause for a moment to refresh his throat. For a bird who is used to standing perfectly still for hours, it was hilarious to see them antsy and aghast.

Murray, chomping on a lily bulb, noted with glee his companion's ability to hold an audience totally rapt.

"Now, gentlemen, please, calm down.", completely relishing every glorious second, "I now recall it precisely this way: as the fox dug into the soil chasing her meal, Mr. Pilsner came up with a plan."

"You know, dear lady, as you are digging away after me, my friends are approaching. If you will but wait a moment or two, they will appear. You can have your whole meal, not just a nibble with me."

"Of course, this was an odd thing, which he was. It stands to reason. Any other critter would simply surrender and wait to be eaten. But not this marvelous madcap marmot. He persuaded the busy burrowing vixen to pause. Ceasing mining, she pulled out a dirt covered muzzle. She took a turn to see if the others were approaching, oblivious to what was in front of them."

"At this precise second, Mr. Pilsner did a truly impetuous thing, he ran straight for the foolish fox's chest. She let out a puff as he hit her fully. He made for the open land and a hole he had spied. It only took a moment. She made for him, letting out a mighty, bellowing screech."

"The whole forest heard her yell, alerting Mr. Beer Nuts and Mr. Poo Poo Platter, not to say every other living thing. He got to the entrance of the hole. He kicked again, as she arrived. The propulsion of the wallop not only took the air out of her lungs, but knocked her against a tree. Life meant a fight. He was not going to meekly lay down."

"At this same instant, a most amazing thing occurred," pausing for another, slower sip of H2O, holding them spellbound, "a skunk came over the knoll. Seeing the situation, she raised her tail at the vixen. The fox froze, blanketed in dirt and shame."

He nodded at the quality of the water's flavor. The stuff was simply marvelous. Soliciting an opinion from Murray and playing out the torture, Mr. Mizithra collected himself. Oh, yes, back to the stories fellas.

"Knowing there was nothing more she could do, the fallen, fervid fox cowered away into the forest, dirty, injured, and beaten, still hungry, never to bother them again. The skunk, one Miss Mulberry, turned to Mr. Pilsner, saying, "You, marmot, you are an impetuous fellow. Any other rodent I know would surrender and become dinner. Kudos to you, sir. You are a marmot of distinction. What, pray tell, is your name?"

One thing about skunks, they were the most formal of all beasts. It is merely the strong odor causing the mystery. Most other animals avoid them. If you can get past the stench, they are quite a decent, cultured bunch. This was a true lady he now met. Like all female skunks, she had an air of royalty, a daintiness, making her one very fascinating polecat.

"Oh, well, please allow me a moment, madam, to collect myself. My name is Mr. Pilsner, the fellows with me are Mr. Beer Nuts and Mr. Poo Poo Platter. We were going around this pond to map it for our colony."

"Were it not for you, gracious lady, that hungry fox would have me for her supper!", raising his voice in fear and terror, "Therefore, I owe you and yours anything, anytime, anywhere. For generations to come, my colony will always be in your debt, dear heroine. Your lovely name will be regaled by marmots from this day. LONG LIVE MISS MULBERRY!"

"With this, he bowed low, raising his right paw to the side of his head in a salute of honor. He arose. His comrades arrived. They witnessed the scene. Mr. Pilsner recounted the episode. Their response was awe, thanking new friend, Miss Mulberry. She was a lovely sow, joining as they made the way around the tarn in which you stand. For you see, my fine, feathery friends, that same foul, fang-laden fox prowls around the area to this very day!"

Mr. Mizithra instantly slapped water with his left paw. The herons, blue and green fell out. They squawked a combination of terror and squealing delight. It was a marvelous little fable. They appreciated Mr. Mizithra's telling it. He mesmerized, each pause accumulating interest. As a seasoned yarn spinner, he successfully made the story memorable, colorful, and instructive. He landed a whopper this time. Heh-heh.

The tension broke when Dmitri called them over. While telling the tale, coach was setting up a net on a handy limb. Herons began stuffing balls in the hoop, passing in a flowing, beautiful way, bouncing them off the water, having a good old time. They loved to play and were excellent. It was a marvelous new thing to watch for our furry twosome. They took the ball in one wing, passed to another player. The same one passed it back to the other wing. Then the sphere goes swish in the basket. Fluidity.

Within minutes, a dashing red dragonfly zipped up to Dmitri and gave the message. He nodded, waving them over. One could plainly see he gave very good news. Practice was over. It was time to head for Sequim to play the game. They flapped, nodding feathery tetes in total delight.

"Ve vant to thank you fellas for ze help. Me und ze boys vould invite you to ze game tonight against the Sequim Schtorm-chasers, but you can't fly. Ze danged seagull may be schmall, but zey are fast und smart as a vip."

One of the green herons approached Mr. Mizithra. While he had seen this green fellow, Mr. Mizithra noted he did not have anything to say. Then, who could get a word in edgewise with Charles bellowing at every opportunity? He liked to suck all the oxygen out of the room. The green one was quieter than the others. An excellent player, he had a way of sailing through the air to the basket with such ease, it was as if he were gliding across the sky.

"Mr. Mizithra, before we go, me and the boys want to give you this as a token of your scary story. You tell it like no one we ever heard. Man, you cool, you are cool, cooler, the coolest. They call me, Clyde the Glide because of the way I fly through the air. And now, we gonna glide on outta here. You keep safe, you and Murray. Thanks again, marmot."

The whole flock flapped frantically in unison, in a moment, gone. Dmitri led them over the trees north, then disappeared fading right. Soon they would find Sequim and have a fine time walloping the hapless seagulls. Mr. Mizithra found a Blue Heron Globetrotters' medallion in his paws. He grinned from one ear to the torn left one.

Chapter 7 Chance Encounter

Murray shook off the episode. By averting attention, he looked ahead. The bright day waned, shadows filling the trees. They needed to find safe shelter for the night. The sun was warm, warm, warm, but evidence was clear: evening would bring wind, water, and worry. Being on the road has uncertainty. Constant guard took it's toll on the energy level. Mr. Mizithra got a handful of violets, delightfully chewing away.

The Olympic Mountains are notorious for instant changing weather. Up here, nature is boss. No one makes it do anything other than what it wants. The warning signs were sensed, seen, and smelled. An animal is very much in touch with nature. It is their only world. For man, it is an escape from civilization. No human can ever be a wild animal. Thumbs eliminated the ability long ago. Try as they might, the detachment caused an irreparable schism.

The pond held great intrigue. Green-backed barn swallows fleeting along water's surface, nabbed insects as they swept. Red breasted nuthatches dove elusively to ambling bugs clinging to camouflage against a tree. Mountain chickadees flitted, seeking succulent seeds for supper.

Frogs croaked, spying flies for a fast feast. Ducks of various types took vacation in these mountains, coming and going as families flocked together. This was a lively, awe inspiring world our trailblazers traversed. The air was full of life. The warmth of the sun embraced a happy mood. Shelter was on the bill of fare.

"Looking for a place for the nighttime, aren't you two?", appeared a subtle, sudden voice out of thin, alpine air.

The fellows froze in place, not daring to so much as turn a head. Murray and Mr. Mizithra's eyes slowly scanned the immediate vicinity for the source of the sound. Their orbs collectively met on a small, blue green feathered figure with an abrupt, very sharp beak and plumage in a short crop, like a Mohawk, on top; he was the punk rocker of the wilderness. Small, he was difficult to see. He sat proudly in front of them, perched on the twig.

He had a noble appearance. A white patch sat in front of a pair of sharp red/black eyes. A white band adorned his neck as well. He had a reddish brown swath for a waistcoat. A white chest and black feet, with a banded tail finished this fine fellow off to a tee.

He was on a lone, branch off a dead little tree in the water. The eyes were locked on them. Since he was the only living thing in sight, they concluded this must be the individual hollering at them.

"You, yeah, you guys. Hey, yew-hew, there's no one else around but you. There aren't any bad dudes here right now. You are safe. I asked you if you are looking for somewhere to stay for the night? The weather is turning as you can feel."

"I'm Boston Charlie, a Banded Kingfisher. We are the true Natives of this land. We know every centimeter, rock, tree, and living thing. The Kingfisher may be just a bird who eats tasty little fishes, but our ancient ways are solely of the forest. We are the birds who have been here the longest. Our ancestors came here eons and eons ago. We have nothing whatsoever to do with the Thumbs, either."

"Some time back, I saw you with those Herons. A big nuisance if you ask me, but there you go. They come up to my mountain and don't even pay homage to our elders. Ah, the young, no respect for the past, roots and history."

With that, he turned, holding up his right wing as a sort of apology, hurling a glob of bones and gunk from inside his belly. It plunked to the ground with a thud and lay. Both rodents were taken aback, disgusted by the sight of a pile of Kingfisher barf. In unison they shuddered in revolt. Boston Charlie saw the horror and did his best to explain the situation.

"Sorry fellas, we don't digest the bones so I gotta hack 'em up. You need a place to stay. I know just the spot for rodents such as yourself. You ever been this way, no. Okay, stop nodding your head, woodrat. Do you see that big lump of wood? Yes, the big pile way over to the right."

"Go up to the front door and tell them Charlie sent you. They will know. Trust me, no one knows nothing about this pond more than old Boston Charlie. Everyone knows this fine feathered friend."

Suddenly, he quit speaking, attention distracted by movement in the water below. He dove in the water, disappearing. In a flash, he was on the branch again sporting a squirming little fishy. The whole process took mere seconds. The boys were impressed by his compact style. To nab a fish with such precision was a thing to behold. Again they learned something new this day and were glad. It made the traveling worthwhile.

"Gotta get this meal home to my wife and kiddies," muttering out of the side of his beak, clenching down on the evening's flopping repast.

"Take care, look out above and each side before you get along. You are nice boys, don't get in trouble. This is a cruel and unforgiving place."

He let out a call like a rapid fire, staccato machine gun, exploding from his perch towards the woods some fifty meters away. Swooping low, the flight pattern was a handful of centimeters above the water. The speed was incredible. His meal gave freedom a wriggly whirl, to no avail. He waved a wing wildly in their general direction, disappearing in the thick greenery with food for his family.

This was one wild character. Still, they were pleased he took the time to help them. The clouds were beginning to swirl, going from fluffy white to a purplish black. They were laden with buckets of water which would soon fall as rain, pelting the rainforest as usual.

With no other knowledge or advice, they surmised Boston Charlie's words as honest. At some point they had to trust someone. It may as well be this vibrant bird. He had no reason to trick them. The fellow seemed to have good intentions. They had little time to dig a new temporary home for the evening. The soggy clouds began to look more ominous with every step. What could it hurt? With little else to go on, they decided to accept Boston Charlie's invitation. It had been a long, tiring day. They made for the pile of logs ahead and uncertainty.

Mr. Mizithra did what marmots do. He paused, rearing up on his hind legs, using the thick, fuzzy tail as the third branch of a tripod. The first rule in the forest is caution, second is safety, and third is to double check the first two several more times. Keen, well-seasoned eyes focused on every detail before, around, and above. His black nose was working overtime. This was virgin country, an unknown place a new friend deemed safe. Rodents tend to stay with rodents, save those vile voles who betray their own kind for a true enemy: the raccoon.

Omnipresent night was fast approaching. Thickening clouds were churning as butter, ever darker gray, bursting at the seams with wet stuff. The knowledge of what was coming prompted them to move forward. Rain would soon fall in sheets. Wind would whip mercilessly upon them were they unable to find shelter from the storm. These facts added up swiftly, making options minimal.

Throwing caution to the wind, they made for the pile. A small, barely visible path of thickly matted wood chips led to a shadowy spot perceived to be a turn. Upon reaching the elbow, they peered around the corner, Murray below and Mr. Mizithra above, to see what they could see.

What was viewed took collective breaths away. The tight corridor opened to a diminutive courtyard. A dim beam of sunlight reflected above the door to an entrance at the end of the path, flickering due to windswept clouds.

Logs, limbs, odd bits of bark, and leaves jammed against the pile to form a stalwart wall a good three meters high.

A small entry way beckoned them forward. A large, heavy wooden door ended the pathway. The threshold was massive, made of fir with a giant Z of cedar holding it in place. A huge cedar handle was bolted on the door. Above the humble, wooden entrance a sign, letters chewed by beavers: Beaver Blues Bungalow. Caution being the word of the day, they crept tentatively closer to the ominous, approach. In the faint light they could make out a handle on the left side of the door about halfway up.

"Excuse me, fellas, comin' through", hurled a hale, hearty voice from behind.

They leaped in the air, grabbing each other in terror. Being keyed up, it was a natural reaction to noise from behind. In a flash, an animal accompanied the deep voice. As they landed, a large, hairy, brown form nudged past, carrying a keg on it's shoulder. He was going through the door. The burnt sienna beast held the cask while tugging on the handle. He nodded towards the two stunned rodents.

"Come on in if you is, cuz the wind is messin' up everything inside the lodge. It's too nasty to be dillydallyin' about lingerin' by the door. What, you wants an invitation? Okay, please, whoever you is, come. Welcome to Buford and Beulah's Beaver Blues Bungalow, Wayside Inn, and Eatery. We da home of the best danged blues music you ever heard"!

With that excellent declaration, the mahogany mass, turned, toting the barrel out of sight through another door left, heavy tail dragging playfully behind. Not bothering to introduce himself, they stood alone and nervous. As the door slowly closed, they were in the midst of a large, dark room. Something else quickly riled the senses. They were not alone. A number of other brown, furry creatures stared at them.

"Who you? What you want here, critters?", came an utterance from the smoky din. It was a strong, deep voice, full and resonant.

"We are travelers, heading for the White Mountain yonder above this pond, past the rapids, and down through the deep pass beyond. We met a Banded Kingfisher who said we could find lodging. His name was Charlie, Boston Charlie.", Mr. Mizithra's words shaky from honest fear.

A few laughs arose from the shady, unseen mass.

"Boston Charlie, well, well, well, it has been a long time since we done heard dat name, huh, folks? Yeah, we knows Charlie, but who is you an' what you want"?

Appearing immediately from the blackness of hidden shadows came a hairy face, ominous, and stern. Shadowy, large eyes narrowed to observe them fully. The mouth was set in a most serious pose. A body next, covered with dense brown fur. This was the same type animal as the one who ushered them inside.

He did not seem quite as friendly. In fact, he seemed downright suspicious and cautious. This was a tense moment. If cast from this place, they would need to locate a safe spot for the evening rather quickly. The swirling sound of violent wind outside increased by the moment.

"My name is Mr. Mizithra. This is my friend, Murray. Boston Charlie said you might have quarters. The storm is upon us. We have no where to go. We are forced to rely on the hospitality of strangers to help our way".

"Okay, I knows what Murray is, a woodrat, but what in da wide, wild world of critters is you"?

"Oh, well, I, sir, am an Olympic Marmot. We live on Hurricane Ridge," he said pointing behind towards the ridge above, "in our colony. Murray lives up the road. We are adventure seekers. By chance we find our old bodies here. You, I take it, are beavers? I never met any of your type of rodent. We have heard of the great work you do with dams".

"Yeah, man, dis is OUR pond. My wife's grandpappy made dis dam wid her grandmammy, her daddy an' his brothers an' sisters. Dis is our dam. We made it. Me an' mine, we tends to it to dis very day".

"At night, at night, man, we get down. We jam, we rock, we go, baby, boogie woogie. Dis is da best danged blues club in da Olympic Mountain Range. By da look on yo' face, dis is news. You kiddin' me, brother, you never heard o' our club? Hey, Beulah, we gots to get mo' posters out so da hayseeds can hear about our club".

"Old man, you get off yo' lazy tail an' make up a batch of stupid posters. I ain't gots time fo' hogwash. I gots me a club to run".

Beulah.

"Now, you boys, Murray an' Mr. what yo' name, pup? Mizithra, what in da sweet world kind of a name is dat? You boys, is you hongry? Oh, now don't you fret none mama Beulah gonna get you up a nice plate o' greens. I think I may have a few choice bugs fo' Murray. You a cutie pie, you little woodsy rat, you.", she said with a bit of sassy sugar in her voice.

Without so much as another word, shaking her wide tail, she disappeared to the back through swinging doors. They looked about. Eyes adjusted to the din. The whole place was clearer. It was warm and quiet. Safety is massive for anyone on the road. They were no exception.

The lodge had a long, straight bar. Behind the bar on the back wall, an assortment of wood, all types, outer casings prominently displayed for customer's to scan for the precise bark to chew for the evening. This was truly an exclusive club for the beaver set. When it came to bark, beavers knew their stuff. They may well be nature's engineers, but a major reason is their affinity for bark. Yummy stuff to them, not so much to we human beings. Bark for supper does not prompt the culinary appetite.

Left, through the haze, a stage, drums to the back. Saxophone and trumpet stands lay smack dab in the middle. They were propped up by stands made of branches and twigs. A piano sat to the far left. A huge stand up bass stood on the distant right. A handful of guitars aligned the entire back area.

"Here's a big mess of greens fo' you, you big vegetarian. Fo' you, my little sweet ratty boy a nice, juicy water root an' a lovely assortment o' bugs. Enjoy, boys, it's always on da house. Now you tell mama if y'all needs any mo' cuz we gots a whole kitchen full jus' fo' you".

"Now, you knows, of course," Beulah said leaning in closer towards the duo, nuzzling closest to Murray, "Since we don't have no money, you gots to earn yo' keep. Here at da Beaver Blues Bungalow, they is only one way to pay yo' way. You gots to get up on stage an' play wid the band or sing or somethin".

"Madam, we marmots, well, unfortunately, are not a musical bunch. All we do is whistle. It's just to warn everybody an eagle or something else nasty is incoming. You really don't want to hear the sound because it is rather loud. Ear splitting. Your ears would never be the same".

"I play a little saxophone.", offered a somber Murray, "Back home, at night ve sit around und make music. Ve do it just to entertain ourselves und ze neighbors. Mama plays ze concertina und Gramma fiddles ze fiddle. Zis is not much in ze vay of talent. If zat is okay mit you, zen I am happy to oblige. If you can use my veak playink, I can pay my vay".

"Und don't let Mr. Mizithra fool you, he can't play a note, that's true, und he can't sing a lick, but he can veave a fabulous fable in such a vay you vill hang on every single vord. He had zose herons ve saw today rapt in terror mit ze schtory he told. He can do ze same for you.", Murray's now stronger voice boasting confidence in his chum's talents.

"Oh lawdy, mama. Alright, boys, you chow down. You wants mo', you gots it. But you, my sweet little ratty baby, you gots to get up on stage tonight. You, Mizzy, you can tell yo' story later. Right now, I gots to get ready fo' tonight. We gonna be reelin' an' rockin', rollin' an' shakin' da very timbers of da dam until she bleeds boogie woogie blues".

With that, Beulah disappeared once again to the back, double doors sashaying and swinging behind. The boys greedily devoured the bowl of greens, vanishing in their gullets. This was quickly replaced by another brimming with clover, flowers, and all kinds of juicy eats.

The waiter was another, smaller beaver with beaming black eyes, a subtle, savvy smile, and red bandana around the top of his head. He was the second son, Tobias, a bass player. He sang back up for the band. He loved to thump that stand up bass.

To add a special twist, Tobias tapped his tail in perfect time to accompany playing. This truly rare talent, Tobias explained, was known to be accomplished by only a handful of beavers. It is sort of like rubbing your belly and scratching your head at the same time, not everyone can do it. It provided a double bass sound, giving the band a richer, deeper tone. Of course, thumping shook the stage constantly. It was named the "Tobias Effect", very unpopular with the aquatic portion of the lake.

Eldest son, who greeted them with the keg, Homer, loved to plink away on the piano, his four digits known to tinkle the ivory with pure soul. Big sister, Bette, was a genuine glittering, gifted, guitar goddess. Youngest brother, Henry played a mighty, mean harmonica. Their family band was known throughout the land, so he said. The new friends would witness this fact tonight.

"You fellas gonna need a place to sack out fo' da night. Come wid me when you is done wid yo' meal. Man, Murray, how can you eat dem damn bugs, yuck. Dang, man, you sick, bro", Tobias said gnawing on a hunk of rare, aged alder bark stashed in his headband.

The delightfully, delectable meal consumed, they hailed Tobias. He led the weary travelers down a long, dark hallway right off of the bar. It wound around under the dam. You could feel water leak a bit here and there. Tobias stopped at these occasional drips. He placed a small mossy flag in the hole. He carried a handful of them as they waddled down the hall. He saw curious looks on their faces.

"Dis is to mark da leaks fo' da repair crews. We on top o' dis, my friends, don't you worry a hair on either layer of fur on yo' head. We has to fix dis thing all da time. It is da way o' water. Okay, here you is, a nice room fo' two. Murray, you hop up on da top bunk. Miz, curl up comfy cozy on dis futon below. See you fellas tonight. I wants to hear dat sax wail, baby, wail".

"What an interesting place we chanced to find. I guess we have to thank Boston Charlie for our good fortune. The fresh storm outside is whipping up rather strongly. I fear we may have to stay here for a couple of days. It would be a grand idea to fatten up anyway, for me. Their greens are the very best".

"Yes, yes, yes, my plate vas ze best. Zey gave me five kinds of bugs. It vas like Purim. Vat a blessing. Und now, I get to play ze saxophone. Ve only play at home. I have no idea if zey vill like ze vay I blow. Ve only play at ze ranger's shed. A moment ago, ve vere on ze road. Schtill, it is exciting".

Weary, they lay down for a much needed nap. This day fulfilled the reason for the journey. They learned new things, saw a world come alive in front of their eyes, discovering new lands. Having eaten a hearty meal and traveled a long way, rest was paramount. Within moments, both were sound asleep. The storm began in earnest outside. Slight smiles smeared furry, fuzzy faces and taut, plump bellies.

Chapter 8 A Wild Night

Murray stirred. Truth be known he was excited at the prospect of playing saxophone for a crowd. This would be a new experience. His only audience was usually forest creatures stopping by to lend an ear. Ernie was a real music fan, for example. Then, a cornucopia (a group of slugs) of Pacific Northwest banana slugs gathered for a woodrat concert is not exactly a thrill, but you already knew this.

He noted most times other animals were impressed and would clap. Occasionally some tossed food or welcome items plucked from the Thumbs into an upturned, old tin cup. They played in front of the cabin. Of course, slugs cannot clap, just gurgle and nod slimy heads in approval.

Hurdling from the upper bunk, Murray playfully poked his catatonic compadre in the colon. He coughed loudly to disturb the sweet, big galoot from his satisfying slumber. Friends can always poke a bit of fun at each other. These two seldom missed such an opportunity. He was filled with the exhilaration of the journey. So far, a real blast. He was going on stage with real musicians to play sax. He could hardly contain the excitement. He wanted to get to the main bar as soon as possible.

Yawning four massive, yellow front teeth jutting from powerful jaws, Mr. Mizithra nodded, acknowledging his friend. He was awake. He shook a hairy head to clear his thoughts, rolling off the bed to the floor. He was all smiles, hoisting his friend in the air with joy, swinging him around twice.

Murray, his arms raised, squealed in sheer delight. As an adventurer, it soon becomes very transparent: food, water, lodging, and safety are of dire need. Once these were secured, the sky was the limit. To find such a glorious oasis forced two seasoned travelers to explode in true feeling. The chance meeting with the Kingfisher reaped instant benefits. The trip to this point was perfect.

Within moments they were down the hall, to the main room of the lodge. Some muskrats were working on a leak as they passed by. The main hall was massive. The place was literally jumping. Two fuzzy Douglas squirrels were hopping a jig on the dance floor. A pair of chipmunks were doing the same. Wriggling fuzzy tails to the beat, they created a beautiful swirl of brown, white, and black fur moving to the rhythm. Two Stellar jays were doing the bird bop, consisting of nodding their head up and down towards their partner, then hopping up and down. Dumb, but they are birds. Hello.

On stage, Tobias waved from his bass. Sister Bette was kicking out a nice lick. Big brother Homer tickled the ivories. Mother Beulah was up front, belting out a low, powerful blues tune laden with pain and soul. A funky looking muskrat was on drums.

At the break, little brother Henry stepped into a blistering mouth organ solo. It reminded one of DeFord Bailey's train song. He brought the house down earning well earned, loud, raucous applause. Henry bowed, leaping off stage, broad tail a nice tripod for landing. He headed for the bar to help his papa.

The transients found a little spot out of the way, making a quick survey of the inhabitants. First, a table full of ducks quacking at the same time. They were vacationing from Oregon; there's just nothing quite like Oregon ducks. They feasted on greens fresh from the bottom of the pond. These were Wigeons. They seemed to be completely self-absorbed and rather noisy.

To one side, a pair of otters. A plate between them held various fish and a lovely, fat eel. They lived, engaged in deep discussion, paying no heed to the music or, anything around them. Each had a large pitcher of water along side. Occasionally, one would grab a ewer, to wet themselves. Otters like to stay damp. It is just more comfortable. The splashes seeped in cracks of the stone floor. This place was prepared for almost anything.

At the next table sat four well-seasoned beavers, heavily sampling the house wares. Each had a number of select, aged bark varieties before them. One would grab a hunk and chew away slowly, looking into space. Apparently, this action assisted in garnering additional flavors, only extracted from this peculiar behavior.

This accomplished, immediate arguments followed of taste, quality, and comparison of bark from days gone by. It was judged, analyzed with a lovely piece some years back, it's flavor still lingering on the palate and increasingly foggy mind. No one was willing to stop talking long enough to appreciate the other's findings. Only their opinion mattered. The seats, well worn, clearly were their familiar digs.

"This cottonwood is the most marvelous stuff, don't you completely agree, brother Abel"?

"Oh yes, brother Seth, I could not agree with you more. Why, it is as good as the cottonwood we chewed two years ago. Probably from the same stand of trees. This place always has the best bark".

"Man, you have to be crazy. This cottonwood is weak compared to the bark we chewed two years ago".

"There goes Cain, always taking the opposing side".

"Hush up, Lincoln, you think this is as good? Yeah, see beavers, even this stubborn old coot agrees. The bark two years ago was better. It had a fitter bite on the teeth, leaving a lingering luster for days. The flavor was ambrosia from the gods, manna from heaven, better than mother's milk".

"Okay, Cain, you know every time you talk we have to move because the beaver droppings pile up so fast. And quit talking about mother".

"Seth, get stuffed", nothing like siblings, huh?

The next table had a pair of Douglas squirrels, so obviously in love it makes you gag. The scene was pure romance. Paws intertwined, as did tails. They were staring into each other's eyes with pure love for each other, oblivious to any other activity. A pawful of aged pine nuts was their only interruption. One thing became very clear, this place was friendly and fun for all.

"Okay, Murray, time to get up an' blow dat sax."

"Tobias, I just play a bit back home und I heard you play. Maybe zere is something else I can do instead. Perhaps, vash ze dishes? Ze last thing I vant to do is embarrass myself in front of your friends."

"Shut up, get up, an' play, rat. Now!".

With discussion for the evening in the books, Murray was tossed on stage. Tobias leaped up, taking his spot at the bass. Henry handed Murray a saxophone. He bit at the reed a time or two, gave it a small blow and nodded. If he was going to have to do this thing, he may as well get it over. They could always retire to their room afterward, defeated, in humiliating shame. What fun. He snagged a deep breath and let it out as calmly as he could muster, supping on another lung load just to be sure. Strength, he thought, strength.

"Fellow, friendly, forest, furry folks, we gots a treat fo' y'all tonight. Let me tell you. All da way from da high country across da west side of da pond, way up on Hurricane Ridge, wid all his little lungs can blow; on saxophone: our favorite little woodrat, MURRAY!", Buford's booming voice bringing the house to it's collective paws in encouraging applause.

The house band began with a good, 12 bar blues riff, easy to follow. Bette hit a sweet groove. On drums, a muskrat named Twigs kept perfect time.

Homer slid in with a solid piano bounce. Tobias' bass thumped, tail bopping in unison. All Murray had to do was blow. Fear made his vocal chords seize up. He began to panic. Thoughts turned to dropping the sax and running for the exit screaming into the dark and stormy night. The poor mountain boy was gripped with stage fright.

"Blow now, rodent, blow.", came Tobias' urgent voice, not joking. To him, music was everything, never taken lightly.

Tossing fear aside, with all his little lungs could muster, Murray let fly. Fingers found levers, pushing into a chord, releasing and finding one more in line with the groove. His right hand moved quickly finding it's place, fingers ready to go. He was determined not to setback a chance. He wanted to play with real professionals. In a moment, he was blowing. It was awesome to simply play along with an actual band.

When it came time for a solo, he was feeling very good. He heaved a lung load into the reed with every gram of force available. Somewhere inside his wee, furry, fuzzy body erupted amazing power. The instrument took on a life of it's own. He wailed away better than ever. Almost stunned, Murray realized he was actually playing music with a real band. He was not bad, either, curious. Instantly one thought of Bobby Keys.

Adrenalin filled him from the toes to the top of his head. It was as if the sight of the crowd and talent of the players made him play better. Encouraging shouts and smiles did something deep inside. After all, a big ego was not an item he toted. Murray was, for everyone who knew him, a timid woodrat. His biggest achievement was accompanying Mr. Mizithra on their adventures.

Now, though, he was on fire, each note stoking burgeoning flames. He felt more excitement at this than anything he had ever done in life. As the number ended, the audience erupted in noise which shook the dam's timbers. In a flash, Buford flew on stage. He waved both paws high back and forth above, shaking his head in amazement. He was grinning from ear to ear. He turned and winked at Murray, then addressed the crowd.

"Well, well, well, my fine forest friends, what does we gots here? Oh my, oh me, oh my, now y'all knows an' I knows dat dis here club has had it's share of true blues rockers comin' in an' out fo'ever an' a day. But animals, I gots to say dat dis here woodrat has it all over dem. Oh lawdy mama, dis is a miracle.

Hand me my ax, we gonna go. Bette, hit it. We gonna soar, babies, we gonna fly like an eagle. Y'all best grab somethin' solid, cuz it's gonna get freaky funky, folks".

With that, he began. The club's patriarch, Buford played old style blues. It was swampy and deep. The guitar was as much a part of his body as his tail. He never bothered to look at frets. The music flowed for hours, the crowd growing as other animals heard the tale. Even with a storm, great music brought them out in the cold, wet night.

Murray did his best to adapt, receiving applause after each song. From time to time, other critters joined in with the band, playing or singing, making the night one of group action, partying, frivolity, and fun. A few errant dragonflies buzzed out to give the word to the herd as it were. The joint was jumping. They had a raging storm inside the club to equal the one outside. The one indoors had more thunder, though.

For his part, Murray's best buddy was all smiles. Mr. Mizithra had never seen his amigo so engaged unless on an adventure, discovering new places and critters. Murray discovered himself in the music shared with others. Beaver, otter, muskrat, or squirrel, it did not matter. His mouth lovingly embraced the reed. His fingers flew over the levers, finding notes and nuances to boggle the mind.

Mr. Mizithra gorged on the finest greens, the clover a particularly delicate flavor. A bouquet of flowers, the centerpiece, was encouraged to be completely consumed. In short, Mr. Mizithra was one happy marmot. The gang was sailing. He even danced with a sweet lady pocket gopher who, sashaying up, flicked her eyes, motioning with her little finger.

As the evening passed, the audience thickened and thinned, but for the most part stayed and grew. The place was packed to the brim with any forest critters who found their way to celebrate a resplendent, raucous, evening of music, dancing, food and fun. The atmosphere was one of true joy. They were having a hopping good time.

The four bark connoisseurs moved to shear delight: aged Quaking Aspen. They were beside themselves in exultation. This select, natural trunk covering is, to beavers, as lovely as the finest truffle, cut of beef, perfect artichoke, sweetest chocolate, finest champagne, or brand new Rolls. A fresh plate arrived, a beautiful, hearty sized, fresh hunk of Aspen bark for each, surrounded by water lily roots and sweet cattails. It was flown in special by swans. They salivated, fidgeting in delight at the prospect of taste, flavor, and joy. These were particular favorites for fat-tailed furry herbivores.

Of course, perfection can never last. The evening seemed to be truly magical, when the front door burst open. Five stark, ominous, terrifying figures moved in, shutting the door with great force. They stood, staring at the crowd in silent judgment. The storm's rain doused this quintet

They stood dripping before the stunned revelers. Everybody parted as the hushed crowd watched. They made angrily for the bar. Buford was still playing guitar on stage. They approached Beulah in a highly bellicose manner. This unsettled everyone in attendance. They moved up quickly to the bar, standing toe to toe with her in clear defiance. These were not happy beavers.

"Bark, now!", demanded the biggest beast of the brutish bunch, banging a soggy, hairy paw on the bar, "The Beaver Brown Boys is here an' we gonna get dis here party started. Joe Joe, Sammy Joe, go take da stage. Jimmy John, Harve, guard da danged door".

The quartet immediately responded. As if by plan, the brutes moved in a flash. The loud one stood commanding at the bar. He reached across the counter, grabbed Beulah by the throat, squeezed hard, causing her to suddenly go weak.

Buford and his offspring were trapped on stage, held by intruders brandishing large pine branches. Each limb had a couple of branches remaining at one end, gnawed down to a sharpened point. These were formidable weapons. The door was blocked by Jimmy John and Harve holding pine branches in their paws. The crowd froze in place, save one.

Displaying a burst of speed hitherto uncharted, Mr. Mizithra exploded from his spot hidden in the shadows of the limelight. Until now he enjoyed Murray's experience. The excitement of being in such a lively setting delighted him. The marmot set on the large one holding Beulah limply in his paw. Four massive front teeth latched on the extended arm.

He planted his feet and tail firmly on the floor. With a grab of arms to assist teeth, Mr. Mizithra pushed with all his might from mighty lower limbs and twisted. The result was the attacker's paw on Beulah's throat came loose, hot blood spurting from a forearm. The aggressor unwillingly flew in the air. He fell to the floor on his back with a solid thud, air in his lungs vomited forth. Mr. Mizithra, not hesitating, dove on top.

Next, he jumped up to his feet, leaping in the air. He came down on the adversary's stomach again and again. He was flipped over by the head, landing with an elbow smacking the enemy in the face. It was a classic wrestler's move reserved only for bad sorts like this nasty fellow. He would never dare do this to his own kind. Mr. Mizithra knew how to scrap. He did not survive in the wild without having a fighter's chance. While a ball of chubbiness, he worked out constantly to remain powerful.

Do not ever mess with a mad marmot when it comes to wrestling because you will lose!

The attacker's stunned companions suddenly faced an angry crowd hostile to their presence. Buford broke loose. Murray bit Joe Joe on the ankle, Tobias, Homer, Bette, and Henry tore into Sammy Joe as he swung his pointed stick in vain. Fur flew and blood hit walls behind the melee. Twigs guarded the disarmed attacker. Those close to the door sprung towards Jimmy John, whose eyes blew up and bulged.

He dropped his pointed stick. He tried to open the cedar door, but no such luck. Harve was trampled by a flock of ducks, geese, and swans, pecking relentlessly, until crumbling in the corner. The pair of otters grabbed Jimmy John by the shoulders, while squirrels and chipmunks descended, teeth bared. Soon on the floor, his life hung in the balance.

Mr. Mizithra pummeled the main invader until he gave no response, blood flowing like a gathering stream from arm, mouth, eyes, ears, and nose. Fur was missing in a patch or two as well. He was possessed with pent up anger at the sight of his new found friends' peril. Twigs stood by in stunned amazement. He hardly noticed this mad marmot from behind his drum kit.

While the bullying beaver was considerably larger than the marmot, his ability to wrestle, quick insight, response to the situation at hand, and strength of body combined to tip the scales in his favor. Shock and awe triumphed. It was over for the loser. His fat, bloody head lay to one side of his limp, unconscious body. He no longer looked so tough, either.

Once the fur stopped flying, Mr. Mizithra rushed to Beulah's side. The rest of her family arrived to an unnecessary rescue. Mr. Mizithra was with her, examining her neck. He called for a red dragonfly, the fastest of all, who darted in a dark and stormy night with dire speed for McGrady's hospital. The dragonflies had their own wee entrance at the back of the barroom. It made it easier to do what they did best.

She was still out cold. Her heart beat, but she was not alert. An interesting fact: a female beaver is the same size as her male counterpart. When attacked, her aggressor had to be extremely strong to so quickly overwhelm her. In many other species, the male is larger while in others the female is bigger. Nature makes it's own choices. We are constantly baffled by the wonder of it. Yup, the out-of-doors is awesome. Go, look.

The rest of the crowd tied and hobbled all five intruders. They lay unconscious where each fell. They were trussed up nicely. As an added insult, the first four were gagged. The leader was not, for he was beaten to a pulp by the most brave, beaver bludgeoning marmot ever. Veneration grasped everyone once the dust settled.

Beulah was laid out on a table, covered with a soft blanket of woven reeds and cattails. Her head had a cushion of clover to rest upon. No one could envision life without her. She was mother to any who ventured in their welcoming door. Her face and wounds were cleansed with water dabbed from a pile of leaves.

To each and every regular at the Beaver Blues Bungalow, Beulah was as much a part of the place as the aged timbers. No one could recall a time she was not tending bar, singing on stage, flirting with her special "critters" (all the while loving Buford with her very soul), and telling everyone how to run their lives.

Now, she lay very quiet, very still, very much in peril, her precious life in limbo. The attack, whose origins were a mystery to Murray and Mr. Mizithra, held curiosity. A hush consumed the place. Our travelers were shocked by the evening's violent turn. One salient question ruled in their collective minds.

Why?

Chapter 9 Cruel, Cruel World

Recovery from the unprovoked and dastardly attack began. They picked up, replacing jumbled furniture. The stage was quickly rebuilt. The bar suffered extensive damage. This incursion tore wood in every direction when the brutes invaded. This, of course being a tavern, was not their first fight. It was common to reassemble the contents of the place. Bette and Mr. Mizithra tended to Beulah as best they could. Murray and Buford saw to the prisoners.

A mess of others headed out to survey for any other companions of the fallen, loser army. They thought more would be needed to defeat the entire bar's attendees. A small reserve force of six other shifty beavers were waiting for the sign to come finish the battle. They were armed, anticipating becoming a part of the fight as well. Hiding in the shadows, they did the best to shelter from constant rain pelting soggy heads.

With a show of force, Beulah's offspring barging up front, the six reservists turned to run. They dropped weapons and beat feet. However, they did not count on the speed of one Bette Beaver. She flew with a fire kindled by seeing her mother laying near death on the floor. She landed in the middle of the crowd as they began to run.

The rest of her siblings struck the aliens as otters, squirrels, beavers, muskrat, chipmunks, and various water fowl vanquished the scattered. Within moments, the entire invasion force lay demeaned, defeated, and demoralized, writhing in pain on the ground with various injuries. The attack was a complete failure.

Bette held a certain beaver, one Beau J. Beaudry, firmly down on the ground. She banged his head on a rock with both hands. Beau had the great misfortune of being Bette's target. Hardened ground, the storm packing it to cement, became his enemy. The impact of Bette's attack flattened him. The concussions kept on coming. He was in trouble of being murdered.

She landed perfectly, flipping him. Bette paid him for unwelcome provocation on they who did no harm. Her emotions got the better of her. She began to wail aloud into the night. Vocalizing let the bad stuff out of a tormented mind. Beau made an enormous, ill-advised gaffe, swinging at her from the ground. Bette let it all go: anger, fear, rage, take your pick.

Unfortunately for Beau, his little attempt to fight back only stoked the fires of acrimony. Hers was a release of vindication, of defense, of protection for her fallen mother. Her screams now took on an eerie tone, haunting everyone. The audacity of the onslaught added to her fervent animosity. This, combined with strength of years in dam building and sweet six string wailing, made Beau's odds at victory slight. Chances for escape evaporated.

The exterior secured, they made for inside. The dampening storm continued to rage, streaming in hard rain. Dragging the new prisoners in the lodge, each was propped roughly against the wall down the hallway. They had a vicious guard staff begging them to try and escape. Restraint was not easy. The downpour added to the soggy drama of the evening.

Ruining an otherwise superior evening earned grave retribution from those harmed. By now, the unidentified leader had revived. Buford held him against the end of the stage, back jammed into the sharp corner. Buford's heavy forearm was shoved against his throat. His head turned side to side in clear pain. Breathing for the prisoner was an issue for some strange reason. Curious.

The bludgeoned Beulah awoke. She lay quietly, well attended to by the two lovey dovey Douglas squirrels who were off duty medics for Dr. McGrady's eastern field office. She was quiescent, forbidden to speak due to the attack. Her wounds were severe. It was up to medical personnel.

One gently wrapped her neck. The other made a full body exam. She was rather beaten, bloody, and bruised. Beulah was cleaned. The wounds were evident. She was dragged violently against the hard wooden bar. Fur was torn out, her neck badly scratched from being gagged and dragged over the counter.

Mr. Mizithra was at Buford's side, a sharpened stick held to the cretin's left eyeball. It was close. A mere sneeze would result in this jerk's blindness. Knowing nothing of marmots, they did not realize the extreme loyalty of these modest rodents, self-sacrifice, and immense power in a critical moment. He did not flinch, did not hesitate, he acted immediately for the salvation of Beulah and all at hand.

"Okay, whoever you are, pathetic cockroach. You who attacks a lady. You are at this very moment on the precipice of death. Tell me your vile name. Speak, you festering boil, why did you come on this festive night to spoil our joy? Dare twitch and I blind you. No funny stuff, no tricks."

"So, Buford," returned the evil adversary, refusing to acknowledge Mr. Mizithra's words, much less pointed stick, "you brought in outsiders to help you? Is you gonna to tell this ugly, scrubby, little mole who I is or does I need to fo' you?"

The tone of voice was demeaning and vicious, arrogance haughty, insults brazen. Mr. Mizithra reacted in kind. He pulled back slightly a hair on the end of his harpoon and quickly sliced above the top of the prisoner's right eye. A look of utter stupefaction flushed his face. Blood flowed from his brow.

He stared in shocked disbelief at this little violent, unseemly thing with massive, yellow teeth bared in defiant rage. This was not expected by the supercilious intruder. Buford, for his part, did not budge. It was apparent so he gladly submitted to Mr. Mizithra's powerful authority.

"I addressed you, low, foul creature. You do not have the right to treat me or my friends in this primitive fashion. You do not have the authority. You cannot come to this place and attack those who are the rightful owners".

The injured beaver scoffed, "You knows so very little, my valiant warrior. This ain't they place, it's mine. I am Jimmy Joe John Brown Jr., true owner o' dis dam an' rightful proprietor o' dis lodge. It is mine, not theirs. Go ahead, ask Buford. He will tell you, it's mine"!

"This sorry boy jus' plain crazy. Has been his whole miserable life. His daddy was Beulah's brother. He died. In his will he lef' dis place, da dam, an' lodge to his sister because he knowed his son was a waste o' space. He even tried to", here Buford hesitated, staring hard at Jimmy Joe John Junior, "well, I am ashamed to admit dis, but he tried to make friends wid the Thumbs"!

This egregious iniquity was, to forest folks, the ultimate betrayal. To seek out and interact with humans was simply not done. It was only right his father cut him off for this travesty. No self-respecting beaver would ever do such a thing. He lost his right to inherit the dam and lodge. He earned his sentence.

For any animal to betray his species, making friends with the creatures who turned past relatives into hats, was unforgivable. His son no longer existed. The shame he bore was heavy, causing grief. He had no choice but to will the dam to his sister. His aunt Beulah was seen as the barrier to a denied ownership.

This schism was due to losing his way. He blamed it on others. This is too common a thing for people. Rather than accept the consequences of actions and choices, they shine a light on innocents. Many, like Jimmy Joe John Jr., believe if they avoid admitting shortcomings, repercussions will never reach them. Sadly, too many human beings do the same.

If you do something wrong, it is best to admit it quickly and accept the consequences rather than bolt. It only makes the results more severe. Holding off or avoiding reality causes more pain in the long run than dealing with a problem early. The avoidance of punishment for sins, no matter how light or severe, often has the by-product of touching the lives of others not directly involved.

Immediate apology, true, humble, contrite, can be the best thing for the soul. Confession really does make it better. Within, you know you did the right thing. Making a wrong right with someone you injured by word or deed delivers a glorious feeling of personal healing. The benefit is relief doing wrong caused in your mind. Sadly, pride keeps most from humility, changing the course of their lives for the better. Guilt, like acid, eats away at the conscience, warping the mind.

Enough was enough, so they took the eleven out along the border of the pond, rain pouring down. They were roughly tossed by the water's edge. Trussed, unable to either move or speak, each was gagged. Buford strode forth, flanked by his children, Mr. Mizithra, Murray, and the rest of the defenders. Anger exuded from the glare in his eyes and vicious tone of his words.

"Now, Jimmy, you can does one o' two things, you can be da fool you is an' end yo' miserable life an' these other worthless traitors. It makes no never mind wid me. You is one stubborn cuss. Or you can apologize, turn tail, an' never return".

"If you decides to flirt wid danger, here's what we gonna do. We gonna leave you an' yo' boys tied up here. Den, we gonna send a bunch of dragonflies to alert da Thumbs they can find eleven easy-to-skin fat, beavers to make pelts out of, boy. Then yo' sorry butt can become a hat. It's up to you, fool. What you wants"?

Knowing he was beat, Jimmy Joe John Brown Jr. slowly nodded his bloodied head to the great relief of a defeated army. He could barely lift his pate. The enemy's lodge was a few ponds south, higher in elevation. They would have to brave the storm, pray no branch fell on them, and dawdle home.

This pathetic blob of demolished, bloodied beavers slithered slowly south. Rain beat down on broken egos. The vain attempt at a coup d'etat failed miserably. They returned maladroit. One look at their feeble leader was enough to see why the plan was a complete dud.

"Well, this evening was one for the books," said Murray.

They turned to call it a night, to sleep a sleep well earned. Still, he mused, playing with a real band was simply the best.

Chapter 10 Retribution

After such a wild, wacky, raucous night, both slept in quite late. Mr. Mizithra's vacant stomach soon got the better of him. He rousted his musical partner. Stimulated, they made for the main room and, hopefully, breakfast. Sadly, everyone remained in various forms of recovery. The mention of food produced only grunts and groans. The medical staff of two were spent from the previous evening.

"You rodents, please go out and wait for the medicine. It should arrive on the pond, on the side we put the beavers we beat up last night.", begged Bette, knuckles bare from wailing on her adversary.

Obedient, and knowing a meal could be had outside, they ventured forth to a new day. Day 4. The sun once again shone in the sky, higher than normal rising so very late in the morning. Not knowing what was arriving, they moseyed along to the loch's rim to wait. They chomped on various forms of abundant lupines, flowers, and greenery. Murray dove in, nabbing a choice root. He snagged a fat, tasty water bug with his left front paw on the way up. Surfacing, a loud noise arose from behind.

Stunned, he let go the water bug, who, grateful for the opportunity, made for the water again. It found sanctuary under lily pads ten meters away. Murray did not dare turn, zipping for the lodge's doorway, some twenty meters away. Fear of safety panicked the poor fellow. The flapping sound increased to a roar.

Finally reaching the entry, quite out of breath, he turned to see a sight. He froze solid. Something in the neighborhood of thirty Northern Pintail

Ducks, flew in tight formation. The flock landed in unison, stopping at the end of the pond. Once ducks land they become a raft, paddling, or, as these fliers preferred, team.

A large quacker, males striking, females spotted and dull to better hide, the Northern Pintail was employed to transport packages due to their reliability and size. Each held a small package in their beak. Not hesitating a moment save to shake water off their tails, they waddled swiftly for the front door.

Mr. Mizithra, mouth agape in stunned amazement, held open the threshold as each delivery fowl nodded politely, doddering in the lodge. Each was stopped by the Douglas squirrels, the package examined. They were next directed to the patient in need of medicine.

The sight of so many birds in flight descending to this spot blew their minds. Murray was eating. This ended his appetite. He went inside. Truth be known, he wanted to play sax. Maybe, just maybe his musical presentation would help those mending wounds. Admittedly a selfish desire, it might quell some of the pain. After all, as everyone knows: music has charms to soothe the savage breast.

Mr. Mizithra, ever hungry and generally not disturbed by the team of delivery ducks, chose to stay outside. The weather was fine. Up on the Olympic Mountains, you must enjoy any and all sunshine whenever it appears. Two thirds of the time, it is cloudy or raining. It's rays, manna from above, showered happily on the furry creatures of the forest.

He spent considerable time gorging a necessary filling of his rodent gullet. The calories packed on were storage for survival in the long, cold, foodless winter of the high mountain country. He ate so much, it was so good, he fell asleep under a small Sitka Spruce tree on the edge of the calm pond. A healthy burp, and out, near the door of the blues club.

Sobbing awoke him from sweet, foggy slumber. Someone was crying in the bushes a short distance from the water. The sound carried to him. No other commotion was outside at that moment. He was intrigued, poking his head out of the dense foliage. Mr. Mizithra rose up on his hind legs supported by a sturdy tail.

A short ways away, over by fallen holly branches from the previous night's gale, sat a swarthy figure shuddering in a fit of tears. It's back was turned away. Identification was impossible. He could make out it was larger than he. So many animals were, he did not regard it with much concern. The sobbing persisted. He was intrigued.

Creeping through the underbrush to avoid detection, he crawled towards the sound of the sobbing someone. Soon the form became a bit clearer. He made out black and white stripes. Now, folks, there is only one animal in alpine country who fits that description, a skunk. Yup, a polecat was leaning against a spruce, propped up, head down, bawling away into a handful of leaves, clenched to a concealed face.

Mistakes in the wild have dire consequences. Now was not the time for one, but here it came. His misfortune was to snap a twig. Breaking the silence, the weeper swung her head towards the sound. Mr. Mizithra was down, silent, still. He chilled. A skunk was larger than he. She could kill him instantly. He froze solid. Hoping the snapping sprig would be discarded, he waited in stone silence until a shadow crept over him, deja vu the story of the fox.

"You, who are you?"

"I am Mr. Mizithra.", he said, rolling around to face her.

"And, why, pray tell, are you spying on me?"

"I came here to eat. After last night's disturbance, there was nothing to eat in the lodge. I came out to get a meal and fell asleep. I awoke and heard your sobbing. Curiosity got the better of me. I ventured over to see if I could be of assistance. After such an unsettling evening, I guess my sensitivity radar is up and running on full."

"Oh, well, I see. So, you were in the lodge last night? What occurred to make you so disgruntled?"

"Well, dear lady, we were having a tremendous evening of frivolity, food, and fun, when invading beavers from a pond to the south hit. We had to fight. Many were injured on our side. We have been up most of the night mending them. I just arose. My friend is inside. Do you not know of the battle?"

"Oh no, I am never allowed inside. I am shunned. They do not like me solely because I am a skunk. They are a bunch of mephitophobics!"

Stunned by the word, Mr. Mizithra paused to take it in his rodent brain and process the linguistics. He was no dummy, but this term was a real winner. Fun with words was one thing, but this? Even running what she said through his head he could not repeat it. She saw the stunned look on his face. She immediately spoke.

"Oh dear, dear marmot, I am so sorry with big words. It means they have an unnatural fear of me and my kind. We have a scent gland. For it alone I am ostracized. They will not let me in to take the stage. I am a singer, a blues singer. In fact, my mother named me, Miss Bluesberry instead of just Blueberry because I sing so well. Since birth, my voice has defined me. Mama said I came out of the womb singing. I am a natural born vocalizer, but they won't let me."

With this, she began to sob again. More tears and louder wailing sparked sympathy. The poor lady was the victim of a form of dishonest partiality so many face or participate in for their own reasons. Prejudice has many forms, existing from ignorance mixed with hate and fear.

Ignorance is, therefore, cured by education. The best way to develop someone is help them face fear with solid reality, not personal bias. For Miss Bluesberry, it was a gland. One little stinkie secreter. The ignorance of the beavers was formed by childish fear, highly exaggerated stories, and her past relatives utilizing that secretory organ to defend themselves.

You see, a skunk uses the stink the gland produces to fight for their children, homes, for their lives. Nature gives us ways to defend ourselves. For the polecat, it is this secreter. The smell is so strong it drives others away, saving both skunk and assailant. In life, even defending yourself in a proper, rightful manner can produce a hatred based on fear and lies.

Once you have an education, prejudices should disappear. You are a better, smarter, more fully rounded person. It is a win-win situation for the entire planet. Mr. Mizithra, from knowledge by lessons taught from ancestors, knew while a skunk can stink to high heavens, they are a very wonderful group of animals. The only problem in a relationship between a marmot and a skunk is one of consumption.

"Well, Miss Bluesberry, if I may be so bold. Once, many generations ago, a relative of mine on my father's side, Mr. Pilsner, was saved from a fox by one of your kind. She rescued him as he was about to be eaten. As he taught us, a skunk is a friend."

"What was the name of the skunk for my family has been here for many generations, too?"

"Her name was Miss Mulberry."

"WHAT? The heroine was my great-great-great-great grandmother. She roamed this land for many years, always telling of the encounter with a marmot. Yes, I know the story well. It has been passed on over the generations. She heard a commotion and headed to see the lady fox going after a marmot in dire peril."

"Things happen for a reason. I must assist you. Marmots are famous for honesty in repaying whom we are indebted. I see this boon as a turn of fate. I am supposed to be here to help you."

With that, Mr. Mizithra turned, marching to the lodge armed with firm determination. He quite forgot Miss Bluesberry, who, composing herself, stumbled eagerly behind. Throwing the door open, he stomped to the stage where Murray and the beavers were busy wailing away. The triage area was modified, just a few remaining down and in pain.

A stunned crowd turned to see the noise from the doorway. No one was up for another round of roughhousing. Mr. Mizithra waddled full force towards the platform. Playing ceased as they witnessed the choler on his face. Angry marmots are a vicious lot, not to be trifled with. They are capable of great mayhem, madcaps that they are. Murray had never seen his chum quite so flustered and irate. Twice in a day he had shown signs of fervor unlike him in any previous escapades.

"I wish to inquire why you refuse to invite a good, decent, lovely gentlewoman in this, this, this *place?* Why are you prejudiced against Miss Bluesberry? She does not stink, as you put it to her. Instead, she is a fine lady. You say you owe me. Okay, this is my payment: give this sow a try. If you think she stinks as a singer only, okay. If you are so narrow-minded you will not give her a chance, you lose all my respect."

A defiant, righteous little fur ball stood before them. As a whole, the entire beaver band broke into laughter. Murray stood still. It was not his fight. The band broke, jumping off stage. They surrounded Mr. Mizithra, Henry rubbing him on his furry head. Next, they came over to a stunned Miss Bluesberry, begging her to come on stage.

Music can cure a lot of problems. It has power no words could ever convey. You would not think so, but it is true. Once notes hit your ears, in your head the soul takes over, producing smiles, nods, and love. Music is groovy, wonderful, soothing, heavenly. It is good for you, better for the soul. Music makes friends out of strangers.

Buford was caring for Beulah in their private apartment behind the kitchen. It gave quiet distance from club noises, providing the privacy married folk need. One can only party so much. She was battered and needed personal attention. He alone loved her in a special way only a good marriage can produce.

Now, though, Buford watched as the two Douglas squirrels tended ardently to his wife. Battle injuries sustained were: pharynx, stomach, and right knee. Jimmy Joe John grabbed his aunt by the throat, hoisting her over the bar. This banged her stomach and right knee on hard wood.

Her knee, sprained, bruised, bloodied, with contusions, swelling, and bald spots was set with bandages provided by the Pintails. They were in the bar, eating and enjoying music, until the commotion Mr. Mizithra caused. Her neck was similarly wrapped. Nothing could be done for the stomach other than rest. Whole bunches of convalescence were needed.

It was dormant in the back. Buford ensured status quo by closing the club for a few days. Nodding his worried head to agree with the attending physician, he heard it. It. His heart did a little pitter-patter. It came slowly at first, but came just the same. It's name was fame. It was a voice from above. It was world shaking, earth quaking, globe shattering.

He left his loving, injured bride with those ministering. He ran like a berserker through the kitchen, to the open dance floor. The house band was cranking out a sweet blues jam. Murray, now a solid member of the group for his dynamite sax, wailed away right next to a skunk. Drawn by the sound, Buford scanned the stage in disbelief. It was the very skunk he dismissed because of personal hatred.

As a young beaver, he chanced across the scent of a skunk blown moments after it's release. The skunk was fighting for his life. The blast of odor staved off his attacker, a hungry, young bobcat. In the wild, no food means death. Starving, the feline fool tried to slay an adult polecat.

The skunk's salvation system of stinky stench caused Buford's nose to go funny and runny. For days he dealt with pain and lingering odor. He did not have any tomato juice to bathe in to rid himself of the stink. While unintentional, the pain nonetheless caused suffering. Since that day, he hated the very thought of a skunk. He did not let them in his club. He forbade his family to contact any skunk. He was a polecat bigot.

A willingly, truly humble person can achieve amazing things. They are freed from the unnecessary, heavy burden of intolerance. This load is borne by so many in life. It is a real pity. Such a closed-minded approach, derived by foolish self-preservation and fear, retards personal growth. One can only see so much wearing blinders. Being wary is one thing, a bigot quite another.

Only by seeing as much as possible in life do we learn what we actually like and do not like. In essence, education, open, unrestrained, free of the dogmatism and narrow-mindedness of instructors, results in becoming a full, well-rounded, whole person of our own making. Clearly, this does not mean violating society's conventions to acquire knowledge. Personal study is self-rewarding.

Choosing not to listen to varying views because you are offended by anyone disagreeing with you will paint you into an intellectual corner. Your ability to learn will be confined to the little spot you allow yourself. Peer pressure can impact this immensely on tender minds. This is wrong. Common sense says fill the mind with information gleaned from multiple, even differing, origins. Single source information is not education, it is indoctrination. Why, you ought to wonder, do they deny you listening to another point of view? What fear drives this notion?

The most beautiful thing a person can possess is themselves, grown by a lifetime of learning, layer by layer as does a tree. We acquire it from others, by observation, application, trial and error. Some of the best teachers never graduated high school. Making mistakes is a part of life. Their purpose is to help see right from wrong. Hopefully, of course, we learn not to repeat the same mistakes over and over and over. That is insanity.

Buford, for his part, was an honest beaver.

He was willing to open his heart. Her voice pried opened his ears. Approaching the stage, rocking and rolling away with Miss Bluesberry's melodious tune melting his soul, he humbly hung his head. They stopped playing. All eyes turned to the floor and patriarchal figure. His contrite body language spoke volumes. Unable to lift his head from pure shame, he spoke.

"Ma'am, years ago, when I was a young beaver, I was hit right in the snoot by skunk smell. Since den, well, I ain't been fair wid any yo' kind. It's me what kept yo' species out o' here."

"My wife is layin' back there, in rough shape, but when I heard you singin', oh ma'am, please forgive an old fool. Since da days when dis club was young, we had us a band. Had us a good one, too. I played guitar, mama did da singin'. Relatives, an' friends made up da rest o' da band."

"An' always, we wanted to find a certain sound. Murray, here, why jus' last night we found us a sax player. Murray, we wants you to get yo' mammy an' grandmammy. Bring 'em here to live. We gots plenty o' room fo' a trio o' woodrats. They can have any room dey wants."

"As for you, why ma'am, I gots to say, yo' voice is extra special. It's da "It" we been searchin' fo' all my born days. When I was out back wid doctors tending to my baby girl, yo' voice tore me up."

"Mama knows, so do not worry. All her life, she wanted to be at the bar, watchin', talkin', teachin', preachin', an' havin' a good ol' time. This is her place. If you stays, yo' life will be happy. Miss Bluesberry, please, from as humble a place in my heart as I can find, forgive my small mind. Music cured me. You da medicine. Please, ma'am, please say yes."

"Of course, good sir, of course. It will be my honor."

Peace, love, and music reigned the day. They played until late, late, late. Eventually, however, slumber claimed it's complete victory. Ha.

Chapter 11 The Journey Continues

They rested two more days to ensure all was well. This gave Murray an opportunity to get things done since life took such a wondrous turn. The unplanned endowment to move, something necessary, was provided by chance. A great pair of lungs on the reed, flying fingers, and sense of rhythm earned this opportunity. He never thought himself a musician, yet every day his talent and confidence grew. This well-earned new found fame did not, thankfully, swell the head under his kippah.

Dragonflies went scurrying home to inform the lady woodrats of this good fortune. Communications went back and forth until his mother was certain it was true. She could finally stop eating the old potato skins. Packing a pack rat is not easy. Grandma Frema was a hoarder. Fear from the loss of her home caused this affliction. She dug out rooms in back of their home to keep things she found. It was ridiculous. Two whole rooms filled to the brim. She did not even know what was in them. They were "treasures". Getting ready took a while.

The time came to leave. They sorted through and got her down to a few, necessary items. They held a big moving sale. After all, explained Murray's sweet mother, Nediva, it was a new life with a home already built. How could they possibly take this with them? The arduous journey would take the ladies longer due to age, resulting in a slow pace. To assist, Murray thought ahead, hiring the armada of Northern Pintails to bring their goods via air in loads ducks could handle. They were paid in grandma's treasures.

"Well, Murray," lisped the lead quacker, "we are off, senor. Pedro is my name and delivery is my game, claim to fame and life's aim. Me and my amigos go to your mamacita's aid. You are the best musician we have ever heard. My flock wishes me to thank you. Gracious. It is, for them you see, an honor to do this for you. For us, music is the greatest of blessings."

With this, he turned, waved a wing, wafting the way west with fully loaded feathered friends. They, in turn, each waved while departing his midst. Murray was impressed. The sight of thirty ducks flying in unison, knowing their mission, made the world a constant source of education.

Nature is it's own text book, each turn, every day, river, or tree provides an opportunity to learn. Nature is very good stuff. It's just outside your door, down the street, outside town. Go find the amazement awaiting. The fresh air will do you good.

He sent word with Pedro for his mother. Continuing his travels with Mr. Mizithra, Murray would stay forever upon his return from the White Mountain. Nothing could deter this duo from the goal. For his part, Mr. Mizithra knew Murray to be a woodrat of his word, never doubting his determination to finish the task. Returning to the lodge, the beaver band and sundry assistants hugged and kissed the boys knowing they had to leave. Even Beulah was able to come thank Mr. Mizithra for his gallantry. Knowing her new boyfriend, Murray, was going to come live with them made her one happy beaver.

"Hubba-bubba, come back to me soon, baby", she cooed to Murray.

Saying good-bye with assurance of a return, our adventurers set off again to the east, different than a few days prior. Life changes on a dime. The experience was unforeseen. Murray found music, Mr. Mizithra a new form of courage. You never know what life will throw at you or when. You only know Why in reflection after an event enters the pages of life's story.

This is what makes life worth living, certainly of getting out of your chair and seeing what the world has to find. Exploring nature means finding yourself. That, my friend, is pretty danged cool stuff. A few days camping out in the great wide open is a tonic to the soul. Strum a guitar and sing under the stars. Cook on an open fire (mmmm, s'mores). Listen to the animal noises. Watch birds fly. These await you.

All you have to do is prepare and go. It is that simple. Campgrounds, hiking trails, scenic road trips make getting out of the city heavenly. One humble request, please clean up when you leave. Thanks.

This being a large pond, more a tarn or mountain lake, it took two days to reach the rapids. Having never seen a rapid or really knowing anything about them, they paused to observe the fast, white, icy water jamming into the calm pond below. They arrived at the base of the last frothy leap downward. They sighted a deep gray bird facing the water. It leaped in, was gone for some time, then arose, a bug impaled on it's beak. A quick flick of the neck and the insect was a meal. They shouted at the feathery friend. Without hesitation, it dove again, not turning around.

They turned to each other in amazement. It was as if the bird was deaf. How rude! They got to the edge of the water. Spray hit their faces, a new experience for both critters. At once, they realized the issue, the sound of rushing water would mute any approaching or even yelling. The bird could not hear a thing. American dipper birds tended to be loners anyway. It is their nature to be distant. Bug impaling is solo stuff.

The feel of fresh, misty liquid gave pause. They turned in unison to see their immediate past. Up the first rapid, the view was dynamic. It provided a distinct leg of the journey. They knew, looking forward, the next stage was a path elevating them to parts unknown by any they knew.

They clasped paws for a moment's reflection, then began, Murray out front to set a good pace. The stay at the lodge added a bit of weight to his bulbous friend. Mr. Mizithra needed it. He was slower as a result. The sun was up and cordial. They could see safety along the water's edge. Anyone diving from above would risk serious, even fatal injury if dashed against rocks in the white water.

The foliage along side changed with the rise in elevation. Sisal bushes were replaced by ferns. No more lilies either, just scrub brush and the odd pine. One other thing, being close to the water made the sun surprisingly hot. It's rays beat against the rocks and shone brightly upon our thrill seekers.

A welcomed break came as the first pool was discovered. Murray came back to a panting, pudgy pal. He informed him the rapids stopped ahead, just a few more meters. He would be in the cool pool. It was a good time to pause. They were spent.

Mr. Mizithra mustered muscle power, making for the top of the hill. Seeing still water, he dove in, uncommon for a marmot. He was extremely hot and tired. Instantly, he was in a state of euphoria. Shaking a furry face, he came on shore, staring at the pond below. It was forever away. Such a long climb was an appreciated accomplishment.

They could make out the pond, lodge, field beyond, and beginning of the path home. Murray could imagine them packing and preparing for the journey to the beaver's welcome. For the older ladies, it would be a comfort to know they would be cared for always. When they returned, a promising future awaited. He felt truly blessed.

Murray and Mr. Mizithra paused a good while, including a nap, to recoup from the long slog up this steep grade. When your body tells you to rest, it is a good idea to listen. If you stub your toe it is a good idea to sit down and tend to the pain. The same is true with rest. After a while, sun still high, they continued, thrilled to ascend the unmapped. Across the stream was a forest. It was thick, filled with ferns and pines.

There was a stillness which seemed unnatural. To the left/north, eventually came a dirt road built by Thumbs. This was expected. They could hug the river bank, hopping stones and hiding under the greenery to avoid detection by any human beings.

The rocky trail alongside the white waterway wended it's way up the hillside ever higher. The occasional pool provided a chance to pause along the way. Without any shelter seen, they pushed on late that summer day. Without warning, common for this land, black clouds filled the sky. The temperature dropped ten degrees Centigrade.

A determined air current appeared out of nowhere. The Strait of Juan De Fuca was back at it. Thanks a lot, Juan. Dust swirled, filling the air. The need to find a place for the night was paramount. Out in the wide open or hovering in the rocks was death. Murray could hide more easily than his best friend. They had to find something fast. The speed of the storm's approach left no time to dilly dally.

With great rapidity, the storm increased. The fresh air filled with all sorts of debris, birds squawking and fluttering about. Terrified, in need of a place to wait out this sudden onslaught, they scattered. In a matter of moments, lightning cracked the gloomy sky, thunder quickly peeled not a second behind. If it hit, they were dead. The rain came down hard, fast, and cold. Raindrops kept falling on their heads. Rocks were no help to shelter from the storm. The ferns easily bowed in humble submission to the power of eventide's tumult. A haven must be found immediately.

Murray spotted something ahead, left by the human road. Way up here in a tempest such as this, no car was going to come by, not that they knew this fact. To them, a road meant one thing, intrusion of the Thumbs in the land. The frightened, mucky, speedy rat was soon completely out of sight. Mr. Mizithra trudged from the stream's edge, through the sloppy foliage, towards his compadre.

In a flash of lightning, a paw urgently waved him towards a water pipe. They were placed by the Thumbs to channel water under the road, to the stream to avoid the road flooding. What was a luxury for visiting people was salvation for animals living there. This pipe was operational. Many were so ravaged by fierce mountain conditions they collapsed after only a few years. This is no place for the feint of heart.

They worked their way into the pipe, whose aged cement sides were cracked, leaking here and there. It was an insecure place. They pushed on north, to the upper end of the pipe. It tilted some to ensure the water flowed in the proper direction. At the finish line, the pathway was blocked by hanging plants, raindrops urgently dripping. Mr. Mizithra pushed these aside with his left arm. He squinted to see what was ahead.

Strained, soggy eyes made out an odd sight on the hill above. Behind this obstruction was a solid wall of rock. It was formidable indeed, well beyond their capabilities. Neither was a climber of note, though Murray had it over a marmot all day long.

To the left, the road dipped to meet the curve. That way was insanity. Right, another steep climb to the edge of the road. It was not an option. The storm pelted down relentlessly. They could not go back. They could not stay in the pipe, left, right, and up were out. This left the odd shape ahead. Since night was now completely soggy black save lightning bolts exploding the stormy night air, straight provided a lone ray of hope of salvation. They moved forward out of the terrible weather.

First, the shape. The difficulty was identifying something unknown in a pouring rain shower in the black of night. It was rather long, perhaps six meters. The form was smooth, low to the ground maybe three dozen centimeters high. It was whitish in color. It did not move in the rain or wind, not one bit. Unless it's a rock, in nature it moves. Even trees sway. This did not, nor did it look remotely like a stone. Curious, they could not wait, shelter was all encompassing.

Right and above a handful of centimeters, sat a pile of red rocks. Lava rocks, how did red lava rocks get here, they pondered. Left grew long grass and a few odd, low bushes. Moving, they made for the red lava rocks. It seemed to give the best opportunity for safety. Mr. Mizithra burrowing the ground would have a fine resting place in mere minutes.

Murray silently nodded in dripping agreement. They made for the anticipated safety. Within a few steps a Stygian figure appeared to block their way, holding a weapon, a barkless pointed stick. It was held firmly, being tapped in the other paw, begging them approach. The rain bounced off a pair of hairy ears and equally furry head. They felt the burn of a pair of glowering eyes glaring.

A mouth next, firm, stern, mad, set hard. Suddenly, the shadowy figure was gone. In a flash it was standing next to them, breathing against Mr. Mizithra's face, the pointed stick up his left nostril. Murray froze as woodrats tend to in conflicts. Knowing this to be a touchy situation, Mr. Mizithra took a deep breath and let it out slowly. He eased his tense body and relaxed.

"Sir, it is a dark and stormy night. We are merely seeking shelter. It is only by chance we came uphill today from the pond and were caught in the storm. We need to find a place in the red lava rocks to wait it out. Our goal is to rest for the night, then push on uphill and White Mountain.", his voice strong against the unknown guard. He only hoped the defender to be a reasonable animal. He knew the intrusion was unexpected, but they were clearly no threat.

"Well…okay, guess I can buy that given the weather tonight. You're a marmot, so am I. Years ago, when the Thumbs built this road, my folks got stuck here with a few other families. We are all that remains. We never did get through the tunnel. Nope, we stayed. We are not like you, oh no, we are settlers."

"My name is Mr. Mizithra, sir. This fellow is my long time traveling companion and best friend, a pack rat named, Murray.", Murray doffed his soggy kippah at their new friend.

"My name is Mr. Gilligan's Island. My wife is Mrs. Maryann. Our children left. We have no desire to move. We are here to stay. She and I are the last of the families who stayed after the road came. Come now, come up to the red rocks. You will be safe, safe indeed."

Chapter 12 Choices

Mr. Gilligan's Island led quickly up a rain soaked path, past the large white thing immediately left, to the red lava rocks. He turned, ushering them to a small hole, left, under a small pine tree. It was good to be out of wind, rain, and uncertainty. They saw a hallway leading straight down, veering off right. The tunnel's left wall was solid rock, black, cold. They spied a lady marmot in a warm chair in the chamber, rocking back and forth. Her appearance was peaceful and charming.

"Oh, my dear, what have we here? Something the storm blew in. It is a terrible one, is it not?", she said smiling, "Please, weary ones, rest here against the fire and warm your bones."

There was, in fact, a small, hot fire in a little fire pit with a chimney whose pipe arose out of sight. A bit of human charcoal which fell out of a truck made a nice fire. They ran for it, hands stuck forth to gain all the warmth they could acquire. They noted small twigs fed a hot, soothing pyre. The charcoal made a nice base. She came back from the stove with a cup of hot clover honey tea for each, including her husband.

On the tray were some grass cakes and a lovely, fat water lily bulb for Murray. His face lit up in shear delight at the care the new friends showed. They ate and drank, then sat back, sharing their story with these two so cordoned off from the rest of the world. Both were exhausted, but wanted to share their experiences. Mr. Mizithra did not go whole hog, either. He was more sedate than he was with the herons. He adapted his delivery for each audience.

The couple's eyes were agog, each new story inciting more and more consternation. They finally called a halt to the conversation. This clearly bothered them. Their lives ceased advancing long ago, voluntarily cutting off the world to dwell quietly alone.

"We are not used to such a life. Here, we are simple, alone, separate. We are happy together, which is most important. You see, fellows, each must find what brings joy, though, not what makes others feel the same.", he said smiling from ear to ear.

They were shown rooms, each his own. It was warm, quiet, and out of the storm continually banging away outside. The colony was old. The walls were lined with old TV Guides and food wrappers from companies, many no longer in business. No new wall paper had been added for a long time, at least not in the two rooms they occupied.

Both lay exhausted. The comfort provided by these solitary marmots was welcomed. Murray was fortunate to spot the opening considering the torrential rain beating constantly upon them. Lightning, thunder, and wind made it rough. Weariness overtook them. Sleep was instantaneous. They slept soundly all night.

Morning's greeting, total silence, eerie, unsettling, odd. In the wild, wooly world of animals, there is always some sort of sound. Deep in the depths of the vacant series of tunnels and rooms, all was quite still. Shrugging off concern, they made for the dining area and a nice morning meal made by Mrs. Maryann. Each had a fresh, steaming, warm cup of flower nectar. Murray had some roots and somehow found him a nibble of a cracker. He squealed a woodrat's little yip, scarfing the wafer down with decadent glee.

This recouping of energy prompted inquiries about departure. The task: head up river towards the goal. While enjoying their company, a chore had to be fulfilled. Urgency meant little time to socialize. A sadness befell the settlers when leaving was imperative. It was plain these were lonely marmots. Mr. Mizithra and Murray had pressing business uphill. Mr. Mizithra approached Mr. Gilligan's Island.

"We must thank you, sir, for your hospitality. Your wife's food and warmth are a true blessing. The rooms were wonderful. Your assistance last night was a life saver. That tunnel is leaky, in danger of falling down, and soon."

"We were wondering if you would be interested in leaving this old home and, provided we find it, moving to a new colony? We have the task of finding fresh residence for our tribe because of the trespass of man. They will never stop putting in more buildings, observatories, and trails. They install cameras to watch constantly in our tunnels. Their infernal dogs run off the trails and terrorize our inhabitants."

"Then we are moved to a green zone or a zoo or another entrapment made by man to control and reduce our numbers. They put tags on you. Tags, attached to your ear by a "device"! Can you imagine such an indignity? How embarrassing and degrading. This way, we can get so far away, they cannot come. The place sought, heard from the ancient times, is a place where you would be welcomed.", Mr. Mizithra sounded a great deal like a salesman. He believed strongly at the prospect of free land for his rodent friends.

Murray had a safe, permanent home for his family. Mr. Mizithra sought the same for his kind, desiring the selfsame for everybody. He found great kindness and goodness in the new, isolated friends. Marmots are very warm and caring about their own.

It is heartwarming to see such love and loyalty. His real hope was Mr. Gilligan's Island and Mrs. Maryann would want to join them. They could live out life with a huge family of furry young ones to help raise. As a communal species it was only natural to want to help these kind souls. They had much to add to any colony.

"Well, brother marmot, that is something we will have to take a long time to discuss. Neither of us has ever even been through the other side of that tunnel. We have seen the road, cars and trucks, we can see the river going by, and small waterfall which feeds our little pool underground."

"It is very hard to think of leaving here. This is home. No one else stayed. They all left. We, we never could. When you return this way, if you stop by, we will give you our answer."

"Please, though, let me show you the whole colony before you go. It is a wonderful spot. We did so much with it when I was a cub. My gramps, Mr. Peter Gunn, and grams, Mrs. Miss Hathaway, built tunnels under the wreck. Come with me, boys, you will be impressed."

Following along a different passageway which turned west, they wound around and around for a bit. A few tunnels veered off of this main one. Suddenly, sunlight shot into view. Then again and yet again, about a meter apart. The holes slanted each at a differing angle. He motioned to follow him. He was grinning.

Up top, they saw. This was an old, off-white colored motor boat, motor removed. The tunnels were from three holes in the bottom created by shotgun blasts. The vessel sat about three dozen centimeters above ground, the remnant underground. The three ascending tunnels came to surface in different parts from the boat's interior.

The edge of the wreck was filled with dirt. There were bushes, undergrowth, and a huge patch of fresh, ripe, clover, glistening from the past evening's deluge. Mr. Mizithra, his stomach aching, ran for a snack, Mr. Gilligan's Island right behind. They munched, making small talk.

Murray, not hungry, made for the bow and spot in the sun. A warm solar shower before going back on the road would do his little heart good. He pulled the kippah down over his eyes and snoozed for a bit. Not a good idea. In moments, his sixth sense forced a roll off the rock just as pair of talons swooped by his head. He dove underground. He was not a marmot, so whistling was not an option.

"HAWK! HAWK! HAWK!", was all he could muster.

The dining duo dove for a handy tunnel. The bombardier made a second swing on bigger prey. Mr. Mizithra's big butt was a tad too tight for the opening. A single talon stung the left cheek. Squeals of pain filled the tunnel as he burst in, rolling to a halt. His rump exploded with fire.

Quickly, the wounds were tended by Mrs. Maryann. She dabbed the open spot, washing away the blood. It was not as bad as one might think. Murray, a smart aleck, spoke first, smirk smeared across his face.

"So, Mizzy, how's zat old tookus? Ha-ha, sorry my friend, but you must admit if you are going to get it, it may as vell be vere it hurts ze most. It is not so bad, all you vill have is a schar und another juicy story to tell ze cubs. Could have been vorse, you know.", ending his joshing with a serious note.

"Thanks, my old chum. Remind me the next time, ow, ow, please madam, not so hard, to be as kind to you, ow, ow, ow. This, ow, ow, this is war. I am going to find that red tailed jerk and fix him. We have to fight back. That raptor must die!"

"Oh, sir, no, please, no, and again, no, sir. This same red-tailed devil torments us constantly. He has taken several from here over the years. That is why many left, it is dangerous. It is our home. We have learned to expect terror from above.", Mrs. Maryann's voice cradling surrender, submission, and sadness. Hers was the tone of one resigned to a universe of fear. This was no way to live, yet they insisted on staying.

"I am resolved to end this bird's reign of terror. Murray, ow, ouch. I am fine now, madam, thank you very much. Murray, let's go see what we can do about tempting him and letting him have it?"

"Vell, Vell, Vell, do you mean to say you vant to schtone him?"

"My thought, old friend, my very thought, indeed."

The tricky thing about trying to fool a feathered, flying killer like a vicious Red-tail Hawk, is evaluating the fantastic speed during descent. They go faster than most ground animals can react until it is too late. The way to get one: provide temptation, Murray. Kippah in shaky, furry paws, he stood near the farthest exit tunnel, to see what he could see. He caught no sign of the hawk. His bravery was overshadowed by devote timidity. He donned the kippah, breathing deep the gathering gloom.

"Murray, we can't very well use you as bait if you stand in the foyer and quiver. Now get out there. Make that stupid aerial idiot think you are his supper. Mr. Gilligan's Island and I will be in the middle hole. Mrs. Maryann is already hiding behind that little bush, yes, see her. She will get things going. Don't worry so much. Go out. Be a hero."

"Danke, you schlub, you putz, you vant my death. You tell mama how I died. Vat a real pal. Let me know ven I can repay zis honor. Okay, I am heading out. You get out zere und vait. If anyone is late, I am an appetizer. Oy veh.", quaked the shaky woodrat voice.

Mr. Mizithra got in perfect position, spotting Mrs. Maryann. Mr. Gilligan's Island was over his right shoulder. He nodded. She returned in kind. Soon, the big, bad red-tail raptor made it's entrance. It came over the top of the ridge, along side the mountain road, massive shadow skulking along side, sporting ominous black.

It came down, swooping towards an oblivious woodrat with a weird little hat on his head. For a hawk, surprise is paramount. The stupid rat was standing without a care in the world. He knew, he just knew in his evil, shriveled up, old, nasty hawk heart meat was on the menu. It was going to be fresh woodrat surprise. Delicious.

Narrowing on the perfect three point landing on the rat's back, his yellow eyes focused on the target. Nothing else mattered. Nada in the way. Nothing could stop him.

Whoa. Not so fast there, big fella.

With one miniscule second to go, Mrs. Maryann tossed a thimble full of boiling water directly in the bird's face. Mr. Mizithra and Mr. Gilligan's Island threw large, solid stones at his blinded beak.

The reverse force of missiles zipping towards the battered bird of prey's downward speed caused a fierce result: a pulverized proboscis. A forceful impact on the rocks gave devastating results. It was so strong, hawk feathers flew all over. The raptor did not move, save a quivering nerve. It would no longer injure, terrorize, or kill. The vindication was overwhelming.

"We will rest for the day and depart on the morrow. My dear friend is very much, truly, and totally done with this day!"

True to form, after a scrumptious meal, Mr. Mizithra shoved a last bit of lupine behind his right ear and told stories until it was time to go to bed. This round, he made certain they were happy ones. Mr. Gilligan's Island and Mrs. Maryann finally yawned, retiring after a full day to say the least. Murray missed his sax and jamming with his new band mates. He was glad to be alive. Lousy, rotten bird.

Chapter 13 Above The Rapids

Night fell. They left the body, waiting on the morrow. To be honest, they were terrified maybe, just maybe, the flying fury faked death. It was awaiting their return to pounce and repay the dastardly deed. Morning found him stiff, done, over, beyond the pale, part of the past, an ex-hawk.

Feathers lay strewn about the stones. Hardened blood, gushing forth on impact, drying, splattering over the crash sight. Eyes shut, his wings were splayed to the side. His mighty, deadly talons curled up, beak crushed, red tail no longer of use. He was deceased. Cool.

Making good-byes, the duo hugged their hosts, assuring of a return once the quest to the White Mountain was completed. They reiterated the need for an answer to the question of moving. It seemed to Mr. Mizithra this prospect for his new friends was ideal. Their current home was a ghost town. They would die here, alone.

Making for the tunnel, both turned to wave good-bye, burning the image in their minds to remember the place on return. This was learned behavior. Getting lost is not fun. Checking their surroundings always made for a wiser journey. Often, relying on solid ancient ways can salvage the future. It is good, therefore, to remember the past. They waded through the tunnel, out the other side, turning left, east up river.

The white water frothed and foamed on the right. They pushed uphill on rocks to the mountain top. The human road rapidly vanished from sight. The woods began to close in left. The temperature cooled. The elevation rose. The previous night's storm left a frigidness uncommon for early summer time. Still, snow was not unheard of this high.

They wended the way up to a solid rock bluff and turned to locate the ground beneath. The pond was nowhere to be seen. Gilligan's Island lay distant as the road turned. It set hundreds of meters down. The white water churned briskly, plummeting to the valley below.

Here grew pine and spruce. Underbrush thinned. The ground was strewn with pine fronds. Small, hard-surfaced rocks tore toughened paws, causing pain. They paused to give the sore feet a good rub and find a less arduous way.

Mr. Mizithra reared up on his hind legs and tail, instantly seeing a better vantage point not far ahead. He told Murray to rest while he got a good look at the area. The stone sat above the little resting spot. It was an easy climb in the shade of a Ponderosa pine trunk, thick enough to hide his approach. Caution was always the first word. Up here, though, animal life was sparse.

On top of the round, red rock, he lay low, scanning carefully what was in front, to each side and above. All clear, he slowly raised up again to see what he could see. A large field of tall, green and yellow grasses, filled with flowers, butterflies, and birds flying carefree. It's size gave pause. It was magnificent.

A light breeze wafted warm scents of the field to marmot nostrils, filling olfactory senses with the promise of fat producing greens for his empty belly. The entire field was ringed by large pines and spruce bushes filled with bush tits, chickadees, and wrens. Blue swallows swooped down diving for flying food, crisscrossing the area with ballet like movements.

At the far southern end, the field gently rose to meet a line of trees, solid, unmoving, firm. This locked in the area. To the west the stream tilted down to the lake below. The north end held a similar treeline. East, the field abruptly ended. A sudden vastness and quick, blue sky dotted with gray, cumulus clouds cruised slowly by. The place was massive, the largest field and pond combination they ever discovered.

"Come up here, Murray, we have found the most glorious place to rest for tonight. All we have to do is find a hole in the ground. I can dig us out a nice room for the evening."

Murray raced to the viewpoint, surveying the bonanza. It was a stupefying sight for a rodent. Eureka. Good, fresh, plentiful food grew easily at hand. Water and an abundance of animal life meant safety. Most assuredly, a nice wayward respite for a well-earned night of rest. In fact, they needed to sit and recount the days since leaving home. It was nearing two weeks on the road.

Their lives would never be the same. The climb wore them down. The rough rock, steep ascent, the thinning air combined to slow them to a near crawl. In short, they were tuckered out. The boys needed a break.

Now, on this bluff, they found an oasis providence made available at the exact moment needed. They were a thankful pair and made downhill for the immediate meal. They dove in the blossoming flowers, gorging until full. Soon, other forest denizens appeared, black-tailed deer, wapiti, skunk, field mice, Douglas squirrels, a short tailed weasel, and a pair of otters. They either merely waved as they passed or stopped for a chat with the new visitors. It was a truly enjoyable, pleasant rest of the day.

Mr. Mizithra soon found a suitable spot under a quiet, hidden lull between two trees, a half meter above the other. It was a cedar. The little crook in between made the ideal location accessible with an easy exit. A second one made it a snap. A large room under the roots of the lower tree finished the temporary abode. Not bad for an hour of hard labor. Murray, for his part, carted off as much as he could muster with his little hands.

They smoothed the front to make a nice entrance, the same to the rear escape. Next, back to the field and another exquisite epicurean episode. It was. It was again for Mr. Mizithra an hour later and then again another hour advanced. This marmot was famished.

Soon night would fall. They kicked back on the large roots of the cedar outside the cabin in the woods basking in the late sunlight. They felt very good about the day, the entire journey, if truth be known. So far, with the hawk and fight in the lodge, it was fairly interesting.

Night came, they slept. All was good with these travelers.

Morning arrived. Being the Sabbath, they rested today for Murray. Warm, wafting early summer breeze greeted them. The sweet, orange sun arose, which it does each and every day. Once again they gorged to the gills with the lovely green stuff about. Murray found a bug or two gave a dash of essential protein on the rough road. Taking in the entire expanse, it was a perfect resting place for the colony, provided they could find somewhere to move. Food was plentiful.

The small passageway was the start of a good weigh station for the thirty odd members making the journey to a new home. The lake was the largest they had ever seen. Murray could swim forever. They hid the hollow entrance for possible future use.

The next day, they made for the narrow gap at the southeast end of the meadow. Here was a very vivid, lively, living place serving a great deal of the wildlife traffic thereabout. Food was abundant. No sight of bad guys nearby or hovering from above.

Once on their merry way, the path was over-trampled. It seems the various critters used it leaving the oasis. Once clear of paw and hoof prints, Mr. Mizithra reared, getting bearings on directions to the White Mountain. The pair moved quickly, but quietly amongst a dense thicket of green bushes. They broke on through to the other side. On occasion, he repeated rearing up on his tripod to determine location. It is good to always know where you are. Common sense, is what animals use.

Continuing for a couple of days, they rested in an old, used burrow, hollowing out some fresh areas to sleep. The next day, it was a hole under a pile of big rocks. The crook between formed a nice, safe resting place for our lads. The time went quickly. They could spy the white tip of the mountain coming closer every step.

On the third day past Grand Meadow, as they came to call it, the journey began to change. They rose in elevation quickly meter by meter. They reached above most of the undergrowth and bushes. Up here, only trees remained. The soil was long void of the necessary nutrients to grow or sustain plant life. The trees' roots could dig deep enough to find the lifesaving water, which is why they survived. Only man, fire, or lightning could fell them.

Suddenly, there it was.

Sunlight shone directly upon the albescent sheen. The snow's reflection was overwhelming. The compadre's joy was palpable, fervent, honest, sincere. The feeling is the same you receive realizing a goal is reached. This fulfilled a major stage in the journey. They felt complete, alive.

This would help the colony, making it more personal. This solemn undertaking was philanthropic. In the case of Mr. Mizithra, giving blood from the scrap with the red-tail. He had flesh in the game. Murray gave up ten years of life being offered as bait.

Making this glorious moment more incredible, reaching the summit of the gap, taking in the full panorama of the opalescent sight before them, a chorus of angelic voices rose above the din. The air filled with the sounds of life joined in exquisite song. It was a composition so lovely, so gracious, so moving, they knew whom it must be.

They heard of these songstresses since cubhood. The vocal tonalities were legend. Here, at this very moment, as if things could not be any grander, the lovely ladies sang a morning serenade, bringing the boys to tears. They were impassioned with emotion. The singing drove hearts to near bursting with happiness and pride in a job well done.

"Mizzy, listen, it's zem. It really is. Listen to zat singing. Wow. Ze legendary Covey of Quail Quartet. Und to hear zem just now, it's, vell, it's a miracle."

"It's a sign, to be sure, my friend. Now, let's sit, enjoy the view, the sound, and some of this lovely fern. Never seen one this delicate in all my days.", claimed a completely content companion, munching ferns.

They rested for some time, appreciating the journey hitherto, the situation at hand, and luscious glory of avian vocalizing of the lady quail choir. Their sound gave the fellas inner calm. It was a soothing tonic to the soul. They spied some berries and helped themselves. When the hens rested from their cheerful chimes, the travelers regaled with unexpected applause. Quail scatter when frightened, so it took a minute or two to gather them back together.

"We thought ourselves quite alone. The quiet of the morning…The quail call for song quickens in early time. We love to sing loud enough to make the leaves quiver. The Covey of Quail Quartet has been singing our songs in this quarter of the mountains for generations."

"Humans pushed us further and further away, quashing our lives up below Grand Meadow. The Thumbs quite like to devour us, our precious eggs especially!", quipped choir director, leader, and great-great-great-great granddaughter of the original quail to form this illustrious alpine choir, Mrs. Queisha Q. Quail.

Her name, Mrs. Quelita Q. Quail. She truly was the queen of joy. Infectious was the term to best describe the aura she exuded to a real love of life. When someone is honestly happy inside, it is best to let it out. There is nothing wrong smiling at the pleasure of singing, breathing, and living. Being alive, appreciating it, is a lot more fun than a life of misery.

Introductions followed: Mrs. Qitarah Q. Q. Quail, the eldest hen of the covey, the essence of good taste, feathers always perfectly quaffed. Fragrant as her name betrayed, she had an impeccably shiny beak, her vocalizations legendary. Next came the heart and soul of the quartet, and only single lady in the covey, Miss Qubilah Quail. Given this name at birth, it seemed she was destined to fulfill its meaning. No quartet can long survive without a good, solid, harmony voice. This, she possessed in ways of no other quail in the choir's long history.

You see, only married women had the middle initial of Q. This is how they designated being wedded in contrast to humans. Human wives take the husband's last name, confusing everyone. Qitarah remarried. Hunters ate her first husband, prepared with a lovely Zinfandel-orange sauce, poached young, white fingerling potatoes, and braised Brussels sprouts in butter and balsamic vinegar with goat cheese, and warm apple strudel!

Last of all, and sweetest, Mrs. Querida Q. Quail. Upon introductions to this lovely quail, she instantly became beloved. It was her manner, gracious soft speech, and sincerity of person. She was the one all quickly grew to love. She made it easy, placing no impediments in one's way.

You see, for love to reach you, your walls must tumble. It is obvious we need protection against evil and harm. Taken to extremes, isolation occurs and death. The only barriers to love, then, we place ourselves. The time spent tarrying with the fine, female, feathered folk was fantasy.

This respite ended quickly for the mission awaited participation. Everybody needs to rest, to have a time out to recoup energy lost. These vagabonds were in need of such a waylaying. The lovely ladies made their first view of the great beyond magical, a good sign for the journey ahead.

Chapter 14 Dramatic News

A small, barely worn trail descended into the valley. This path found it's way to a babbling brook at the bottom. To the right, south, sat another ridge. It's path disappeared over the rocks. This was the correct way. Others skirted the canyon edge. Any raptor would be easily spotted. They had to be wary at all times. The sun kept them company all day. They meandered here and there, pausing at times to eat and gauge progress.

Exploring is exhaustive work because you have to double back. Poor Murray's paws wore out from clinging to the stones for fear of falling. Basophobia is real, anxiety heightened every time a stiff breeze pushed him around. This environment was not conducive for woodrats.

The plunge downward made it warmer. Sun rays roasted the rocks. Good fortune was found again. They located a small cave in the stones as they arrived to an obvious little clearing. It was too late to make a secure resting place. Spelunking was the order of the day. Poor Murray became nervous. No one knows what is lurking in the dark waiting to pounce.

For a little woodrat, this was a very real possibility. Cautiously, they explored every nook and cranny until certain the place was animal free. A bit of grass, moving of a clump of dirt or two, and they had the most incredible view for the evening. Hunger got the best of them. Feasting began in earnest. A small pond provided water. Mr. Mizithra wrote in his journal. They had been on the road twenty-four days.

The quail story would excite everyone back home. He relished the opportunity to put his particular twist on the retelling. He gained a bit of weight, but not nearly enough to sustain him through hibernation. Instantly, he grabbed a clump of flowers growing on the edge of the cave and chewed away with fervor.

All was quiet until a movement came from the same direction they earlier traversed. Both of our travelers were gripped in fear. A soft noise broke the evening's stillness. At once a mass of brown moved closer, again quietly and slowly. Whatever it was, it was very cautious. And, this shadowy creature was considerably larger than Mr. Mizithra.

"You there, in the cave, whatever you are, I smell you. Make yourself known. My name is Jon, a black-tailed deer. You can call me Jon Deer if you like. Come out of the cave, you will suffer no harm".

The voice was strong, not intimidating, rather confident and sure. A mouse's voice, a chickadee, or any small critter would be higher. That of a large stag such as Jon Deer was deep, resonant, strong. At the same time, it was tender. Mr. Mizithra and Murray ventured into the light.

"Ha-ha, well, well, well, a marmot and a woodrat. Are you the two who come from up on Hurricane Ridge? Ah, we have heard of you small ones and the journey to the White Mountain. A big fight at the Beaver Blues Bungalow? Wow. You slayed a red-tailed hawk? Very impressive for rodents. Word gets around, mountain gossip. Okay now to your needs. You see it there, the way is long and you cannot reach it with your size".

"I will tell you though, for your plight is known amongst those of us in the forest who care. As man approaches, we die. They hunt my kind. My own brother was killed last year. The shot hit him in the heart. He was dead before he knew it. Sad, sad, so very sad".

"Since my travels chanced to intertwine with yours, this I will do for you. There is a place not far from here for you. You see where the ridge rises slowly, disappearing at the peak? Yes, there, exactly. Okay, well once you reach a little clearing, do not follow the path which leads to the immediate right".

"Instead, push on left a bit further, there is another smaller, almost hidden trail. You must pay attention for the way is narrow, the other, wide road, leads you to certain doom. A little slit in the undergrowth will show the path to safety for you and yours."

"Once on this bottleneck path, you journey a crooked, rocky road. At the end is the place you will discover. I feel it is the best place for a colony of marmots. It is called, the Valley of the Stones. Once a mighty glacier lay there for many centuries. Over time it thawed enough to break off a ledge and land there, down the trail."

"It is this place you must find. There you will locate the new marmot homeland. Having been there before and knowing something of the ways of your kind, it is an ideal location. This haven will be, for you, one of perpetual safety, food, shelter, and growth."

Jon stayed a while longer, then insisted he had work to do. His task was not over for the day. At home wife, My, awaited his return. He bid them adieu. Instantly a bobbing tail was the last thing they saw. The help he rendered was invaluable. He was a kindly sort. Deer had a reputation for kindness. Wapiti, not so much.

Morning brought fog below creating a surreal effect; the mountains were floating on a fluffy sea of white. Murray especially enjoyed the view. It bestowed a feeling of safety: hawks cannot find you in the fog. A quick bite and they were off, checking behind and ahead for bearings. They would recall it for the journey the entire colony had to endure. Getting lost at that point would spell disaster.

In his rodent mind, Mr. Mizithra's internal gears ground and churned, working out a time schedule for the colony. Were they to find this Eden, it was two days ahead according to the deer, which would be 26 days. Word could be sent back. Leaders could organize the tribe to move. Murray's family was busy getting ready to relocate to the Beaver Blues Bungalow. It would be beneficial if they were to coordinate. This would provide safety for Murray's mother and grandmother.

All day they strode with purpose towards the little rock ridge on the canyon's rim. The path was not always easy. Little gaps appeared to impede passage. For a deer no problem, for a fat little marmot, creativity was required. Luckily, he was full of that particular necessity. They traversed the narrow way with few problems. Night came. Sabbath the next day and rest. They found a pine outcropping near a small rivulet behind the cliff side. It's roots provided a quiet shelter for the night and a chance to recant the day's activity.

Progress was excellent. The next day would provide an opportunity at a new home for Mr. Mizithra's family. Meeting Jon Deer was extremely fortuitous for all concerned. Life is interesting that way, you simply never know when good things will pop up. You must, therefore, keep aware of being lest the best stuff passes by whilst you nod.

So filled with anticipation he could not sleep, our marmot was up and rearing to go before dawn. Murray knew his chum was anxious. He did not object to being stirred rather early, though could easily sleep a bit longer. Never mind, he mused, time to rest once they got to the valley. At a quicker gait set by Mr. Mizithra, the day soon focused on one thing, one goal: the glacier valley, the Valley of the Stones. Murray did his best to keep up with much longer legs, eventually grabbing a handy tail, hanging on for dear life.

Near noon, they could feel a shift in the wind. A parting in the trees betrayed the necessary evidence: they were near the edge. As if it were not fast enough, Mr. Mizithra picked up the pace and sped along. Pleas to halt were completely ignored. He was driven. Suddenly, the opening.

Majestic is perhaps the word to best describe it. Yes, majestic. Jon Deer was a deer of his word. The valley was perfect; East, the cliff fell a thousand meters to the valley. Any creature wanting to get at them could not do so from below. West, a lofty wall rose gently, then perfectly up and down, built of solid stone. Firs and snow banks topped the rampart.

To the far, southern end, trees rose to fill the sky, then clearly were gone. This meant the space beyond them was short, additional good news. This would be a nice ditch out spot to avoid any intruders. Then, the valley floor lay before them.

The path down was strewn with large rocks, remnants of the late glacier. Cleared, this would make an excellent way up and down. It was extremely narrow, good for them, bad for predators. The rest of the old ice mass lay on the hillside, this portion having broken off as the glacier moseyed down the mountain over the centuries.

Glaciers traverse downward due to gravity. Over time, they melt enough to move. An earthquake or avalanche can cause the same effect. When a glacier is in the way of a route in the Himalayas, they cut a new road through it. Some glaciers have five or six such alleys sliced across them. In this case, about ten percent broke off, landing on the platform below. The rocks remained when the ice thawed. A pattern was formed, culminating as they dropped over the side to the unknown abyss.

This was absolutely perfect.

The possibilities were infinite. A massive village of marmots could exist here. They could spread out so multiple families could co-exist without eager males getting their dander up come mating season. No sign of man was anywhere found. The only way to find this slice of heaven would be by flight. It would be extremely tight for any raptor in this high an elevation. By the end of this glorious day, shadows crept away replaced with the cool of the evening.

It mattered little. Our sojourners found a marvelous place to begin at the bottom. To descend, it took a bit of hopping. They knew a way back would have to be provided. Once down, they could cut off the world, reachable only by those truly desiring to say hello to a bunch of marmots. He could see an excellent way to build a protective wall, concealed to provide safety from predators.

Any sort of hunter would have great difficulty trying to go through the maze of rock in a pattern left by the thawing process. No cougar, lynx, coyote, or fox could pierce such an impenetrable fort. Mr. Mizithra knew in his heart this valley was the ideal place to rid themselves of man. Only a bear could make it were it to descend.

Utilizing wits, the next day dragonflies were sent with haste back to the beavers, then on to the colony on Hurricane Ridge. As fast as these zipping messengers were, it took a full day to reach the lodge, arriving at Mr. Mizithra's homeland at noon the next day. As one dragonfly tired, another took it's place, carrying the precise message. They really are amazing creatures of nature, aren't they?

The news sent a ripple of excitement throughout the colony. Mrs. Lip Gloss made the joyous announcement and prescribed instant moves to evacuate. Get rid of everything you absolutely do not need was the charge. This would not be an easy journey. Some were old, some young, and a few injured. All were going regardless of condition. Man won, they had to skedaddle. Tag your own ears, humans.

Dragonflies informed her of the beaver lodge. The ability they had to rest there made the initial leg less arduous. Their joy filled message included news of the rapids, Gilligan's Island, Grand Meadow, and the Valley of Stones. They need only arrive at the Beaver Blues Bungalow. The vagabonds would come back for them once the quest up the mountain was fulfilled.

All agreed their marmot brother found a suitable, if not superior new homeland. They had only to move an entire colony. In a flash, the lady woodrats were contacted, providing them relief. They worried about Murray and a journey at their ages.

Once mama's hoarding larder was empty, the load was light enough to carry. The Northern pintail ducks benefited. Spirits were high, high enough a massive party broke out. They celebrated good fortune. Mr. Mizithra was more than a mere yarn spinner, he was a lifesaver. All hail, Mr. Mizithra.

Now they had a new home. No longer did they need to fear man or his cameras intruding, dogs hunting, guns killing, or his fires burning. Salvation was available, they just had to go and get it. They were pioneers, following a trail blazed by Mr. Mizithra to the promised land. Within a day, they were ready.

Marmots do not have much to pack. They carry most of it around with them. They are the only pioneers to gain weight along the journey. Mrs. Lip Gloss stopped in to get Murray's family. Off they went towards the Beaver Blues Bungalow, the old home a memory. The tribe lived there for untold generations. It was a good home, but change is a part of life. Often adaptation is dramatic; this was. It would leave a mystery for the Thumbs. The marmots could care less.

Chapter 15 On To White Mountain

Having secured this oasis and knowing the colony was heading towards them, Mr. Mizithra and Murray burrowed for a couple of days to get some nice tunnel action going. When the marmot nation arrived, all would be exhausted. Murray would head back to the lodge and direct them back while Mr. Mizithra worked on the digging alone.

Dragonflies whizzed back, providing information from woodrats, marmots, and beaver alike. All looked good. They decided to make for the White Mountain. The goal lay before them. It had heightened purpose, as the other aim of a permanent abode was successful.

Sending a last dragonfly to impart their intention, they wove up out of the Valley of Stones, to the pathway which led south. They had to walk a long way. Finally they spied the alabaster peak at dusk. Food and sleep followed. Two more days found them in darkened shadows, unfrozen snow packed about as they steadily rose in elevation.

It was colder all day, brutal once it got black. Stars did not even shine in this place. Once this high, the environment changes dramatically. Still, with a goal at hand, there was no time to worry on small problems. They slept next to each other, back to back for warmth. The small hovel kept out the cold and wind, but little else. With far fewer of the enemy about, the fear was cold and a lack of food.

An odd precipice faced them the next day when cresting a small, rocky rise. Pines were sparse. Frost covered rocks and cold, hard dirt were the predominant features. During this whole day they saw no other animals on the ground or in the sky. Landing on the cliff of the precipice, they were close.

However, deep, white stuff was plentiful. They needed to get clear of it to stay warm, dry, and on track to the mountain.

Murray hit a patch of snow ahead of his larger companion. It was icy. He slid off the edge of the cliff, disappearing in the fog. His screams soon dissipated. Mr. Mizithra was suddenly alone. The only thing left was his tattered kippah. It fell off as he plunged. Murray vanished in the cotton like white, pea soup. Life is like that, you are here, you are gone. Life, then, is measured by how you fill time between the birth and death.

Distraught, he sat on the stone, head in paws, sobbing, clutching the ragged kippah. The shock of such a dramatic event occurring right before his eyes was devastating. Losing someone dear with no warning leaves an emptiness which can never be refilled. In his heart, he could not push on any longer. Mr. Mizithra sat down, pondering what to do. A myriad of thoughts swelled his mind. He was alone, friend gone. He was frightened, cold. He stayed that night on the same spot, begging to hear something rise from below. He hoped the hope of the hopeless. Murray was lost.

Morning broke. Mr. Mizithra was unmoved. Mourning and grief are very powerful emotions. The loss of loved ones, of friends, of pets takes a toll on bereavement. Lamentation wears you down. By holding in pain caused by death, a cork plugs emotions. They must be exposed, exorcised, exuded in order to continue life in a productive fashion. The pain never fully goes, impossible. You can diminish it by grieving and mourning.

Here we may learn from Mr. Mizithra. He did not destroy anything. He did not hit or hurt either himself or any other creature. He sat down, had a good cry, reflecting all night about life with his dear friend. He resolved with Murray gone, he must continue the objective to White Mountain. This deflected inner emotions well hidden. It stirred his heart. He would forever bear the burden of this passing, as is perfectly normal.

When we lose someone close, there is no way to "get past" it. In fact, it is impractical. To give yourself in love to a spouse, child, parent, friend, or a pet, is to impart part of the inner you. There is only so much of you to give. When someone passes away, your spirit is marred. The hurting reverberates throughout body, mind, and spirit. It is natural and reasonable to experience internal pain from such a traumatic event.

"Hallo?", came a marmot's trill whistle of identification.

The sound broke concentration on Murray's kippah. He clutched it all night. The oddity of the whistle shocked him. What in the world was this call? He turned, eyes puffy from the tears and a lack of sleep. A male marmot stood a dozen meters from him, on a little stump of a fallen spruce tree. Mr. Mizithra returned a like whistle. The stranger's signal bore a different tonality. Humans would say he had an accent.

"I am, Mr. Mizithra. You startled me, sir. Sorry. I lost my traveling companion last night. He slipped off this precipice on that patch of ice. He was gone in an instant. We traveled from Hurricane Ridge, heading for the White Mountain over there.", he said, pointing towards the bleak obelisk across the divide.

"Ah, sorry for your loss. Was he a marmot? No, hmm. Well, no mind. It is mournful to lose one so close to you. My name is Mr. Milk Chocolate. My village was over the side of this mound, down the hill, up the other side, a good five kilometers south."

"We were attacked by Ralston the Rancid Raccoon and his Vile Vole Vanguard. That rotten raccoon tore our homes to shreds. The voles hit us from every side, burrowing in, attacking women, cubs, elders, and the injured. We scattered. It was a total annihilation. We had no chance. No time to even give a warning whistle."

"For some time, I sought other marmots to no avail. Alone, roaming, I have seen many things. All I want is to find a place. The devastation our colony suffered is not unheard of in these times. I must find a new home. Afterward, I tell you, my sole purpose in life is to see the diabolical raccoon and his vile voles destroyed."

"Dear friend, my task for the colony was to find a new home. Before my companion fell to his death, we found such a place. It is vast. You are welcome to our new settlement. There you can build a new life. Together, perhaps, we can defeat this evil source of pain and sorrow. I know him."

They spoke for some time. It eased Mr. Mizithra's pain. He focused on a specific problem. By so doing, it temporarily averted his mourning, providing purpose. This is an excellent way to deal with loss; not to hide from it, rather accept the reality of death and prove you can move on without falling apart.

Time, this takes time. It is very difficult, more so for some than others, but the method does work. This can ease the pain of loss as you see life continue. Sadly, we all lose someone eventually. The need is there to help those who suffer by those who suffered first.

Soon, Mr. Mizithra felt at ease as did his new friend. Their chance meeting was positive for both. Alone, we struggle. With sympathetic and understanding help, we recover better. Suffering alone is rarely the best answer. He would never forget Murray. He knew life had to go on, just vastly much more vacant. He knew for his wife and girls deaths how much it could hurt and for how long.

Mr. Milk Chocolate sent a dragonfly to let the other survivors know what he was going to do, and where to hook up, then agreed to go with Mr. Mizithra on his quest to White Mountain.

Chapter 16 Not So Fast

Actually, death was not the case. While Murray did fall, it was most amazing. With spread out furry arms and legs, he slowed. He glanced a pine bough on the edge, softening landing on a soft bed of deep snow. He sunk very deeply into the white stuff. Due to the way he fell, impact was minor. He bonked his head and fell asleep. Time passed as he lay still, sheltered from further harm.

Having no idea where he was upon waking, the first thought was of his friend. The tumble was so far down his little voice did not carry far. He wanted to gain his bearings and head back up. Mr. Mizithra would be distraught. Murray wanted to reach him to relieve the mourning.

Murray was on a small, snowy outcropping. It was daylight. The tiny, ice-laden shelf was three meters wide by five long. The shape was similar to an eye. It was half open to sunlight, half dark in shadows of a shallow cave. He moved off of the snow to the dry portion. Rocks from above lay where they landed, shards of stone strewn about. The wall was straight up, a few minor, small outcroppings dotting the journey upward.

Murray inched towards the edge, nervously peering over the side. It was a sheer drop into oblivion. Fog lay thick below. Another parachuting episode was impossible. He was stuck. Climbing up top, he would have to work step by step up a wet, cold, windy rock wall. He would have to do it without any assistance.

Resignation is when you give up. Surrender, capitulation, defeat are terms which describe his immediate emotional state. To a little woodrat, it was as if a human had to scale an icy cliff a kilometer high solo. It was insurmountable. He knew it. The acceptance of the situation sunk him into depression and despair. Death was all that seemed possible. Food was nowhere in sight.

"Say, little fellow," broke the stillness.

The voice shook with the heftiest tones Murray ever heard. He quivered from a combination of fear, cold, and shock. It stirred the terra firma upon which he quaked. For a moment he froze. There was no way up. Down meant certain death. No escape. He was trapped. All he knew was an omnipotent sound called from the heart of darkness. An immense thing was in the cave.

"I must say, I never expected anyone to visit. Come closer, little rat, I promise not to hurt you."

Now, for some reason, while he should not have thrown caution to the wind, Murray trusted the booming bass emanating from the black cave. It was the tone, one of comfort, goodness, and kindness. Slowly, he ventured toward the shadows. A large, light brown nose surrounded by lighter, curry colored fur and long, white whiskers prodded from darkness to light. The mouth, while once a dense brown, now lay rimmed with white hairs. Four immense, dull, yellowed fangs, the lower right one heavily chipped, pushed out as the old, worn teeth showed.

Murray knew it to be only one thing: a cougar. Bears, raccoons, bald eagles, hawks, dogs, or any other hunter paled in comparison to this big cat. They were solemn, silent, and stealthy. Oft times, victims were dead before they felt the attack. No other forest denizen could beat them, not even bears. Murray knew he was not much of a meal for one as large as he. This led him to wonder what a cougar was doing in a place so remote.

"You know, Mr. Mouse, it is my good fortune you fell my way. I ask some assistance. The reason for my presence on this cliff is this is my place to pass away. I am old, tired, no longer able to feed myself properly. As you know, the law of the forest, the code of the mountains, the credo of the cougar is when you become a burden, you must go off to die alone."

"My name is Einar," he growled with distinct pride, "I am he who fights alone. Those who knew me in my prime called me the ruler of this forest. I hold to the old ways of my kind. Here I was born, above this precipice in the cave beyond, now here I will die. My wife was slain by man many years ago. My young ones grew and left. My days are finished."

"Some time back, I spotted this place and made it down. Here I am and here I will pass. It is simple. No reason to get in a tizzy. Don't get your fur in a wad. We are born, live, die. It is the cycle of life. I do not object. No, little fellow, no, the purpose you may serve me is one of relief."

"You see, as I leaped down, I hit an unseen branch hanging out from the wall. A shard of it lodged up here, ow, here on my back. If you would please be so kind as to remove it, I will get you up the cliff side to safety. What do you say? I solemnly promise not to eat you. Food will never enter my aged, dying body again."

So sincere, so grave were his words, Murray at once climbed aboard the massive Goliath. He located the offending, imbedded, bloody wooden stake, planted his back paws, and yanked with all his strength. With a loud pop, the woody javelin burst free. Murray fell back, tumbling off the great beast to the cave floor, at his gargantuan paws. Blood instantly flowed as Einar let go a deep roar of comfort.

"AHHHHH. That was so painful. I don't mind dying, but that was too much. Thank you, tiny Mr. Mouse."

"Murray, oy vey, Murray is ze name. Of course, I am more zan happy to help any creature in need. Even vone as immense as you, sir. I fail to see any vay to get out of zis und, pardon me, sir, but you cannot get out of here eizer."

"Oh ye of little faith," Einar mused.

A sound came bellowing from deep inside to his orifice, shaking the landing. A few pebbles fell vibrating beside Murray who leaped from fear. Snow dropped in large clumps over the side to oblivion. The explosion of sound took a great deal out of Einar. He sunk against the rocks in the cave in utter exhaustion. He was so close to death, this final shout about finished him. He panted and wheezed, seeking comfort from the cold cave wall. His breathing dragged. His eyes slowly, softly shut.

Energies spent, Einar lay with a slight smile painted across his face. Pain subsided, easing the pathway to death. Within moments, noise traveled from above. It was sharp. The rocks clacked together. More snow fell. Sound reverberated off the canyon ramparts. Again, pebbles came crashing around poor Murray's head. Snow shimmered lightly down the innocent ravine.

In an instant, standing before him were eight white, hairy legs, eight hard, black hooves, and two big, white, bearded faces. Mountain goats are able to climb up and down where no other animal can. A little known fact: they are nature's attendants for those passing away. They are like noble humans who tend the dying, yet are not related. These wonderful people do society an important, necessary service. They see us pass from life to the next in as comfortable a fashion possible. Bravo hospice.

It is said no one knows where animals go to die. This is true for the Thumbs. Pet cats often wander off to pass away, leaving loving masters to crossover alone. Then, there is a saying about elephants. In the wild, animals reach the point when the end nears. They contact goats to let them know.

At this point, these marvelous bovids go to work, ensuring the one passing does so quietly, undisturbed. In regular times, a mountain goat stays far from cougar for obvious reasons. When death approaches, things change. The mountain lion no longer eats. The goats are quite safe. Theirs was a grave task. Species prejudice was not a thought. They kept silent for all others. The dead should not be disturbed.

Einar's bellow caused them to believe, quite correctly, it was time for Einar, quickly arriving to carry out final rites. He was a noble beast, feared as a mighty animal. This made all creation his family. The loud death call was for them to witness the end. In a gesture proving him not to be heartless, he saved his little doctor. The pain appeased, Einar's entire body relaxed, breathing quite slowly. He had just enough strength to summon the goats. It was done. The final effort drained him.

"What have we here? A woodrat, now? No one is supposed to be here, ever."

"He, he fell. Oh, oh. Pulled the branch, pulled it out. Saved me. You, you, Gottfried, you take him up. Gerta, please stay.", Einar weakened.

"He is traveling. Slipped on the ice last night. Ohhhhhhhhhh, the pain. He fell. Landed here. Brave. Brave, I cannot thank him enough. Now, I can go. I can rest. I, I, I am finished", Einar's death rattle sounded. He twitched a bit, then lay still, spirit departing now vacant eyes.

Einar was gone. Aristocratic, strong, fearless in life, now still, lost, lifeless. In his final moment, he saved Murray's life. For us, the end of our days come down how we lived life, treated others, and served our planet. Toys and possessions are seldom remembered. It is the love we show, the goodness we share that make life. It was so for this stately mountain lion.

He was a cougar, but his soul was pure. He killed other animals to survive, just as herbivores kill plants to exist. In the deepest depths of his soul, though, he was a good beast. Saving Murray proved that to him, Gottfried and Gerta.

"Let us thank this time. We are here to see a fine feline pass. He is gone. We must perform our task for him. We have for a good many of his relatives. This is our duty in these canyons, mountains, and forests. Gerta will take care of Einar, do not worry, my friend."

"I see you are touched. What he did was incredible. With his dying utterance, he asked not for himself, but another. Fine, noble, an excellent end to a magnificent life. We shall solely inform his final, surviving son."

Murray stood before the ex-cougar, paws together in prayer. A tear flooded his eye. He wiped it from his nose. A short time prior, he was bounding about with Mr. Mizithra, headed for the White Mountain. Now, he stood in total permutation from events of the past hours. The incident forever altered his entire existence.

Never again would he take life for granted, or the good someone did him. He would, whenever possible, do right for others regardless of who they were. This gave an inner calm hitherto unknown. Murray knew he was blessed. He grew.

"Climb up here on my back. Hold on to the long hairs between my horns. It is the safest place to be. Do not let go. If you do, there is no way to save you. No one can go down a valley and come out. No one ever has."

"Now, grab tight. Up we go. I jump stone to stone in ways only Mountain goats can! Nothing can follow us up and down. Only the Thumbs stop us, the evil beasts they are. They flew out friends of ours last month via helicopter."

"Okay, here we go!", with that, Gottfried reared up on his back legs and sprang to the heavens, landing on a miniscule spot. One after another followed, terrifying poor Murray.

After a few choice leaps, to his amazement and joy, Gottfried landed on the exact spot of Murray's slip spot. He climbed off, shinnying down the front left fetlock. He stood in front of his transporter, looking up with an odd grin. Warped, he thought of asking for another go. It was fun.

"Gottfried, I am humbled by zis, all of it. To see Einar die, zat vas so sad. In falling, I thought I vas dead. Oh, it vas terrifying. Und vat he did for me, vell I am truly very thankful. Thank you, Gottfried und please thank Gerta for me, too."

"Now, I must find Mr. Mizithra. I see marmot tracks. I shall follow zem. Again, thank you. I vill tell no one, of zis you can be certain."

With a blink, Gottfried nodded, plunging back down the cliff wall, gone in a breath. Murray, shuddering, looked over the drop-off, clinging to a handful of grass. His full mind raged vividly with thoughts of what happened. His first goal was reconnecting with his friend. The canyon's trail led in the same direction they were heading when he fell.

He quickly deduced Mr. Mizithra headed towards their aim after considering what course to take. He spotted an additional set of male marmot tracks. Interesting. Where did he come from? Knowing his friend well, his mind was set to reach a goal, nothing would deter. Mr. Mizithra was possessed with resolve. Resolve is a most powerful emotion.

Murray ran.

Chapter 17 So Very Close

While his burden lay internally, it was nonetheless there, heavy. Mr. Mizithra sucked it up his mind, pushing on with his new friend, Mr. Milk Chocolate. Since both were mature marmots, they made a brisk pace, only pausing to eat. Of course, dining took quite a while. If they came across some lovely patch of flowers, clover, or soft ferns, it was time for a feast. Food up here was sparse. Pine nuts were plentiful.

He still needed to gain weight, like a clock constantly ticking. He had to add tonnage to survive. It was only during these lengthy meal breaks Murray made up distance. Nabbing a wad of insects, he could fill with protein and keep moving. Greens took more time to consume.

Daytime muddled along. The marmot duo found the canyon's end, turning east. Bam, the mountain began. Hard summer snow blanketed everything everywhere in sight. Odd massive, black rocks or small green pine trees struck out of the alabaster canvas. Before them lay the White Mountain (Mt. Olympus). Mr. Mizithra's heart swelled with joy. Mr. Milk Chocolate did not fully comprehend. They made camp for the night in the hollow between some fallen trees.

Gathering leaves and pine fronds, they made a bed. It was so dense, they could rest safely without worry. Nothing was small enough to get through or strong enough to tear it apart. A bear could, but the marmots would simply escape out the back side. They felt safe.

"In the morning light, we ascend the White Mountain. This is the baby toe. We arrive on the morrow. My only sadness is Murray will not see it. We traveled many times together, so many adventures. We were good friends. It is difficult to accept he will not be here on our most important journey."

"Not so fast, my friend,", quipped a familiar Yiddish tone, "I sink I have somethink to say about it."

In a flash, Mr. Mizithra burst from the nest in the tree and found his friend. Eyes bulging, furry mouth agape in a broad smile, yellow teeth gleaming, he fell upon Murray, tossing him lovingly in the air. Upon landing, the kippah went on his head by a beaming buddy. Murray told of the fall and return, omitting entirely the episode with Einar. His tale was the mountain goats helped.

No mention of a dying cougar, nor would there ever be. If someone asks you to keep a secret for a righteous cause, it is your duty to oblige. This is the decent and honorable thing to do. After all, you would want someone to do the same for you, right? Trust, my friends, is a precious and rare commodity.

Though he only knew Einar for a moment, it mattered little. The obligation fell to Murray's own character and conscience.

With joyous reconciliation, plans for ascent revived. Knowing snow would be the main hindrance, they employed an old tactic which served well in the past. Murray came behind a quick digging marmot to relieve his snow tunneling long enough for Mr. Mizithra's nose to warm. During the period behind his large friend, Murray hard packed the sides, thus securing the exit when the time was right.

With the addition of another marmot digger of equal ability, they could shoot through the path and climb steadily upward. By aiming for a higher, rocky outcrop, they could take a breather long enough to plan the next pile of stones and appreciate the view. Poking a head up through the white stuff, would ensure reaching the goal on a steady path. It is very easy to get lost if you cannot see where you are going.

"I have to ask", queried their new found friend, "why do you scale these peaks? I never heard of mountain climbing marmots or, for that matter, woodrats."

"Because it's there", they acclaimed in boisterous unison, grinning from ear to ear. It was a joy to see the chums together again, it truly was.

Dumbfounded, Mr. Milk Chocolate shook his head. He prepared to sleep, anticipating morning's incline up White Mountain. The lodgings secure, the trio slept deeply. It is amazing how fresh mountain air works on a tired body. This well earned slumber was a tonic to weary bones. A good night's sleep healed. Morning found them ready and rearing to go. Excited, a quick meal was consumed. Time to climb!

Agreeing to make for a large outcropping two hundred meters ahead and above, Mr. Mizithra shoved his furry snoot in the hardened snow and began. He dug with a furry fury of one used to the exercise. While it is easy to get through snow, keeping direction is not.

By sticking a head straight out of the new, snowy tunnel, it would often collapse. The solution, from long learned lessons, was making a minor caveat to the side. Then, stick a noggin up through the white stuff to sunlight above. The sight of a snow covered marmot head shooting out of a snowbank was most humorous.

The little tunnel off to the side could cave in because it would not effect the main route towards the goal. Mr. Milk Chocolate was hip to the common sense of experience, quickly adapting to their action. Every twenty meters or so, a brown, fuzzy head would pop up, turn around 360 degrees a couple of times, then disappear only to reappear minutes later ever closer to the rocky goal.

Employing this well seasoned method, they successfully zigzagged up the monster. A worthy goal takes time, effort, help, or cooperation from others. In this case, three made the two better. The goal was a noble one. Joint, selfless focus made the acclivity possible. Utilizing this well tried system, they ascended the frigid berg at a furious pace. At each outcropping or tree, they paused to size up the next task and appreciate the scene. There is something truly breathtaking about the view from the top of the Olympic range. Unspoiled by humans, it is a humbling force of shear, raw beauty.

"See, Mr. Milk Chocolate, the whole world lays before you. You have never seen it like this, have you?"

"Oh my, this is surreal. Being this high, you can see so much of the land. It is as if we are a bird able to see the whole panorama instead of our little peak from the lip of the passageways."

"Yes, yes, yes, my friend, precisely. The best education we gleaned from our journeys upward is the vantage point of the enemy. Those who hunt from above can see better, move faster, and kill quicker. Many a marmot are caught unawares, snatched out of the wild blue yonder by a plummeting bald eagle."

"The knowledge acquired by traveling have serious benefits for our colony. The education we impart helps marmots and his woodrats live a safer life. In the wilderness, refuge is of paramount necessity. There are so many adversaries. Any edge, advantage, method to improve alerts, any facts about the enemy added are tools for our whole nation."

"Truth be told, our mission is not solely for adventure, though we honestly admit they do make life worth living. No, no, no, we are charged by our colony's leader, Mrs. Lip Gloss, to go, discover, learn, and share. Once my dear wife and family were taken from me, I turned to this way of life to cure my pain."

"Now we found a new home for our people. Mr. Milk Chocolate, you are welcome to be a part of our new family. Once we return from this climb, my thought is go seek the remnants of your colony. Come live with us in the Valley of the Stones".

"There is enough room to live in peace and harmony, removed from man. It is structured in such a way as no raptor can safely fly in and out with one of us in it's talons. As for coyotes, raccoons, and the like, it is so large they would never be able to penetrate. A bear would find the going extremely difficult. The path is thin and narrow. It would be so loud all of us could safely hide. We plan to build sturdy walls to protect us up high on the wall ledge."

"Honestly, my new friend, I sink zis is ze right vay for you. Mizzy is my best friend. He vould not schteer you wrong. He may be a goyim, but he is a true mentsh. Let me tell you, if you vant a real home, go mit Mizzy. You can't go wrong, Mr. Milky."

"Yes, yes, yes it appears the best solution for me. The invitation is accepted. Now, let's climb."

Up, up, up the trio slogged, jogging side to side, scaling the hard mountainside. Evening painted it's journey across a snowy canvas. This cast a panoply of red, pink, yellow, orange, and gold against the soft, steel blue sky, bursting with plump, white clouds, green, sturdy pines, and black, stoic rocks dotting below. Collective sensory organs were aglow with colorful life. They rested far above the start of the day.

It was clear to Mr. Milk Chocolate why they did so well at the task. Experience guided by knowledge acquired through survival made expert climbers. Once on a fine spot with a suitable cave, they stopped for the night to rest. A bit of food was found by a nearby handful of Ponderosa pines in the form of ferns and a few pines nut. It was not much, but enough. A good day done.

Chapter 18 Climb, Climb, Climb

For three solid days, they ascended the object. Utilizing the same system, hopping from place to place, sticking a head out of the snow to find bearings, they made progress. Mr. Mizithra's past had him at the advantage over their new friend. Some time was spent instructing him on the reason they did as they did. In learning how to make the little detour in a hole in the snow and head for an outcropping or clump of trees, Mr. Milk Chocolate acquired necessary tools to be an asset.

If you have a task to perform, learn from those who know how, and soon you will become a good, reliable component. It is easiest to witness this in practice versus reading it in a book. Get out and live life. It passes far too quickly. You are soon too old to do what you crave. Of course, go only after you finish this lovely, little book, naturally.

After this triad of days, Mr. Milk Chocolate saw the objective. Upon arrival to the final crest on high, the entire valley, forest surrounding above the canyon, to the ocean below, lay in all it's splendor. Yes, they could see the mighty, majestic Pacific, constant waves crashing vivid white upon the shore. At this sight, he squealed with delight, grasping his companions on either side, leaping up and down in shear joy.

"Mr. Mizithra, Murray, how can I ever thank you for this, this view of the world? I am moved beyond words to describe the wondrous, vast, beauty before me. I realize the added purpose. You can perceive the safest places from raptors, coyote, bobcat, raccoon, or even bear. The mission is a success. What an amazing sight. Oh, thank you, thank you, thank you".

"Our pleasure. We rest here tonight, then head down come morning. Some of our caves will be fully intact. We go down, so the time will go quickly. If done properly, we will be back at the base by nightfall".

They stayed for a time, appreciating the panoramic view, pointing out various sights. Murray spied the outcropping from which he was saved, Einar's tomb. He humbly hung his head. He held the melancholy inside, cradling the pain and joy time benevolently provided.

A life lesson he must keep to himself, surely, but it was the sweetness of it, possessing personal knowledge of such import. He managed to find the good in the bad. This should be the goal when we face turmoil. It was a silver lining, often so difficult to find. True to his prediction, the path Mr. Mizithra carved up the slope stayed intact for the most part. They were soon at the first spot they stopped days prior. One last exquisite scanning of the area and they made for the warmth of the forest floor.

They located the place in the log jam. Eating as much as they could stuff in the mouth, they fell into a well-earned, deep slumber. It was the sleep of they who succeed a mission. A happier bunch of rodents would be hard to find. The memory of what they saw served better than any dream could. They took Sabbath, the next day, to rest.

Dawn the day after found them in good spirits, even better weather. It was warm from the time they arose. A sweet, welcomed thermal draft rose from the valley. Wafting breezes provided a pleasing blast of heat to begin the day. Eating took longer than normal, especially the marmots. They needed to get back to packing on the pounds.

The warmth did a lot to slow them as did the massive meal. No one wants to see a skinny marmot. Moving back went swiftly. They knew the route. No detours increased the distance traversed. They made for the spot Mr. Mizithra and Mr. Milk Chocolate met, where Murray's fall occurred.

"Here is my thought, go back to what's left of your colony and see who you can find. Bring them to the Valley of the Stones. I told you how to get there in extreme detail. We are making for Grand Meadow, then the Beaver Blues Bungalow. Murray convinced me we can do the work in time. It is best for me to guide marmots to the safety of their new home."

"There are two other marmots we must gather, too. It is fate I must return to the beaver's home. They are lovely folks. It is of no trouble. The saying goes: Life is what happens when you are busy making other plans."

With this agreement, Mr. Milk Chocolate gave Mr. Mizithra a marmot's hug, kiss on each hairy cheek, and a slap on the back to Murray. He was off in the direction from whence he came. Armed with most wonderful news, he gathered those devastated survivors of the raccoon and vole invasion to establish a new home.

Murray and Mr. Mizithra headed back the way they came, pausing at the edge where he slipped. The impact for both was strong, unique to each. It is interesting how both experienced the same event, yet came away with vastly contrasting impressions. This shows we can witness the same thing, yet see it completely different. No two people experience the same event in an identical way.

"I must confess, I thought you were a goner. I sat here all night, holding your kippah and weeping. It was so sudden. You, you are my best friend, Murray. I never say it, but you make my life worth living. Our adventures are what fuels my energies. I am truly grateful for you and your friendship."

"Once we get back to the bungalow, you will stay. I must return to the Valley of the Stones. And…well, we may never see each other again. I want you to know in the future my stories will glorify you to the utmost. No grander woodrat will be regaled in my tales.", Mr. Mizithra's voice tapering off softly with the utter acquiescence.

It is a good thing to let those closest to you know how you feel. How can they if you do not inform them? Telling people they are special, that you love them or care deeply, strengthens the bond between you. Frankly, telling someone you love them is a marvelous gift to bestow. Try it, you will see. Showing real, honest emotion accompanies speaking love. The two feelings walk hand in hand. Cool stuff.

"Oy vey, vat? Now mit ze emotions. A shower of tears for a tattered old voodrat. It is touching, but not necessary. You are my best friend, too, you old Mizzy, you. Big galoot. Now let's get goink before I schtart to tear up, too."

Secretly, Murray had the same mushy feelings, but was not going on a crying jag alone with his friend up in the woods. They had things to do. Sitting around sobbing would do no good. Duty called. They headed back along the same route, the scent still relatively fresh.

One way animals ensure the way is to mark it was a scent gland or some other method. In so doing, the fellows made excellent time, pausing so Mr. Mizithra could gorge adequately. Two swift days and they were at Grand Meadow. It was a familiar and welcomed sight for our weary boys, it's abundance scrumptious. The hovel previously dug was unearthed. They spent a truly pleasant evening filled to the brim with good stuff to eat. A cordial breeze lulled them to slumber.

The travelers spent another entire day, resting, eating, drinking, and eating, basking in the salubrious weather. Theirs was a sense of genuine achievement. The White Mountain was assailed to shear delight and wonder. Grand Meadow was a boon, an oasis for all seeking safety. Finally, discovering the Valley of the Stones, new home for Mr. Mizithra's colony, made it worthwhile.

Morning would bring them to the river, down to where they met Mr. Gilligan's Island and Mrs. Maryann. Then it would simply be a short jaunt down the rapids to the pond, the Beaver Blues Bungalow, reuniting with family and friends.

Murray, in particular, was anxious. He missed mother and grumpy, old grandmother. Even more, he was chomping at the bit to reunite with that saxophone and wail out the blues. Music grabs the soul, carrying it to boundless places of joy. This is not just for you, but others who listen and are moved by it's tones. Ah, music, nothing is quite like it, you know?

They made the pathway swiftly downward.

It was easier descending with a sense of achievement. They beamed. The sun radiated in crisp, summer sunshine. The air swelled with tepid temperatures. Clouds were on vacation, wind a mere breath. Bugs swirled about as birds filled the skies. They were happy in their travels. As always, in their world, caution was key.

They followed the same trail taken in ascending. Occasionally, they would come across another critter. Nodding hello and exchanging any information was a common courtesy. They availed themselves heartily. No one wants to get caught off guard. The news: a coyote nearby. Be very aware. She was a mother looking to feed her family. Two young mouths to feed placed a severe burden. The father had died by puma attack.

They walked all day, resting for the night in a small hollow of an old cedar. Adventures were the fuel of combined joy. A goal you have which involves others always has an added reward: the ability to share passion and success with those in it for the same reason. Like a sports team, if you win, all win. Feelings of camaraderie are wonderful to experience.

The next day found them in range of the cascading river once again. Water sped up, turned frothy white, bouncing off large stones midstream. In a breath, they were downhill to the familiar pipe which led to safety a short stint prior. They turned and walked through the tunnel.

Midway, an earsplitting noise deafened them. One of the Thumbs was driving a massive truck uphill. The interior of the pipe quaked. They scattered, spat out near the boat. Dust from the road spread like a brown cloud, settling in a dirty layer, covering everything. Through the haze, they made out the form of Mr. Gilligan's Island.

He waved them along. They headed for the comfort of the lodging. Mrs. Maryann greeted with a hug and kiss. Mr. Mizithra and Murray shared discovering the Valley of the Stones. While an arduous journey, they tried to make it sound like they would not suffer.

"Our whole colony will be going. We met another marmot, one Mr. Milk Chocolate. He is going to round up the survivors of their invasion and bring them. You will have a mass of marmots with whom to live. You need this. Both of you have been lonely long enough. Please, please do join us."

"We have discussed this at great length. The pull to remain is strong. This is home. My three sons left to find their way in life. The rest died or moved along. With what few days remain, we are in harmony, it is time to go. We will follow when you come back. We will pack our meager stores and be ready when you return."

Mr. Gilligan's Island nodded in agreement with his sagacious wife. Her tender words showed passion. It was a wonderful life they had, but lonely. The loss of sons and friends was irreplaceable. Now, late in life, they were offered an opportunity to change destiny. It was time to go. They knew it. Human intrusion was a fact. The passing truck was a reminder of man's power.

They spent a glorious night together, talking about the journey up the White Mountain, Grand Meadow, and the Valley of the Stones. To a marmot, the thought of a place so ideal was very tempting. In his avid descriptions, Mr. Mizithra was cautious not to oversell the place.

It was difficult to contain himself, though. It was simply the perfect home. From the protection, remoteness, and size, it was as good as could be. Sleepy time found our quartet happy, sated with good food and company. The next day would find them down at the pond, reuniting with the beavers. Marvelous.

Chapter 19 Back To The Rapids

With a hug and kiss, they left their friends, who assured all would be ready. The colony would come back in approximately a week. The move was a massive undertaking: what to carry, plus emotions of leaving a lifetime home.

Both were born in this remote colony, but the dirt road was used more and more. Human contact was dangerous to survival. One errant idiot with a gun and they were trophies. It was a very safe place to live, save raptors from above. Now it was time to move, as it was for Mr. Mizithra's colony.

Change is part of life. We advance as we learn. It causes growth, adds a layer of knowledge like skin. We retain what we can of past education, hopefully the best parts, progressing through life, constantly growing, like ear hair. We pass wisdom to our young, as those in the wild do.

The warm summer sun shone, birds flit about. The stream sparkled, a frothy, happy white, tumbling quickly towards friends. It would be an all day affair. At it's end: the promise of a warm belly and well soul. Our travelers were up to the task, heading at a good clip, caution the rule.

One disadvantage was rushing water. It was deafening. They were shaded by ferns growing heartily along the shore. After, the only safety were low pines, useful as cover as a last resort. Descending from last night's lodging and friends, their hearts filled with joy borne with grins ear to ear. Nothing is so humorous as seeing an Olympic marmot smile.

Reverie vanished in an instant.

Without warning, from behind an approaching predator's sound froze them. The coyote. The hunter's chances were good. She had two from which to choose. They were hard against the rushing water. The only cover, a few meager trees. She was hungry for marmot.

Fear gripped. Coyotes are marmot's biggest killers. They can easily run one down in the open. If there is no tunnel in which to hide, they are usually goners. Time and time again coyotes achieved victory over the smaller, slower rodent. This was going to be the case unless something dramatic and immediate occurred.

It did.

As they ran, Murray spied salvation: a log. In a heartbeat, he dove, paws in front. Mr. Mizithra saw the move, joining with zeal. They reached the log and gave it their all. Force loosened it from sand and rocks. They headed towards water. She was closing quickly. Fear engulfed, completely possessing them.

Hitting liquid, Murray leaped, landing square on the widest part of the log. He scurried forward. With no thought of geometry: speed of the log in rushing water, his weight, distance to the log, and height necessary to reach it, Mr. Mizithra urgently tossed his hefty physical structure into space. Should he land in water: drowning, death.

Mr. Mizithra hurled a chubby furry frame into the wild blue yonder towards the log, his friend, and safety from the pursuing coyote. He did not so much soar as simply heave an immense weight in the air, falling like a shotput. His aim was pretty good, not quite perfect. His back paws did not make the raft. He quickly scrambled forward, eyes bugging out of his head in terror.

Murray, seafaring at a fantastic rate, terrified, navigating a bumpy torrent on a measly stick, avoiding becoming lunch, praying they did not hit a rock, could not help but laugh out loud at the sight of his friend's face at that moment. It was one for the books.

Now, to the present. They were on the scrimpy stick. It was hurtling downstream at a fantastic velocity. Within a second, a huge shadow came from above. It was the widowed mother, desperately searching for food for her pups. They were hungry. She had to kill meat soon.

Murray shoved his little body to the beam's edge, swung his left leg over the side, steering the stick starboard, avoiding the diving assailant. She hit water ahead of them with an enormous splash. She immediately swam for the opposite shore. The coyote was zipping downstream so quickly, it was impossible to get to the side.

They were behind, watching as she crashed headlong into a stone. She went limp in the water, disappearing beneath the rapids, a trail of red behind. Her young would not survive. This is the harsh reality of the wild. One mistake and death quickly follows. Our travelers had good fortune.

Murray's swift mind spotted the stick, developed an instant, albeit madcap plan, and put it into action. His boon companion knew precisely what his pack rat buddy had in mind. He joined in as the only option. The sailors continued towards the pond, leaping off as it slowed to a pool.

They made for the same side of the river they were on before being so rudely interrupted. Both shook off the water and wet once on shore. They made it a few meters up the side of the bank away from the liquid, to the safety of some dense grass, plopping down for a well earned rest.

"Quick thinking, my friend, quick thinking indeed.", heaving heavy breath, "Spotting that log saved our lives. Once again, Murray, I owe you my life. How can I ever repay you?"

"It vas nothink, my friend, nothink. I saw ze log und knew ve could make it if ve could push it. But ven you yumped on the log, I am sorry, but your eyes vas buggin' out of your skull. I had to laugh. Zen zat rotten coyote....vell, she got vat she had cominck."

"Now, Mr, Mizithra, I sink ve must rest for a schpell. Let us find a suitable schpot to hide out und grab a some shuteye."

They found a nice little cradle in some rocks. A previous critter made a minor start at a hole. They dug a bit and had a neat hovel perfect for their immediate needs. Sometimes plans alter. We must modify and adjust to change. This is never simple or easy. Being prepared means if disaster befalls us, be ready with food, water, heat, shelter, etc.

They slept for hours. Rising, they ate. After a bit of reconnoitering, consisting of Murray climbing a handy alder for a good view down to the pond, they ascertained it would take a couple more hours to arrive at the destination. This buoyed soggy spirits, placing the goal within grasp.

The pace Murray set was brisk. They felt a bit safer having survived the onslaught by the coyote, but were conscious to watch above. Raptors are excellent at following waterways for prey. Everything has to drink. One swipe from a hawk or eagle, and it was over. This kept them moving swiftly, safety in mind. Each step brought them closer. The pace increased.

His estimate was off, gait quicker than reckoned. In less than an hour they were at the bottom, waddling along the pond. They saw dozens of birds, bugs, fish, and animals. The place was abuzz with stories of the arrival of the marmots and old lady woodrats. Gossip increased when Mr. Mizithra and Murray were sighted.

Dragonflies zipped off in a mad race taking news to the beavers. A short time later, exhausted and delighted, our searchers saw the turning point to the blues bungalow. They paused to grasp each other by the arms and stare in each other's eyes. The look was one of mutual admiration and knowledge of their success. For both, death was a distinct possibility. Mr. Mizithra rubbed his scarred butt as a reminder. The goal was successful, beyond the dreams of avarice. Murray recalled his fall and Einar.

They turned the corner, opened the door, and strode into the club

Usually, this time of day before happy hour, it was dark inside, hazy, and dead quiet. The door shut. There was nada ahead save blackness. It was eerie, the place apparently empty. With a collective sigh, they turned around, heading for the door and light. The room exploded in shouts of joy and peals of laughter. Lights blazed. The band jammed.

Mrs. Lip Gloss started things with a massive and loving hug for our heroes. Then came Nediva and Frema to embrace Murray and pinch Mr. Mizithra's chubby cheeks, their big galoot. The entire colony engulfed them in love, followed by beavers. The rest who crowded the Beaver Blues Bungalow that special night came next.

A WELCOME HOME banner hung in front of the stage for returning traveling champions.

After a quick hug, Murray was chucked on stage, not a moment to rest after such an arduous journey. He snagged the sax with a happy little smirk, coaxing the reed into submission. The band hit the break. He blew with all his might. His mother and gramma wore identical, open mouthed looks of amazement, hugging in glee at Murray's skill.

"Such a boy."

"Ja, such a boy. Our boy, eh, Frema? No longer a boy. Now a mentsh."

The night danced along. They ate, drank, and played until it was truly late in twilight. The young ones were in bed as was Frema. Beulah, now fully recovered, tended bar and sang harmony with Miss Bluesberry. Bette was happy Murray came back so she could wail along side him. They were musical best buddies. They played until the wee hours, then made for sleepy time.

In the morning, plans were unveiled for the journey to the Valley of the Stones. It would take longer. They had the elderly, young, and a few injured. The estimation was ten days. Thankfully, they were at the lodge instead of back on the ridge. Murray and his family were safe. The ladies were given lovely quarters around back, far from the din. It was roomy, a nice exit to the left under a small fir stand. Perfection was the word.

Murray's room sat across the hall. He had a nice spot. The beavers spared no expense preparing rooms. Their welcome was affable and genuine. This beaver family was golden, full of honest love and affection, bound by music, and joy it brought the world.

He had a brand new, woodrat sized sax, music stand for sheet music, extra reeds, and a branch and twig stand to hold it. It glistened in the dull light afforded in this wonderful room Murray would live his days with family and beavers. The smile was genuine, ear to ear.

"The first stage is to break in five groups. I will lead the front group, Murray agreed to accompany us. He will head the rear guard. We ascend the rapids to the tunnel. I will get Mr. Gilligan's Island and Mrs. Maryann. We camp at a spot called, Grand Meadow. We stay for two days."

"Once there, I will immediately take a group of strong, young ones to join with Mr. Milk Chocolate and remnants of his colony. They will follow him to the Valley of the Stones and begin the chore of digging the initial tunnels and rooms. We started the process for two days upon arrival. They will make progress to ensure the others have rooms upon arrival. We can expand after a time."

"I will then return, leading you to the Valley of the Stones, our new home. The journey to the spot we descend will take about eight days with this group. The additional two days of rest will equate to ten full days to reach the destination. They move slowly. It is vital we exercise extreme caution along the journey. We were almost taken out by a red-tail hawk and a coyote. And we are hale and able."

"With so many on the path, we have to work harder to provide food for the elderly and young. Those injured will not be a burden, but will be slow. Grand Meadow is a larder of great abundance. Afterward, substance becomes sparse. Upon reaching our goal, the food supply is endless. We will gain weight yearly far from the deadly paws of foes we always face. No more MAN."

"Tonight, though, we are throwing a massive party to have fun with our dear friends. I see a few strangers have come since my arrival. The beavers found room for all of us. It is a bit tight. Each has a comfy bed. Now, I must speak with Beulah and Buford. Questions? Good, we depart at first light."

The most important tidbit on the pond was a buzz of the huge brawl the night Mr. Mizithra and Murray arrived, and latter's amazing talent on saxophone. A rat with his ability was a real find. He was going to live in comfort under the dam. Once others knew, they descended on the blues club, cunning in greedy eyes. Music was essential to Olympic creatures. His newly discovered virtuoso gift was big news.

Some might persuade him to perform at their clubs. The muskrats had a nice one. The squirrels had the sweetest loft party center not far in a huge stand of old cedar. There was no way he would consider the owl's place. Buford saw his potential and "signed" him, offering a home for his family. How could he turn down such an offer? Buford knew every musically inclined animal would descend like a plague of locusts.

To wit, most notorious were two crows. Oh yes, the most wise, wily, cunning of the bird world is a crow, no exception. They are renowned for an ability to out think any critter on the planet, covering six continents. In the Olympics, the two most celebrated were best buds from way, way back: Tutti Frutti and Mon. Fame came from an innate ability to hear what no other creature could: perfect musical tones.

Oh to be sure, an owl or two and a few select beaver had this talent, but no others had what these two did. They knew it. No one would ever ask an owl for anything lest they wished to become supper. Thus, their opinion was never considered. Ever casual, always laid back, this crow duo scoured mountains and valleys. They flew along Thumbs' roadways to save time traveling to find the best talent up in them thar hills.

Of course, humans do not know this, but the denizens of the forest love music. Heck, get out a guitar, flute, or violin and play to a herd of cows. What happens; you will be surrounded. The only thing they cannot do is clap. Hooves do not do that sort of thing. They could "high hoof", similar to a high five. They moo and wag tails in approval.

Word got out. Our feathered fellas flew to the familiar haunt of the Blues Bungalow. News included Miss Bluesberry. They had doubts about others utilizing her dulcet tones. The Muskrat club was too small, Squirrels too high, the Wapiti club too snobby. Still, one never knew.

If this little woodrat was all everyone insisted, the ebony duet would have a ready contract to sign. A rat who wailed on sax would be an easy commodity to peddle. For the record, no one, I mean no one, could hawk like these achromatic, feather clad, silver-tongued devils.

They arrived sans fanfare, sneaking in quietly, seeking the very back in a small booth entirely out of light. In fact, no one saw them. They were not even served. The cloak and dagger routine's express purpose: should anyone know they were present, the club would change it's show to expose the "star" versus the usual. They got the real jam session, jointly blown away by Murray's skill. Then, then there was Miss Bluesberry; it was a jolt, a bolt. If you missed it, you were a dolt. Her tonality was spot on. No one could miss Bette's chops on guitar.

Tutti Frutti glared, emblazoned, glistening black eyes focused stone at his chum. Mon nodded in a hypnotized glaze. This woman could sing with more soul than any they ever heard. The rat on sax, oh please. They could see him booked throughout the season here, then there, back home and then around again. They could count the loot they would receive for a job well done. Something shiny, shimmery in the light, would do nicely.

The fellas had a dandy collection of all sorts of glossy baubles and fishing lures, not to mention soda cans, etc. This was payment for being so adept at their job. Once talent was spotted, it was signed. Once signed, it went on tour, making benefactors wealthy. No one knew the benefactors. The crows were not blabbing. The dynamic duo got loot as commission. They did rather well.

Mon motioned. They slunk outside to confer. They flew to a remote, high branch on a huge fir. Once they were certain nothing else was in the area, they discussed the talent and how to proceed. A plan must be worked out to a tee.

"Mon, dese two, oh mon, Mon. We gots to sign bot. Dat skunk, she so full o' soul an' dat rat, oh Mon, you know we needs dis one fo' sho.'"

"Tutti Frutti, yeah, mon. We gots to have dat lady on guitar. We need to rap wit' Bufo' da beaver and his mama, Beulah. Dey gots to let dat boy come out an' play. So, what we offer dem? What dey want?"

"Don' know, bro. Let's go fine' out, Mon."

They flew back, strolling in with pride, noise, and fanfare due stars such as they. The black coats shimmied and shone in the light of the club. They strode to the stage and awaiting frowning beavers.

"Y'all cain'ts have 'em," spat Beulah before the crows so much as squawked, "My little ratty ain't leavin' this here club. Mizz Bluesberry don' wants to go neither, so git yo' black-tailed butts outta my club."

"Oh now, come on, Beulah, you gots to listen, gyal. We only wants to "borrow" him fo' a while. You can have him back, Jack."

"My wife said, No, so y'all gots yo' answer. Now git.", shouted Buford in defense of his wife's position. He knew they would not see Murray for a long time were they to get their sleazy wings on him.

Disheveled, heads bent, beaks held low, they sauntered slowly out of the bungalow, up to confer in the fir. They did not like to lose. They came up with a drastic scheme. The one sure way to rile pesky beavers, impress Murray and Miss Bluesberry, and get a reward. The duo cawed in greedy, utter delight at their juicy plan.

Chapter 20 A Murder Of Crows

No dragonflies for this message, it was too important. Tutti Frutti flapped due east. Mon spirited north. They knew whom to contact. Oh to be sure, this was not something done lightly, no indeed. In fact, it was not considered for quite ages. Not since the case a good while back when they applied the same acrimonious measure. As the party swelled that evening for farewell to the marmots, a sudden wind blew, taking the crowd's attention towards the entrance.

The music was acquiring heat. Everyone was getting in the groove of festivities. The threshold burst open, rays of summer solstice sunlight shot a shard direct to the stage. In stepped a sight. Tutti Frutti and Mon strode in, followed by twelve other, older crows. It was the famous "Murderers' Row". This touted twelve, distinguished dozen, were famous throughout the Olympics as the best judges of talent in the mountain range, coast, and Puget Sound.

"Why, it's old Winston and Beverly from the west coast near Willapa Bay. You can't get further south with this group. There's Delroy, Neville and Clive from Neah Bay, far north where the ocean meets Puget Sound at the

Strait of Juan De Fuca. Oh wow, Lloyd Joseph, I thought he was no longer alive. He is an old, old crow. For him to come from Cedar Crescent at Whiskey Bend on Elwha river is something special."

"Darla, Vivienne and Simone, the Ladies of the Lake, up from Lake Crescent….they rarely leave. Derrick, he's most critical of them, out of Seattle across Puget Sound. If he is here, this will be good to watch. And naturally, finishing off this "murder" of crows (a group of animals is given a distinction like a "covey" of quail or a "gaggle" of geese. For crows, a flock is a "murder") are Everton and Clifton from down Port Angeles way. You knew if the others would come so would those sleaze balls.", spouted Lincoln, aged beaver of the quartet. He practically lived in the bungalow.

The dusky, eminent dozen moved without a word towards the front row and four tables nearest the stage. Every eye in the place was on them. They collectively fluffed their feathers in the chairs and peered straight ahead. They came for one purpose, one alone: to see two new discoveries whose early reviews lit up the Olympics from south in Willapa Bay to the top in Neah Bay, clear across Puget Sound, and east to mighty Seattle.

Seattle alone had so much talent none of Buford and Beulah's children had ever seen any of these twelve. Save Everton and Clifton, who visited a few years back. The rest were far too important and busy to bother to come in the mountains to see hill animals for pity's sake. To these snobs, it was nothing but hillbilly stuff: no front teeth, married to a cousin, and a banjo. For any decent music to be found would be shocking.

Buford called the band for a quick quorum. The cabal huddled, nodding noggins. Buford made his way on stage. He raised his paws overhead to gain the attention of every critter in the club. This was his lodge, his band, and no one, not even this famed band of judges, was going to put them down, no way, no sir. Bette loosened up her ax behind papa, twisting a knob or two to ensure perfect tone.

Homer plopped down on the piano bench, rattling keys. Henry did a light run on harmonica getting it moist and ready. Tobias thumped a note or two on his beloved bass, slapping tail. Twigs laid down a tight drum roll in back. Buford had subdued accompaniment to emphasize his proclamation.

"Well now lookie here, folks, we gots amongst us tonight da mos' magisterial list o' visitors from all over da Olympics! These here twelve eminent crows is here to revel in da sweetest sax sounds from brother rat, MURRAY!"

The crowd erupted in a roar causing water in the pond to ripple in response. Trout were not sleeping tonight. They moved to the far end of the pool. Shouts, roars, marmot whistles, caws, and squawks rose in support of a humble looking woodrat wearing an odd hat, clutching a glistening saxophone. Crows love, I mean love, any glossy instrument, piccolo to tuba. Shiny objects are a crow's weakness, as you already knew.

"An' let's don' forgit da lovely, darlin' lady, wid da voice to melt a puma's pumpin' heart, Mizz Bluesberry!"

Again the crowd exploded in approval and adoration for this find, true talent, diva in the making. She bowed to the audience, revealing a lovely red bow tied on back of her head extenuating the white stripes beautifully. It lay in a field of black fur. The crows, The Murderers' Row, seemed entirely bored with this back water display. Derrick appeared totally languid, primed to depart, having not so much as heard a note.

City crows demand something at the slightest caw. The birds were served a small plate of various edibles. They are omnivores. Each had a bowl of water to wash it down. Crows, as most know, are notoriously cheap. The servers knew not to bother with a menu or to expect a tip. They ate everything in sight. The reason for such an arduous trek was to hear some good sounds. It was time to put up or shut up.

Go.

Twigs hit. He hit hard. Really hard. Tight. Real. Tobias picked up the beat, honing it to a fine point, tail thumping in perfect rhythm with bass. Henry took a low path. He threw heart and soul in his harmonious harmonica. Add with absolute perfection, Homer's piano, tinkling ivory with flawless rhythm. Bette joined, guitar more a part of her body than not. Daddy taught her very well. Father/daughter night with them was dueling guitar solos. Mama and the siblings judged.

The unmistakable sound of blues and soul combined flowing from heart to fingers to strings, pouring out a sound, loosening hard liners immediately. Music was the voice of her inner being. She had flesh in the game. Her guitar rode hard and mean over the backs of other musicians accustomed to this power. Her sound lit up the bungalow, moving all who heard it's painful, simultaneous joyful moan.

Then, oh then, it came. Miss Bluesberry's voice tore a hole in the night. Twelve pairs of feathers ruffled and rustled. Sensuous, melodious tones filled the room with a shower of love and sweetness. She hit the chorus, warbling it through. She had so much gumption, so much spirit, Lloyd Joseph, seated with Winston and Beverly, the latter Lloyd Joseph's younger sister, leaned forward, whispering in their ears. To their credit, they nodded and turned back to absorb more.

She finished the second verse and chorus. Her voice trailed off to the bridge, allowing Murray to step up and do his thing. Heretofore, he had not so much as blown a note. He strode forth, grasping the golden sax, kippah pulled low over his right eye, light providing an enhanced, dangerous appearance. The band was waiting. He delivered. He served it hot in their faces. He strutted up and down the stage, pausing at each table to display his funkified wares.

You want to judge this rat, he thought, okay here you go. If you do not like it, that's your problem. He was not haughty, not proud, no not at all. He was simply confident in his ability. He could play. Everyone said so. He lived it in his heart. It was the only goal. Traveling days were past. He secured a place for the family to live. He found his true love: music.

He sashayed across the stage back and forth. He handed it to Homer who plunked away, passing to brother Henry who stole the limelight. He was on fire, wailing on his mouth organ for all his worth. Yet, as he hit the peak, a sound arose from way out back. Bette came armed with a truly blistering lead. It sprang from her furry beaver toes, soared fully to her heart, essence, soul, coming to light in her fuzzy fingers. She laid it down. She slammed it down hard. She set it down clean. Sweetness personified.

She made that guitar sing. Her fingers strummed faster and faster, the band happily going along for a ride. Bette handed it back to Murray, who leaned up against her back, screaming away. They united in a sound sent roaring through the timberland, shaking trees. No animal in a ten kilometer radius could miss the sound emanating from the Beaver Blues Bungalow that night, no way, no sir. Every turtle, bug, and frog in the pond joined the fish at the far end to escape the pounding.

They combined to a true mountain top of sound. The band threw caution to the wind, ending on a solid, monstrous note. The crowd was dazed and confused, storming the stage, zigzagging around Murderer's Row. They surrounded their friends in stunned amazement and love. It was the best, the very top-grade anyone ever heard.

Even those few who hesitated to get too near Miss Bluesberry were there to give her hugs and hearty congratulations. As the crowd swelled, Buford rose to the stage, voice bellowing over the din. Beulah's face was one giant beaming smile of beaver satisfaction.

"Okay, y'all there, you Murderers' Row, what y'all think? You tell me who gots dis much soul, dis much sound, dis much talent anywhere? They ain't gots dis on da coast. Ain't gots dis in da Bay, an' dey damn sho ain't gots dis in no Seattle!", the final point made with impact.

There no question of talent, not only of the two they came to hear, but the whole band. Bette alone was worth the price of a ticket.

"Let us concede and confer. Thank you, Tutti Frutti and Mon, you are to await our verdict." said Clive in a sullen tone.

The twelve went out-of-doors to the same handy fir branch Tutti Frutti and Mon employed. Upon arrival, the Olympics showed their stuff, soon pelting rain with a steady wind. They huddled tightly, cawing and squawking muffled by wind and darkness. Out of shadows, a young female crow flew up, awaiting commands. She handed an object to Simone. In time, Lloyd Joseph and Simone cawed at this unknown courier. She nodded and flew off, vanishing in the black of night, south.

After what seemed like forever, they descended from on high. As a whole, they entered the bungalow, making for the center of the club to render their promulgation. Every bird, dragonfly, every mammal was on pins and needles in expectancy of judgment. Buford hushed the crowd. Simone, most distinguished Lady of the Lake, would render the verdict. Her word held ultimate sway in the Olympic Mountain range.

"Forest friends," words spoken in regal tones, "tonight you provided a most enjoyable sample of mountain sounds. My goodness, to see and hear what Buford and Beulah have done is astounding and so very wonderful to witness."

"This whole band impressed us. For a "house" band, using the term loosely, we never heard better. The proposed stars were made aware to us by Tutti Frutti and Mon. The boys earned their commission. My my, these musicians are talent to be shared."

With that pronouncement, she threw down the ultimate prize, the hallowed "Boxcar". It was an old black domino with twelve dulled white dots. This was the object the young messenger handed Simone. It was the highest number of crows to approve of a particular talent. It was for three: Bette, Miss Bluesberry, and Murray. Each was pointed out by Simone's beak in cold silence. This reward was rarely bestowed. Only a handful were so favored. The honor was a sign of the abilities of the triad.

"It is clear Miss Bluesberry cannot travel far, but she could be a star in this mountain circuit throughout the year, spending, say, half of her time here and half on the road?"

"Ho ho ho right dere, now Lady Simone. First, we gots to ax her if dat's what she wants. Next, we ain't about to have her gone dat long. Da most I could say we could allow is………oh, say six weeks."

"Well, most certainly, Buford, my friend," as Everton's slithery voice arose from amidst the mob, "you are able to give a bit more time to let others enjoy her dulcet tones. Surely, you cannot deny her public, which will most clearly arise once they hear her sing with a most lovely set of vocal chords?"

Everton was from a low class crow family who lived exclusively out of human garbage cans and campgrounds. His family made it's way up the ladder having so much to trade in shiny objects gathered from massive amounts of refuse people leave everywhere. Humans are a messy species. Everton was a manipulator. He slithered up to Buford's side, continuing vile attempts to wrangle a better deal for the crow side.

"Now, you know, I can provide you other entertainment, mon, when dey away. We could arrange it so dey not gone at da same time. Would dat do, my furry friend?"

"Man, da las' time you sent us "talent", you had a squirrel juggler who couldn't juggle. You had a muskrat singer who couldn't sing. You had a beaver magician who dropped his cards…an' his wife was his assistant, who came out o' his hat. Terrible."

Bickering back and forth put an end to the music. Both sides tossed words like kids with mud pies. Feathers were fluffing, beaks cawing. In no way were the beavers and entourage going to back to down to this revered mob. The only thing resolved was both sides were capable of making a lot of noise, ending an otherwise incredible evening.

"May I offer a bit of assistance, please?', came a clarion voice from deep at the back of the joint.

"Who dat?"

"Tis I, Boston Charlie. I have no ally in a battle. I am neutral. Seems to me there is middle ground you can both see, but both sides are too proud to concede any valid point."

"Now, here is the easy solution. Bette does not want to travel. She has duties to family, dam, and band. Perhaps she might do the big show at the end of the season at Grand Meadow? Miss Bluesberry will work, at most ten weeks, no more. She prefers seven. I propose eight by assent. Yes, from the skunk. Yes, from crows."

"Last, but not least, my friend, Murray. We met on the road. I steered him here with Mr. Mizithra. I have a vested interest in this little fellow. It is clear he is in high demand. There is the huge festival at Willapa Bay. He can fly via Trumpeter Swan Air Lines. That is a clear must, but after that it is really up to him. He says he only wants the same amount of time. The festival is three weeks. So, let's say five more weeks on the circuit?"

A moment of calm reflection and both sides concurred. The band hopped back on stage and wailed away for a couple more hours. Then all concerned lay down. On the morrow, the marmot colony was leaving on a journey to the promise land. Boston Charlie simply beamed.

Chapter 21 Journey to a New Home

Morning offered dryness and sunshine. By the time the marmots arose and ate, the birds and Boston Charlie were long gone. The crows said something about urgent business. It was hard to figure out what it might be. The issues were amicably ironed out by an excellent Kingfisher. The twelve rarely did anything together. Truth be known, the group was not united individually. They felt Derrick, Clive, and Everton lower class. The Ladies of the Lake preferred the solitude of Lake Crescent and it's magical waters. Lloyd Joseph was a sea of mystery.

Nonetheless, they took off as one at the crack of dawn, squawking as they arose, scattering a scold of Stellar jays who hollered from the safety of the trees in retaliation. The entire group clearly were on a path south. One could imagine they were preparing for big goings on at Willapa Bay.

"Well, it is sadly time to bid farewell, my dearest beaver friends. I can only humbly say, Thank you for all you have done. My best friend has a place for he and his family. You are a warm, loving group. Like us, you are a colony, knowing the power of strong family. With deep affection, sadness, and a smile we must take our leave.", his speech severed by a choking voice and tears.

Hugs and kisses abounded. Beulah gave the best hug he ever had. He paused in front of Nediva and Frema, hanging his head. Nediva lifted the sagging pate, staring lovingly. Their eyes met. He saw her love for him.

"Vat? Such a big galoot. Get out before I schtart to cry. I love ya, ya big old marmot, you. Go on, und take my boy. He can help you. Ve vill see you again, Mr. Mizithra, ve vill, mark my vord. Now give me a hug und go or I vill never be able to let you leave."

Mr. Mizithra hugged and kissed them. He stood staring at this wonderful family. He left the Beaver Blues Bungalow to head for his new home in the Valley of the Stones. This new abode was far from danger, man, bears, eagles, hawks, coyotes, and the like.

Leaving loved ones knowing you may never see them again deposits emptiness in the pit of your stomach hopeless to quell. It makes a hole in a heart never truly refilled. The power of love is strong, so deep nothing can erase the pain when removed. Sadness was understandable.

Assembling thirty odd marmots in five groups with Murray in back, Mr. Mizithra up front was awkward, like a train pulling out of the station. They organized. Each group of six would travel a distance, wait for the next group, then proceed. This way, the chances of someone getting hurt was greatly reduced. It kept the lines of communication open, which was vital. They hugged the shoreline of the stream. None of them had seen rapids. It was quite a distraction at first.

By staying against the water and hiding under ferns growing in abundance, they were kept virtually hidden. The only fear was if some big critter assaulted from the side. The system worked. In a short two days they assembled to the top of the hill. The road lay left, river right, Grand Meadow straight ahead.

Spotting a familiar pipe, Mr. Mizithra called a halt. He directed the second in command, Mr. Opie, an up and coming young marmot three years of age, to tend the others. He could see this one as a potential new leader for the colony. This was something they were always seeking to sustain good leadership that they might thrive.

Word passed to the next group, eventually reaching the finale` directed by Murray and another three year old with promise, Mr. Beef Stew. He was a burly marmot whose prowess as a wrestler was already renowned in the Olympics. It was ascertained he would be best to anchor the family. His strength would give a good, safe feeling for the rest. Mrs. Lip Gloss selected well. Everyone made it safely uphill to the pipe. The elders were spread amongst squads to give more stability.

Mr. Mizithra did not wait, going straight through to retrieve Mr. Gilligan's Island and Mrs. Maryann. They had little to carry. It was now a matter of departing the place they always known as home. Even though a dangerous location in which to dwell, it was all they knew. Leaving this relative comfort was not easy.

"Okay, dearest, one last look and we simply must go", urged Mrs. Maryann to her forlorn husband.

That small chore complete, rough for him, they went under man's road to the other side for the first time. They found thirty odd marmots of various ages staring. They had never seen this many. They recognized Murray and gave a holler. Mr. Mizithra would escort the timid couple to Grand Meadow. They would stay two days to eat. Scouts would hook up with Mr. Milk Chocolate and the remnants of his colony.

He coordinated how things were to proceed. They made straight for Grand Meadow, as day waned. The old ones were not as able to keep up, so provision was made. Each group gave up a young male who went to the resting place Murray and Mr. Mizithra slept when they first reached this lovely spot. They easily dug several large rooms in a short time in preparation for others. All they had to do was make the rest stop.

Food, water, and safety lay in sweet abundance. This bundle of big boys trundled eagerly up the pathway towards the goal. One interesting fact about Olympic marmots is the difference in size between male and female is the greatest of any rodent. Where the beavers are the same size, not so here. The disparity of male/female is extreme, called: dimorphism. The variation is more than 20%, making size a dramatic differentiation.

The big guys went over the hill, through the dale, and outta sight. Mr. Mizithra got his group going with two new traveling companions. It was contrastive for them. They had never gone anywhere. Once their sons left and others departed, it was just them. The elders passed away. Now older, no one reproduced.

Predators got some, a couple were hit by trucks and cars going on the road. Mr. Mizithra was sensitive to the fact they were experiencing a brand-new world. He hoped they would see beauty in the Valley of the Stones and be happy. He knew the trek would be especially arduous due to isolation and lack of travel. He stuck to them like glue.

Light began to fade in midsummer's eve. They got over the crest of the last hill, spying compadres at the knoll of trees they stayed in hiking to the White Mountain. On approach, Mr. Opie spotted them, whistling in the awaiting hole. They paused at the entrance, hearing digging.

The first began to eat the most handy greens, They were famished and stuffed faces, grazing in hearty joy, shoving as much as they could grab. A few made it to the pond to drink, snagging some lovely roots to chomp on for the short journey back to the temporary home.

Sated, they entered, finding a massive room, five tunnels branching off of it. One went straight ahead, clearly an escape route. It shot between some roots, under a sisal bush. Ideal. The smell of freshly dug dirt filled happy nostrils.

The other passageways led to four large rooms, laid with fresh dry leaves, a stack of lupines and greens in each room for snacking. With six or seven to a room, all would be warm, snug, and cozy, perfect for a hostel on the way to the new home.

As it darkened, Murray arrived with the last group. They were likewise tuckered. Knowing they could rest for a couple of days brought desires to eat, drink, and sleep. This is precisely what the entire colony did, sealing the entrance with a few choice stones.

It was an excellent beginning to this most important journey.

Morning found the colony up early to stretch and eat, eat, eat. Guards were placed on vital spots to ensure safety. Once one marmot was filled, a guard was relieved to gorge on the cornucopia of food awaiting. For children it was a playground. For old ones, it was time to get as much fat on as possible. They felt winter's swift, serious approach. Autumn is brief in the Olympics.

The lodgings were fine. It was thought the young men ought to stay for the day, enjoy the surroundings and family, then head for the Valley of the Stones come morning, hooking up with Mr. Milk Chocolate. Once contact was made, they dug permanent homes. The valley was huge, massive. Space was not a concern. They could build an excellent colony for winter, expanding in spring once things thawed. Since the cliff upon which the homes would be built faced the sun, the melting process would progress quickly.

This day was a holiday. Marmots lay in the sun, ate until their bellies were ready to burst, and drank sweet pond water. Birds and bugs flew about, other animals, species new to the young, meandered through their midst, ignoring them for the most part. It was difficult not to do a bit of rubber necking. It was not common to see a mass of marmots waddling about shoving every green thing edible in sight in their gobs. It looked like a brown mass of obese fuzzballs on the prowl.

As the afternoon nap approached, Mr. Mizithra saw something sad to witness, yet not unexpected: Mr. Gilligan's Island and Mrs. Maryann turning back. They were clearly uncomfortable with new friends. While nice to see their kind, so much food and water, it was alien.

It is odd to see people's reaction coming out of a place, a situation, clearly detrimental physically, spiritually, or psychologically. Even with a much better and safer condition, many often return to misery. The shock of a better existence is too much to accept. At times, a bevy of benefits causes fear and rebellion. It overwhelmed them.

There is something in the way people and animals behave. It was no different with these stick-in-the-muds. Eyes on the prize, Mr. Mizithra snagged Murray and Mrs. Lip Gloss. She saw the problem and asked Mr. Marathon Bar and Mrs. Ellie Mae to help. This clique made a beeline for the vacillating duet. They were skipping out, back downhill to old haunts.

"Mr. Gilligan's Island, Mrs. Maryann, please, let us step over to the shade of this Noble fir and have a chat."

With a soft-toe approach, he, Mrs. Lip Gloss, especially Mr. Marathon Bar and Mrs. Ellie Mae, tried to soothe the frightened ones. Mr. Marathon Bar and Mrs. Ellie Mae joined the colony a year before, theirs a tale of woe. The old family was decimated by man. A bulldozer dug up the home for a road. Forced to flee, they lost touch with all they ever knew for a new life in Mr. Mizithra's colony.

While very happy to join another solid family, they harkened back to the former abode. Love of a particular colony kept their eyes turned to the past. It was easy to see comfort in bygone days, albeit an unsafe place to live. The red-tail hawk was the latest. They experienced so many others.

Yes, the Thumbs stuck their noses in from time to time and let kids jamb sticks down in the holes to scare the marmots. A dog came once, a long-haired Dachshund, let out of a passing minivan. He stuck a snoot in the nearest hole and began to dig with fervor.

Only when a chubby lady with plaid shorts, loud, flowery blouse, and a camera came over did she shoo away the mutt back to the vehicle. She giggled as she put the pup away, telling her traveling pals how cute it was for Fritz to dig up gophers. People, what can you do with them?

The island was, nevertheless, home. Parting was not sweet sorrow. It was downright misery. They were resigned to return. Nothing anyone said made a dent in their steely resolve. They knew what the potential was going back, but was all they knew. Change was too painful.

As the discussion continued, a loud whistle broke the conference. It emanated from over the hillock. The whistle came from Mr. Beef Stew. It was the warning call of danger. It was loud. Three distinct times, then a pause, then twice more. This was the sign of serious trouble from the ground. Were it above the three would have been followed by three. All concerned made for the new, temporary hovel.

Murray was last. The room was full and snug. The entire colony held it's collective breath waiting for Mr. Beef Stew to show his face to explain the warning. Mr. Opie and Mr. Diet Cola stood vigil at the front entrance. What could it be? Coyote, fox, cougar, bear, or man? Prospects were endless. Fear gripped them in darkness. Eyes turned to see what was next.

A furry face stuck itself in the entrance, grinning a toothy smile. It was Mr. Beef Stew. He waved to come out, massive grin on his big mug. What greeted them was a handful of other marmots. Instantly, Mr. Mizithra recognized Mr. Milk Chocolate as head of the group. He smiled and whistled to come ahead. Mr. Milk Chocolate did so with zeal.

However, this was not the news. Next to him were three young male marmots, remnants of his lost tribe. There were two females and three young ones, two of which appeared to be mere babies. This ragtag bunch were standing to the side with Mr. Milk Chocolate. A howl came from Mr. Mizithra's group.

Mrs. Maryann exploded from the crowd, waddling pellmell towards the others. The three males screamed shouts of delight. Mr. Gilligan's Island was immediately behind his wife, running for the trio, too. The five joined together in shouts of glee. They linked in a tight circle around Mrs. Maryann, whistling a happy tune.

It was their sons. The three departed to find wives. They fell in with Mr. Milk Chocolate. Before able to wed, Ralston the Rancid Raccoon demolished the colony. The fact they were safe, alive, and reunited with their parents was a wondrous turn of events. The boys were more than happy to head for the Valley of the Stones. The decision for their parents was now easy: they would go to the valley and live with their sons, future wives, and children.

"Mr. Mizithra, Murray, these are our sons: Mr. Hoss, Mr. Little Joe, and Mr. Hop Sing", beamed Mr. Gilligan's Island.

The mother and father in them emerged, heretofore unseen. People are who they are. In various situations they may seem different than whom you assume them to be. Parenting is such a thing; one never knows what kind of a childrearer one will be until opportunity presents itself. To see these five Olympic marmots reunite was a beautiful thing.

Beholding it led those observing the entire affair to tear up with joy. Love poured out, the entire colony, who knew none of the marmots, burst in one huge hug. A spontaneous song of felicity erupted from the midst. They rejoiced at this marvelous turn of events. Mr. Gilligan's Island and Mrs. Maryann would stay and head to the Valley of the Stones.

The addition of Mr. Milk Chocolate, Mr. Hoss, Mr. Little Joe, and Mr. Hop Sing, with the others, made it clear: Murray was no longer needed. He could head downstream tomorrow to begin life as a touring musician and member of the Beaver Blues Bungalow Band. Dreams had not only been realized, but superseded. The future would be a happy one.

However, one sad item remained. These adventurers, madcaps, best buds, had to part. In fact, it may be they would never meet again. Once at the Valley of the Stones, he would have no reason to come back to the club. In similar fashion, Murray's life was never going to take him up that way, unless perhaps….

"Say, Murray, I have a thought, why don't we hire you once a year to come play at our new home? We would be more than happy to have you stay with us even for just an evening's entertainment."

"Mizzy, you old marmot, zat is a capital idea, but I vould insist on three days und nights."

That resolved, they turned in for the night. It had been a very full day. The colony rested and replenished. A family reunited. A new friend came to help lead them to the new home. Ardent friends found a way to perpetuate friendship. Come dawn, forest creatures came to eat and drink. Robins came bob bob bobbing along, yanking unwilling worms from their early morning beds, gobbling them with zeal.

They were early birds. They got the worm. Then came everyone else. The marmots knew today was one of complete rest. They went right for the handiest food, then spread along the shoreline to eat and drink with gusto. They had to add more tonnage before winter snows buried them in tight for nine months.

Along about noon, Mr. Mizithra and Murray snuck off alone.

"Well, well, my friend. It is time to say farewell. Let us ensure this is not the last time. I will contact the crows and get on your schedule. I am so delighted you will play music. When we first went in the bungalow neither of us knew your talent. It is a most wonderful one. You are very good on sax."

"Murray, you are a friend I can only admire. You have saved my life many times. You have kidded me, played jokes on me, listened to my pain about my wife and children passing. You have been here for my travels. All I can most humbly say is: Thank you, sir."

"Oy veh, not now. Vat? Now ze vater virks. Mizzy, you old marmot, you. Ja, ve have a very vell built friendship. Ve vill see each other again. Zis you can count on, zis you can trust."

They hugged a long time. Murray began the jaunt downhill to a new life. He did not look back. Mr. Mizithra watched a bit, hanging his rodent head. He went to the edge of the pond amongst cattails and gave it a long think. A good friend is a valuable asset. A true ally is someone to help shoulder the load, bounce ideas off, and share escapades. Best friends look out for your foremost interests, even if you do not.

This relationship is one people hope for and need. Someone on your side in thick and thin is valuable. Murray and Mr. Mizithra had it. Yes, they were varied species with adverse habits. Murray did not hibernate. Mr. Mizithra did. If you accept people as unique you bypass things which separate. This permits you to see people for who they are, not what you want them to be. Open eyes, hold your values, do not close your mind.

Murray made it to where Mr. Gilligan's Island and Mrs. Maryann lived. He paused by the water's edge to drink and check the area. This was where the coyote tried to nab them days ago. While he was resting, a park ranger's truck zipped past. He knew it was. It had the same logo on the side as the ranger who stayed at his old home under the woodshed.

It stirred a pile of dust, an opportune time to head downstream to his new home. Two days later, he arrived, making straight for his room. He stayed all night, alone, commiserating his loss, counting good fortune, and excited about the future as a sax idol.

The marmots at Grand Meadow spent the day resting, eating, and getting acquainted with new members. Mr Milk Chocolate they knew. The sons of Mr. Gilligan's Island and Mrs. Maryann were known, so only had to introduce themselves to Mrs. Lip Gloss as head of the clan.

The two young ladies, both by the way eligible for marriage, were: Miss Mascara and Miss Saffron. The young ones were: Mr. Garlic Powder, Mr. Salt and Pepper, and Miss Perfume. The last young lady was almost three. Their families were murdered in the Battle of Queets River. It was at the headwaters of Mount Meany. They were ravaged, torn asunder. The army invaded, ruthlessly killing, no whisper of sympathy. They wanted marmots dead and did everything to accomplish the goal.

The result was success for the enemy. The complete surprise of their prey ensured victory. It was swift. The marmot leader, one elder of great praise, Mr. Mickey's Bigmouth, was first to die. Obviously the invaders knew what they were about. A mere nine made it alive, a third babies.

Mr. Milk Chocolate finding Mr. Mizithra was a real boon for he and his colony. The survivors held him in high regard. He was viewed as a hero. Miss Saffron thought he was cute, scars and all. Mr. Opie found interest in Miss Perfume.

Mr. Hoss, Mr. Little Joe, and Mr. Hop Sing were eying three lovely, eligible young ladies. They were the daughters of Mr. Marathon Bar and Mrs. Ellie Mae. Respectively: Miss Wilma, Miss Betty, and Miss Pebbles. Marmot weddings occur prior to dormancy. Provisions are made during the dog days of summer. With new additions, the future for Mr. Egg Flower Soup, the colony's wedding performer, was going to be busy. With so much room to spread out, the future looked bright.

All they had to do was reach the goal.

Chapter 22 The Valley of the Stones

Early the next day, the wagon train turned southeast. A large group of marmots went ahead. They kept plenty of able bodies to safeguard slower groups. Mr. Milk Chocolate led the trailblazers. Mr. Beef Stew, Mr. Egg Flower Soup, Mr. Hoss, Mr. Little Joe, and Mr. Hop Sing, and others took off at a brisk pace, aimed directly towards the objective. It was a six day trek for Mr. Mizithra and slow moving convoy anchored by Mr. Opie. He made sure Miss Perfume was in his group.

The scout party arrived in two days. Next day, digging in earnest. First, they carved out a large hall. It was over eight meters long, a meter and a half tall, and more than four meters wide. There were six tunnels for escapes. They built room after room until they were spent. Mr. Beef Stew, an engineer, proved invaluable. Dirt was piled around the rocks. They ate evenings until they dropped.

One thing Mr. Mizithra missed in the initial discovery was a small stream trickling down the far south side. A dash of water traipsing off the side simply fizzled to mist, plunging a thousand meters to the valley floor. The males built a small pond pushing rocks around a modest hole. It was less than a meter deep, more than ten in radius. Around the stones, they packed dirt hard against it to seal the deal.

This way, they would have a constant source of fresh water all one hundred days. Digging was difficult in some areas. Weight from the glacier shoved solid bits of rock in soft earth. The Olympics are the only mountain range in the "Ring of Fire" (from the bottom of Chile north, over Alaska east to Asia, south along the coast to New Zealand) with no volcanoes whatsoever.

This means no granite. The rock was basalt. They simply dug under or around impediments. Mature males were adept at tunneling. Face it, when it comes to digging a hole, your first option is marmot. They love excavation, are experts, and constantly work to improve living quarters. Besides, moles, voles, and gophers are ugly. By the end of the fifth day, the first scouts appeared ahead of the colony expected the next day.

They rested that night, vigil kept on top of the hill in a safe hollow. Mr. Burrito stood watch, his family amongst those on the way. He did not want to miss their arrival. His mother, Mrs. Jazz, father, Mr. Perry Mason, and sister, Miss Ballerina were among those still on the road.

He knew food was scarce and his mother not strong. She was hurt avoiding a near death experience with an eagle a week before they departed. Her shoulder was still painful where she struck a stone while entering an emergency hole. This would slow them down, not to mention the elderly and cubs.

All night he waited. He ate food at hand. He denied relief by Mr. Kato. He wanted to see his family. This is a common reaction of those who are ahead. They look back to ensure those not so fleet of foot are safe. This concern is admirable and shows the passion Mr. Burrito felt for his loved ones. Long about noon, there came Mr. Marathon Bar, Mr. Mizithra, and Mrs. Ellie Mae.

Then came the next group. Another came, then another. He greeted each with a hug and kiss. He was happy the colony was safe. He could not rest until seeing his people again. One more group passed. They let him know two groups remained on the road. They would by there soon. For some reason there was a holdup in the group immediately behind theirs.

Mr. Burrito could not leave his post. The safety of the whole was of more consideration than his family. It was simply a matter of numbers. This caused more anxiety. He sent ahead to Mr. Kato for relief. In minutes the change of guard took place. Mr. Burrito hugged his friend, dashing northeast to his missing family. It did not take long.

His sister wandered off, something she was prone to do. It took a while for the group to locate her. Mrs. Jazz berated little, naughty Miss. Ballerina for her lack of attention. A young animal separated from it's parents is fodder for any looking for an easy meal. The babes are tender, too, so are very much sought for by any hunter.

He wrapped his right arm around little sister, gave a loving squeeze, then led her down the garden path to the Promised Land. In due course, Mr. Opie, Miss Perfume, and last pioneers arrived at the final destination. In the morning, they examined this immense, safe area.

They were very happy at the results of Mr. Mizithra's discovery. This was thanks to a chance meeting with Jon Deer. Mr. Mizithra reminded himself to send a dragonfly with a message of humble gratitude for that valuable information provided by the generous stag. His help was of the utmost gravity. They were now in the safest place to dwell.

The next day, Mrs. Lip Gloss took Mr. Mizithra off to the side.

"You, sir, are a marmot of the highest order. We are blessed to have you as a member of this colony. I can only say: Thank you. This family loves you from oldest to youngest. Thank you, Mr. Mizithra, you have ensured the survival of our people. Thank you, forever."

Mr. Mizithra humbly bowed his head and smiled.

BOOK 2: HIBERNATION

Chapters

1. Preparation 146
2. Meanwhile 149
3. Miss Bluesberry Scores 152
4. Club Club 157
5. The Grand Meadow Festival 165
6. Terry, Chris, and Lee 174
7. Willapa Bay 176
8. Carnival Island 182
9. The Carnival Continues 200
10. Reality 208
11. Aftermath 215
12. Mr. Mizithra Arises 218
13. Kai the Kestrel 219
14. The Four Whistles 224
15. On Meeting Murray 230
16. The Day Of Terror And Pain 233
17. Ollie 240
18. The Folks 244
19. Murray And The Fox 247
20. Rayen And The Night Ninjas 253
21. A Nice Slice Of Marmot Anyone? 258
22. The Exception 265
23. Springtime 271

Chapter 1 Preparation

Once the Olympic marmots arrived at their new home, the colony prepared for impending weather changes. Even now in late summer, they felt coolness arrive earlier come morning and materialize sooner come twilight. Days shortened, sun disappearing over the cliff more quickly.

Hibernation is a most important thing for them. You see, certain species "sleep" all winter. Why is not fully known, though a portion has to do with the lack of food. However, prepare they must. Marmots are a part of this lot. The goal on arrival was one of preparation.

The most immediate need was for everybody to gorge on food. In a very short few weeks, they would descend in the new digs and spend the first hibernation's dormancy underground. In order to survive rugged Olympic Mountain Range winters they had to be as fat as could be.

Imagine being told to gain as much weight as you could by eating as much as possible to be huge for the next nine months. By the time most adult marmots were ready for "Sleepy Time" as they called it, they were breathlessly wedged down in the tunnel to sleep chambers to settle down for a long winter's nap.

The urge was to find a mate as well. Males dug new rooms for such needs. Everything had to be just so for newlyweds. Young couples would have autumn, winter, and early spring to begin their new lives together. Mr. Egg Flower Soup kept very busy during wedding time.

They had to dig alternative escape passageways, though it became obvious it would be nearly impossible to do much. The stones were impenetrable. They began a second colony a good forty meters away, so the young families could establish their own village.

This was fostered by Mr. Milk Chocolate. It was an agreed to notion, for the last thing any marmot desired was for fighting to occur. This would come about were there too much competition for mates. Since land was plentiful, Mrs. Lip Gloss authorized the second colony after conferring with other elders. The new spot was perfect and had a great view of the mountains across the great divide of the valley.

The entire group marveled at the abundance of food, the quiet, the security, and stunning vista of the valley a thousand meters below. It was a perfectly idyllic place. Each day, Mr. Mizithra was stopped by others so they might thank him for finding this marvelous homeland. The body as a whole felt certain no human would ever find them.

Predators would have a very difficult time. The only fear was of raptors from above. After he dug his own nice room, Mr. Mizithra began to edit his journal and line up the days. Once he had an opportunity to review his notes, he realized how arduous and wondrous this journey had been.

He was pleased at what occurred on this latest, and more than likely last, great journey. Certainly, it was for Murray, who expressed his desire to live with beavers the rest of his days. Any further sojourns would require a new marmot partner. A good rat for traveling is hard to find.

He wooed Miss Saffron. She agreed to unite with him as one. He spoke with Mr. Egg Flower Soup. A time for the ceremony was placed. She approved of his large room with two smaller ones for sleep and storage. They could cozy up in one of the lesser rooms and be very comfortable. Since she was so much smaller than he, his chubby body would keep her diminutive one warm. They would form a marmot cocoon of love in their comfy nest.

Mr. Milk Chocolate found his true love in the lady Miss Mascara. He quickly sought out Mr. Egg Flower Soup to set the date for their wedding. It was going to be a speedy affair as he had to now wed Mr. Mizithra and Miss Saffron, Mr. Little Joe to Miss Pebbles, Mr. Hoss to Miss Wilma, and Mr Hop Sing to Miss Betty. Mr. Opie and Miss Perfume were added with abundant joy.

Marriage season gave a sense of pride at solid beginnings for their new and permanent colonies. There were to be more as well, so he got ready. He called the intended. Mr. Egg Flower Soup stood amongst them and asked all who wished to be wedded to gather before him.

Ten males and ten respective females stood in a phalanx, males on one side, the females on the other. Each had a garland of Flett's violets around their necks and a small bouquet for each lady. Mr. Mizithra, true to form, was the only one with a small sprig in his mouth. He was the oldest groom in the gang.

Smiles gleamed from all concerned. Forty odd Olympic marmots stood grinning, yellow teeth glistening in the bright sunshine, at the spectacle of so many joining. It was an odd sight to see ten large males and ten smaller females beaming at each other in the joyous prospect of pending nuptials. The colonies could see how this would give grit to their actions. It is with family societies thrive. A breakdown of the family unit is no way to advance or sustain a species, including human beings.

Mr. Egg Flower Soup raised his arms above his head, "To all here, you are witness to the joining of these pairs who have chosen to be as one from here unto the end. As is our custom, males will turn to ladies and pledge undying devotion to their lives. They will defend them, help raise the young, and live in harmony one with the other.".

This they did.

"Now ladies, you will agree to their pledge and agree to adhere to the lives you have now openly chosen".

This the brides did with glee.

"Finally, you are to exchange garlands and kiss. Lovely, perfect. You are now united as mates. Please, will the whole colony come and embrace these newlyweds and welcome them to the family of bonded marmots".

This everyone did. They spent the rest of the day congratulating the new couples and wishing upon them nothing but the very best. Gifts of food were pretty much it for all that anyone wanted or gave. Wrestling was the favorite way to release energy. Newlywed wrestling tag teamed the new couples against others. Mr. Mizithra and Mrs. Saffron won three matches to become the victorious pair.

Olympic marmots mate for life, there are no divorces or any such thing in their world. It was not any honest way to maintain their culture. To preserve it they know the bonds of male to female are lifetime; this is a thing they find great pleasure in living.

For some in the human world marriage is not so highly valued. Mr. Gilligan's Island and Mrs. Maryann were both especially happy. They had rediscovered their sons, had three new, lovely daughter-in-laws, plus in-laws, Mr. Marathon Bar and Mrs. Ellie Mae. These eight would go with Mr. Milk Chocolate and Mrs. Mascara to initiate the new colony. It was a wonderful way to end "Above Time" or "Sun Days", as they called the period above ground.

The rest of the days above terra firma were spent eating, grappling, and loving each other. Both colonies thrived. All were concentrating on the upcoming change in weather. The cold could be felt up here much more quickly than down below along the coast or back at the pond.

In the Olympics, winds come directly off the cooling Pacific. These strike the range with a vigorous and devastating blow at high elevations, while down on the coastline it is calm as can be. The contrast in height is a change in lifestyles. However, this was home. To them it was a wonderful place to live.

Mr. Mizithra approached Mrs. Saffron with an altogether different look in his eye than she had ever seen. She was intrigued.

"It is time, my dear, to go in our chamber and ready ourselves for Sleepy Time. We must conceive young ones for our future".

He oozed love for her. As far as female Olympic marmots go she was exceedingly lovely and desirable. He was pleased at his good fortune to survive long enough to find another mate. He would always remember his first wife. It was time to preserve the memory.

She saw the reason for his different countenance: she had never seen the romantic side of this stalwart adventurer. A shy, coy smirk crossed her face, exposing her yellowed front four.

"Oh yes, my love, let us go to the chamber for our long nap".

With a moving, randy smile the lovebirds entered their warm and loving enclosure to explore what awaited them for the next 265 days.

Chapter 2 Meanwhile

The marmots were busy preparing for changes in weather. Down at the pond, blues seeped from beneath the dam, with sweetness hitherto unknown in this humble beaver club. Adding Miss Bluesberry and Murray, upped their game. Beulah recovered nicely. Buford was in his usual spot jamming on stage or basting bark out back.

Beulah was up to old tricks dishing out what made her everybody's darlin'. In no time the place was rockin' every day and night. Frema and Nediva were happy ladies. They could play with the band if they wanted, rest, eat, sleep, whatever. Woodrats are not hard to please. Aches and pains were the only concern. They had all the food and safety any animal could ask for. The sounds of beautiful music serenaded them day and night. After a lifetime of struggle, this was heaven on earth.

Murray was floating above the clouds. He knew he would be going other places to play. For the immediate future, he was having the time of his life. He would climb on stage, grab his sax, lean against his new best friend Bette, and wail away. Miss Bluesberry, for her part, began to really get in the groove.

She was quite comfortable and accepted by others. Occasionally a newcomer would balk at the sight of a polecat. In time, her soothing, mellisonant tones eased the fear of stench, replaced with veneration at her talent. Quality overcame qualms. Miss Bluesberry became a hit.

The joy of this period was interrupted by two now familiar crows: Tutti Frutti and Mon. They first went up to the bar and spoke with Beulah and Buford, then made for the stage. They waved their wings to stop the music. They motioned for Murray and Miss Bluesberry to come have a little chat.

It was time to plan club dates agreed to when Murderers' Row saw them a few weeks back. Since she was so large, Miss Bluesberry could not travel as far as Murray. He was transportable via trumpeter swan. She would go to clubs closest to the bungalow. She could visit other places, spread the wealth of her immense talent, and gain a solid reputation in the Olympic mountain music community. Her career readied to blossom. Murray, on the other hand, was slated to head very soon to the greatest festival in the Northwest: Willapa Bay.

"Hey, Mon".

"Yo, Tutti Frutti, what you want?"

"Where, Mon, does Mizz Bluesberry go first"?

"To da Hart Den, braddah, den to da Muskrat Cavern. After dat she go to da Crow's Nest. Den she go to dat big Pond Gatherin' up at Grand Meadow, den back to da Beaver Dam. After dat, she make one last trip to da Squirrel Square. Oh wait, dat one is canceled due to weather. By den it gonna be snowin', braddah, an' we no go in da snow. NO".

"Okay, Missy, we gonna go wid you on dese trips, cuz we want you to have a good time an' be safe. Each spot has it's own house band, so you can sing your songs or join in wid dem. Dey all wants to see you, it's true. We have been puttin' out da word, you da bomb, lady, da bomb"!

Miss Bluesberry was overwhelmed and began to sob into a handy handful of leaves. Emotional anyway, this was simply too much. It was her dream come true, to be able to sing songs to animals. The clubs and other venues were where she had to go to share her love of songs. A happier skunk simply did not exist at that exact moment. She had actual engagements.

She had to work on her arrangements. She had to get her bow ties ready. Tutti Frutti and Mon were going to go with her. They would guide her to each show. Her head swam with ideas, emotions, smiles, because of the work of one wonderful fellow. She thought of the day and how Mr. Mizithra went to bat for her. Since then, her life was one of joy and love. He was a marmot for all seasons.

She reached out, gently touching Murray, then the crows, sobbing in shear joy. She ran back to her small cave not far from the beaver club to recover and pack. Clutching her leaves tightly, she wandered off, simply inconsolable, enraptured in passion and wonderment. As she meandered along, shards of sunlight shone upon her head, glistening beautifully off her dome. As autumn's early days began, she was happier now than ever. The future was golden.

"Now, Murray, you gonna be headin' to Willapa Bay soon enuf. We gonna send you to Neah Bay fo' a week, den you gonna get ready to fly down to Willapa Bay. In Neah Bay, Delroy, Neville, an' Clive are da kings. Dey got you a sweet spot. Dey gots a place out on Waadah Island dat is da bomb. You can go dere an' jus' jam away. Da band dey gots is da Waadah Widow-Makers. Dey a tight blues band wid a real sound. You gonna love dem, Murray".

Over the next few days he prepared to fly for the first time on the back of a trumpeter swan and begin his career as a professional musician. His knowledge of geography was limited to the area around which he lived. Otherwise, it was the numerous trips he and Mr. Mizithra made throughout the Olympic Mountains, all on terra firma.

Yes, he saw the Pacific Ocean afar off in the hazy distance up on the ridge, but he had not ventured far. His excitement was piqued. He could not sleep, tossing and turning the night away. Murray did this with an unbreakable smile on his furry mug. Telling his mother and grandmother presented it's own problem.

"VAT?, screamed his mother, "You vill not go to zis. Ven you go up in ze air, vat is to keep you on zis narish Schwan"?

"Mama, it's okay. They do zis all za time. It vill be fun".

"Oy vey, do you have to do zis? I vill be a bundle of nerves until you return. Oh, Murray, do not do zis, I beg you. I am out of potato schkins".

Though Murray tried and tried, he could not calm them down. He resolved to allow them to fret. They were protective anyway. He was their son and grandson. He bore a wry grin walking away, daring not to shake his head. Were he to dismiss them, they would be hurt.

He spoke with Beulah and Buford. They would watch them. They had come to love the old ladies. Their room at the back was a great place to set and worry. Beulah would supply the spuds. They could eat moldy old potato skins in the corner and enjoy their misery. The truth is, these ladies were very happy with the new digs. It was warm, quiet, and most importantly, risk-free. Safety is key, once food is found for woodrats.

Murray approached the band members and gave each a hug. Bette was his best friend. The hours spent jamming, leaning up against each other, established an undeniable bond. He loved Tobias. He was the first welcoming him aboard. Murray had a blessed life, with Mr. Mizithra, his mother and grandmother, Beulah and Buford, the kids, Miss Bluesberry, Twigs, Tutti Frutti and Mon. The rest served to make his life joyful.

He did what perhaps we should do to improve life: he took time to sit back and count his blessings. He saw the danger of life, near misses, the Thumbs and desire to kill him and his kind. He knew this, but saw the overwhelming positives. He admitted it was far better a fate than anyone else he knew. He was a humble old woodrat.

The day would come soon. He would begin a new chapter in his life: traveling musician.

Chapter 3 Miss Bluesberry Scores

To understand what happened, you need some prospective of the Olympic animal world. Deer are fairly lone critters. Mama will keep and raise her young. They often have two or three fawns, but males are usually off to themselves. When they mate the stag will hang around more. It is the way of nature, not of man, who marries and stays with his family until the young are ready to move out. One just hopes the kids do not think of home as a permanent solution.

However, here, up high, far from man, they had a club. It was called the Hart Den. It consisted of a pseudo circle of fir trees up against a large chunk of hard, solid basalt. At the back, the rock gave way to a large opening. It was not a cave, more of a hollowed out space in stone. They could get under it and be safe in the stand of trees. Up to fifty of these big creatures could assemble there and go wild.

Deer absolutely love to party. It is one of those things man simply does not know. The presence of Thumbs is quickly detected and they disperse. Up here though, this far from any trails high in elevation, no human being had ever been seen. This Den was fabled in the hills for late nights and delirious antics. They could not play any instruments, so they had a local band perform.

Tutti Frutti and Mon found as good a group as they could given the limits of the area. The drummer was a muskrat named, Boz. They had a muskrat bass player named, Delmer, who was not related to Boz. They had a fisher (a type of local weasel solely found in the Olympics) banjo player named, Zippy, and his brother, Sweet Siggy, who played guitar. Finally, they had a woodrat named, Moshe, who knew Murray, playing trumpet.

This band did not regularly play together, so they were not "tight" as musicians call it when individuals play as one. They were professionals, so would find a way to mesh. It was the reason the dam band was the best. They were so tight it was scary. For Murray to fit in so well was proof positive of his talent to all. The same was said of Miss Bluesberry.

She was given guidance on how to get there with the crows. They were talent scouts. While merely agents, they cared for Miss Bluesberry deeply. She was a blessing. Her voice was so enveloping and deep, they would stay to hear her sing when they ought to be finding new talent. Tutti Frutti and Mon were the very best at it in this mountain range. No one could doubt the results. Success was virtually guaranteed with this lovely lady; she was a raw talent to be honed.

They ascended the mountains. She soon huffed and puffed, slowing considerably. Winded, she halted several times, finally pausing to catch her breath. She turned, saw the heights and smiled at the view. Face it folks, having eyesight is a blessing, seeing vistas offered by rock ledges of the Olympics was icing on the cake. She sat simply to catch her breath, but found herself singing to the magnificent world.

Lilting tones echoed through the hills. These descended beautifully to the valley, filling the air with a most fantastic sound. Miss Bluesberry's voice awoke all who heard it. When she ceased, a roaring applause arose about her. The forest denizens came alive in appreciation of the song from on high. It enraptured them as evening pushed it's way across the world.

Miss Bluesberry was humbled to her toes at the response. She stood and bowed. Then she turned and continued the journey to the Hart Den. Tutti Frutti and Mon led her along a narrow trail. It hugged the wall of an odd, slender ridge of hard rock with a long line of fir trees dotting it's path. No human could hike this trail. It was simply too slim.

They flew from tree to tree, waiting on her as best they could. They led her to a large, flat plateau with a huge circle of fir trees lining it. Once inside the arena, trees gave way to an opening, a tall cave-like structure, forming the stark back of the club. The basalt rampart rose hundreds of meters. It was a perfect barrier to any intruders who might consider deer for dinner.

The deer hired beavers to come chew out the few trees marring a cleared clearing. Accomplished, they had an excellent place to unwind and hang out. It was stags right, does left. Once the music began, they danced the night away as only deer can. Truly, they are a wild bunch, but not rowdy. It was always fun. Deer are not the most serious of creatures. If you want serious, one thinks instantly of wapiti. They are the most sullen in the forest world.

"Tonight", began the emcee, a stag named, Big Horn, because of his huge rack, "we have a treat provided by our feathered friends, Tutti Frutti and Mon. Give 'em a rattle of your racks, folks. Thank you, and what they have special for us tonight, all the way from the famous Beaver Blues Bungalow, down at the big lake, is Miss Bluesberry. Come on, give her a rattle of your racks. Thank you. They tell me she can sing up a storm. So without further adieu, Miss Bluesberry"!

She strode out as the band started the song she told them was her best to open with, "I Gots Dem Polecat Blues". She had a beautiful red ribbon on each side of her head, highlighting her black and white stripes. She was combed out with a pine cone brush, looking sleek and shiny. The crows had a few buddies carry her show things up ahead of time. Miss Bluesberry had but to don her accoutrements and belt out the tunes.

With total unity, the audience gasped. The sight of this black and white creature on stage shocked them. No one thought a skunk would ever enter the Hart Den. Then, oh yes, then she sang. As happened all her life, Miss Bluesberry knew prejudice because of what she was. She was judged not for what was inside, but solely because of her fur. Them seeing this dreaded creature gave a quick pause. So to combat this, she opened her mouth and let her singing do her talking.

"Listen up all you dear deer to what Miss Bluesberry gonna say, cuz nobody gonna give it to you better, baby, nobody."

"I gots dem polecat blues, I gots dem polecat blues, what you gonna, what you gonna do, when you gots dem polecat blues? Dem people runs away, dat polecat stinks they say, oh yes, I gots dem polecat blues".

By the time she finished the first verse, they shed all fear and rushed the stage to the left side near the ledge. It was a perfect place as the acoustics were the best. Once she began to sing, even the band was held spellbound. All endeavored to play their very best. Sweet Siggy, Boz, and Delmer got really into the song.

At the break, Sweet Siggy proved why he had his name for no fisher in the Olympics could play guitar like he. He laid down a sweet, lazy lead which built to a crescendo by the power of the muskrats, whose bass and drums became louder and louder. They hit the top and Sweet Siggy hit his highest note, Miss Bluesberry retaliated with her powerful lungs.

"What you wants to get, what you wants to have, what you need to be is a happy ol' skunk. What you gonna get, it's dem ol' polecat blues".

She carried the final note long after the band ceased play. She held it, strong, lilting, powerful, perfect. Defiant, she wanted them to know who she was and appreciate her offering in spite of bigotry. The moment her singing ended, the Hart Den heard the loudest response to a song in it's long history.

Stags stomped, does danced, the band stood and clapped for all they were worth. Creatures emerged from shadows, squirrels, tiny Olympic chipmunks, crows, and rodents. The reverie lived for some time.

"Whoa Nelly and fill my belly," shouted Big Horn, "Please, dear lady, please continue. We are humbled. Tutti Frutti and Mon were truthful when they said you were better than could be believed. Stamp your hooves, my mule deer brethren, let her know what you think".

With this the din was raised further. The band returned to their instruments and tore into an old mountain favorite. It featured banjo, for which Zippy was famous, the "Olympics Stomp". Zippy and Sweet Siggy got going, the lyrics were simple, country, and a perfect followup. They alternated between mountain music and blues. It seemed there was no song she did not know completely. Miss Bluesberry wowed them.

She owned the club's audience. Her point proven, she now plowed in fully impressing with her vocals. It was late, late, late. She was tired, worn out, and frazzled They begged for one more song. The evening's event was emotionally overwhelming.

With a smile and soft shake of her head back in forth in joyous love, she turned to the band They gathered about her. Since this was another blues song, she did not need the banjo. Zippy picked up another guitar. Moshe was not needed. He yanked out his trusty harmonica.

This was a slow, sexy, strong song, "Oh Baby". They began quietly, harmonica melding in with the guitar Zippy laid down. Boz and Delmer fell in rhythm which left Sweet Siggy and Miss Bluesberry with nothing to do. That is until he let go the most sultry lead any of them ever heard. They played in what could only be described in a "musky" way. They jammed for a while, then, only when it was absolutely perfect, she stepped forward, her voice almost obscene in it's sensuality.

"Oh-h-h, oh oh yes, b-b-baby", the stags shuffled uncomfortably from the lascivious, salacious, simply filthy way she sang, "oh baby, what, oh what, oh what I need is you. Touch me, baby, touch me there. Touch my soft fur, rub it, oh like that, just like that, just, oh, oh, just like that, baby".

Well, needless to say, by the time this burner was over, so were they. Stags were eying does. Does were coyly returning stares with huge, coquettish lady deer orbs. The band arose, stood in line on either side of this songstress, locked arms, and bowed down in thanks for the audience and invitation to play.

Then, they embraced Miss Bluesberry, fading to the background. It was her moment alone in the spotlight. Her performance that evening was baptism under fire She emerged from the baptismal pool a star. With strength of spirit and immense, amazing talent, she won over a prejudice crowd.

They hailed her, calling her down from the stage. Once amongst them, being much taller, they led her to a stump upon which to stand. She was lauded with praise, and nuzzled which is what sufficed for a hug in the deer world. Well wishes, praises, and accolades surrounded her.

It was overwhelming. She was proud of herself, humbled, too. She had never envisioned this sort of a life. A short time prior, she was a lonely skunk sobbing into a handful of leaves. Now, now she was a hit and had a career which was off to a flying start.

Happy is hardly the appropriate word, but it would have to suffice.

Chapter 4 Club Club

While Miss Bluesberry was conquering the Hart Den, Murray got ready for his gig in Neah Bay. Tutti Frutti and Mon said Delroy, Neville, and Clive would have everything ready. It was a first rate club with an excellent band. They were very excited to have a sax player. He was the element they did not possess. He felt it would be a challenge.

They had two lead guitar players, otters, the Mackenzie brothers, Paisley and Ness, north coast legends. Why, sea otters were known to come ashore to listen to their sound. Their father, old Angus Mackenzie, had his own guitar school. Graduates from his academy were spread over the peninsula, east to Gig Harbor, south to the Naselle River.

His sons proved to be a bite off the same apple. They ruled Neah Bay and the club on Waadah Island. No Thumbs lived on that spot of dry land. The ait was shaped like a club, naturally they called the place: Club Club.

The rest of the band was a drummer, Pockets, a short tailed weasel of ill repute for his past no one dared inquire about. The bass player was a beaver named, Hammer. He was a solitary type whose family were killed by this and that, he would not say, leaving him an orphan.

He drifted into the band when he swam out to the island to become a hermit. With no family, he felt it best to simply disappear from the rest of the cruel, cruel world. Instead, he found a home with the bass and lived with the Waadah Widow-makers, as they were known. The sound was so tight, you could bounce a quarter off it.

Their singer was a Douglas squirrel, Chicky Baby. Say what you will about a squirrel, dude had pipes. He was a hard core blues singer with a deep voice, certainly deeper than one would imagine could come from such a little guy. The other thing about him was he could climb into the higher highs vocally, even falsetto. He had a girlfriend Douglas squirrel, Mamiko, daughter of the sea, who was his agent. She was vicious and demanded the best for her boys as she called the band.

They anticipated Murray's arrival. The crows worked so everything was perfect. Bringing in outside talent was not a common occurrence on this soggy spot. Mamiko worked with the three and got a piano player, an otter who they really did not know, Graham "Ivory" O'Shea of limited reputation. It was a gamble, but no one else could be found in time. Everyone with any semblance of talent was booked down for the big digs at Willapa Bay. Delroy, Clive, and Neville were sure Murray's talent would make any weak parts of the band moot.

Back at the pond, Murray saw his two saxes loaded on the backs of trumpeter swans. The formation was: one in front with two more rows directly behind and to the side a smidgen. Inside were three others, one for each sax. The one in the very heart, the largest by far, carried Murray. He realized he was going to ride on the back of a massive bird. A tad bit of fear grappled his confidence to the ground.

"My name is, Theodor, and my wife over there, yes just there, she is Terese, which means Summer Harvest. In front, we have, Torbjorn, or Bear of Thor, for he is the strongest. He will lead us to Waadah Island. We have Taavi, the adored one, Trygve, the trusted one, Thrya, who is strong like thunder, Torhild, the fighter. Finally, Torvald, Rule of Thor, our leader."

"Our Nordic names are from out-lands and northern affiliations. We came and found it best for us to work while we are here. We know every place to go and are very strong. We chose to begin the line. Trumpeter Swan Airlines is a success story. We have a perfect record".

"In a minute, we take off, heading due north by northwest. So nestle down in my down. Enjoy the flight, it will take about two hours with tailwinds. It takes over 100 meters to get airborne. We will be at an altitude of over 1800 meters, flying at over 80 kilometers an hour, so hang on tight. The view is incredible. Make sure and take a look or two on the way. Relax, my furry Murray, we will have you there in one piece"

Murray pulled off his kippah and hugged on the massive neck. The swan weighed over 11 kilos. He spread his three meter wide mighty wings and flapped. In an instant, the other seven swans began. They moved off, making a flying V soaring into the sky. It was a lovely day. The sun shone as they cleared the pond, heading north. Everybody came to wave good-bye. Murray saw none of it. His head was buried in Theodor's feathers.

Soon they were wending their way to the end of the United States of America, literally. He finally raised his little rodent noggin and saw the wonder of it: the Pacific Ocean lay before him in the distance. He could see the forest below and activity. Lives of animals were being lived out downstairs from him. He could see the action. He saw coyotes, deer, squirrels, crows, all the wildlife he knew moving about below. This view caused him to shrink back in the safe warmth of the swan's gentle down.

Murray was a rodent, a ground dweller. He was no flier. Yet, with the scene before him, he changed. He saw the world as a much bigger place and he a miniscule part thereof. This was different that the view from the White Mountain. It humbled, causing a feeling of inner quiet. He was envious of any flying creature. Then, Murray realized he was so good on saxophone, he was being flown to play.

He was being lofted far away from home because the talent was that huge. No ego followed this revelation, but he did regain confidence. He stuck his head out of the feathers once again, reveling in the ride. Murray dug flying, oh yeah.

Sure enough, within two hours time, north by northwest for 140 kilometers, they descended over 1700 meters in elevation. All arrived in late afternoon at Waadah Island off the shoreline of the tiny town of Neah Bay, native

Makah tribe headquarters. They were a proud, simple tribe whose fight to stay a viable people was hard and long. They were now coming back strong. In spite of the odds they endured, standing powerful after so many years of oppression and sadness.

They set down along the saline shore. Murray hopped off onto dry land. Two otters came out and got the saxophones. Theodor waved a wing and took off again. They would return in two days to take him to his next gig. With a few flaps of powerful wings, they ascended the heavens and were gone. Apparently, swans are not blues lovers, preferring opera and classical. In fact, he was certain he heard them humming The Ride of the Valkyries as they wended the way south. Each to his own, eh?

"Now Murray, dese are your band-mates for da next two days. We done told dem dat you da best. All you has to do is play wid 'em", Clive explained. To him, it was just entertainment they were providing for the wildlife in the area.

"Dis is da band: Paisley and Ness on guitars, Pockets drums, Hammer bass, Graham piano, and, finally, Chicky Baby is your singer. As usual he don't show till show time. Now, you fellas welcome Murray. He's da best sax player da tree of us ever heard. Da first show is tonight, so you best jam quick".

Clive, Delroy, and Neville winged off to set up the rest of the night. Murray stood before them, a humble little woodrat amongst critters much larger than he. Graham did not garner affection from the coastal rockers either. They had an attitude problem which seemed to be fairly common amongst maritime folks. Without a word, they motioned him forward. He saw the club. It was a hollow rise in the ground, a cave of sorts. Once you passed in the dark entrance, an immediate turn left.

They descended several meters in a deep cavern. This was out of the sight of man. It was dimly lit by a few large holes dug in the ceiling. It did not leak water. The place was massive, ceiling three meters high. A large round area of about fifteen meters from one side of the circle made the hall. There was a small stage on one side, well lit by slits, with a bar and tables and chairs strewn about. Then it opened up to a space to dance.

Guitars were on stage, awaiting nimble fingers. Drums were to the back, piano off to the side, reminiscent of the Bungalow. He saw his two saxophones on stands made from twigs and branches. The Mackenzie brothers leaped on stage and tuned up, ditto the bass. Hammer being a beaver received a dragonfly note about Murray from his cousins up in the hills. They told him while this may be a mere woodrat, he was a part of their family.

Hammer had better mind his P's and Q's and treat this fellow right unless he wanted their wrath. While it was clear the Mackenzie brothers were not throwing out the red carpet, Hammer did his level best to show Murray around. Pockets found his spot and began to tap about getting ready. Graham, also new, sat at the piano and began plunking the keys. He was an odd otter, distant from the other otters. It seemed he preferred kinship with Murray, smiling at him constantly.

They did not so much practice as jam. The newbies were expected to join. The brothers did not have time to converse; their playing did the talking. They assumed the rest could catch up or get lost. They were hard otters, guitar gods, professionals. They did not have time to waste on amateurs. Add pride, 'We don't need outsiders coming to show us up'. You had an attitude problem. Once "practice" was over, they laid down their instruments and made for the bar.

The brothers ordered fish, sitting down to await the evening's jam session. They were not going to be any more cordial than necessary. Pockets fell asleep at his kit. Hammer hung with them and let Murray alone. He was stuck with Graham the odd one.

They had not seen Chicky Baby. His girlfriend, Mamiko, kept him in silence to "prepare" for the show. He was an egotistical maniac, but he was a Douglas squirrel. They lean that way anyway. A prima donna if there ever was, good old Chicky Baby.

The crowd began to assemble loosely, meandering in pellmell, eying the stage. This club was the island's only source of entertainment. They heard the ruckus Clive and the fellas stirred, so were curious about the new talent. Saxophone was not a common sound up north. They were used to the Mackenzie brothers hard blues music and deep wailing tones of Chicky Baby. A brass instrument was novel.

The place filled, bar jammed with otters demanding a varied variety of fish. The popular ones were sardines and mackerel, the unusual choice of octopus, grayling, and salmon arose as well. A few local squirrels and chipmunks made it, filling the corner nearest to where Chicky Baby usually hung to sing. They were given bowls of nuts to nibble on during the show. The lady chipmunks had a fan club for their "hunk". The chairs were packed with those who did not wish to stand or dance.

The club owners were an old otter couple: Haggis and Heather MacTavish. Their bevy owned this club for ten generations. Their children were going to continue the family business. They were busy, offspring as well, filling orders and overseeing events. At one point, Haggis looked up to the stage and nodded to the awaiting emcee, a short tailed weasel named, Beaumont St. James, but everyone called him, Jimmy.

"Hey there everybody, it's old Jimmy with tonight's entertainment. Without further adieu, you islanders", he shouted above the din, "tonight the Mackenzie brother's Waadah Widow-makers has a pal, visiting from on high, from far in the Olympics, up above the clouds, a woodrat of great distinction, MURRAY!"

The boys began a hard, really angry riff. Pockets picked the beat up, followed by Hammer who hammered his bass with a thump so hard it rattled the room. The quartet were a powerhouse of sound. Both guitars were actually different models, Paisley's a flying V and Ness a semi-hollow body. The sound they produced was something new to Murray. It had the direct effect of raising his crushed ego.

He at once understood the standoffishness. He had to prove himself first before any respect would be rendered. Unproven reputations were a dime a dozen, especially from boastful crows as they well knew. While Graham tried to plink and plunk along, Murray held himself in the background. The brothers wondered where in the wild world of sporting animals he was. They cared not as the jam grew into a monster. Chicky Baby was no where to be seen as he always made an entrance after the first number for "dramatic effect", as Mamiko put it.

How droll.

The brothers wailed away in utter love with their axes and sound they produced together. Brotherly love was not simply fur deep. They stood facing each other, strumming violently away, smirking at each other on stage. Proud, yeah. As they hit their first real break, in time for a bridge, a sound exploded in between them, forcing it's way to the top, above their "noise".

It was Murray and his tenor sax. Just as he had with the crows of Murderers' Row, he wanted them to know with whom they were dealing; he was no mere woodrat, this was Murray sax master, the rest be hanged.

He did not wait for them to catch up. He did not care what they thought of him. Murray would show precisely why the ebony, feathered freaks brought him this way. He was going to prove it was worth the effort and would never forget him. Not looking about, Murray stomped out in front of the band. He assumed they would follow his lead.

He dove further into his instrument, finding sweet keys, throwing caution to the wind. After making a bold statement, at a most critical point he bowed out, the custom for blues players so the next guy could have a go. Graham's time to shine. Suddenly, they were attacked by 88 black and white keys plinking with a powerful drive, howling the room.

Graham nodded back to the boys, who were in a state of shock. They wailed away, closing with the entire band jamming one chord until a nod from Paisley. It ended abruptly and froze. The audience went into frantic applause. The Mackenzie brothers were stunned. Haggis and Heather froze in unison, mouths agape. Murray was the real deal. Graham was over the top. They knew the club was going to explode in musical mastery. Paisley and Ness turned to Murray and bowed.

"Dang nab it, Murray, and grab me beard. Laddie, you got it. Now, me bucko, let's have some fun", screamed Ness over the crowd.

"Graham, oh my, my, my, you laddie, you just found a home, so let's roam", hollered Paisley.

Not waiting on Chicky Baby, the gang followed up the initial jam with an old Olympic Peninsula blues standard, "Damn The Strait of Juan De Fuca". It was a hard pounding number made to imitate the sound of driving rain famous in the area. The best line of the song was, *'Hundred days of rain, just about all of my pain, love gone down the drain, hundred days of rain'.* Chicky Baby bounded on stage as the first verse ended. He picked it up as a pro and began entertaining the crowd with his crooning.

Dragonflies were sent northwest to Cape Flattery, east to Cape Clallam Bay, and south to Ozette Lake to let others know what was going at Club Club. Knowing Murray would be there the next eventide was all they needed. It was something special.

Critters left lairs in droves to enjoy jams and entertainment. No one was going to miss an opportunity to rock out. This island's reputation was that of going above and beyond expectations. This new sound with Murray and Graham was doing the job nicely, very nicely indeed.

Haggis sent his own personal messenger, a dove named Sweetness, to the Blues Bungalow. He needed Murray for one additional night. After only four songs he could tell this was a genius. A humble, little, moth-eaten woodrat with an even more moth-eaten kippah had given a lesson on saxophone the Waadah Island crowd would never forget.

The request was expected. An extra day was granted. No sooner had the return note come, than Haggis told Jimmy the good news. He rushed to the stage, waved his arms over his head and made the super fantastic proclamation.

"Folks, this is your loving emcee Beaumont St. James, giving you the word, news, hot off the presses: Murray is being held over an extra night, so go out and tell your friends. Now, back to the music"!

By the end of the evening, moved by the talent displayed, Paisley and Ness hoisted Murray on their shoulders between them and waved to the crowd for silence. It was common knowledge these two were not given to talking. They let their guitars do theirs. For them to even take the precious time to do this was completely unprecedented. Ness raised his fist to quell the crowd's noise and make their announcement.

"Now ye know us. We be simple otters who love music above it all. Aye, tis true, we were hard on our new friends. We were because you can talk the talk only if you can walk the walk. And my friends, tis a bit of a shock, pleasant though it be, to find a woodrat of any value. Yet this fellow here has humbled our pride and we honor him. Here's to our new friend, Murray", shouted both above the crowd.

The audience surrounded he and Graham, welcoming them to the Club Club and Waadah Island. Later Heather led Murray to a nice room behind the bar to a narrow passageway. He collapsed on a bed of dried grasses and leaves and fell fast asleep. This day was one he would have to share with his best friend. Mr. Mizithra would not believe it. As he slipped to a deep, well-earned repose, a warm smile crossed his lips. He considered himself blessed.

Chapter 5 The Grand Meadow Festival

She returned to the bungalow. The next gig was Muskrat Cavern, then the Crow's Nest, both Miss Bluesberry knocked off in true diva fashion. After the work she had done at the Blues Bungalow and the first show up at Hart's Den, her reputation quickly built and spread. Word traveled them thar hills and valleys of a lady skunk with pipes of gold.

She was now called, the "Darling Doe" by crows who spread news of her talent. They were proud of the fact their discoveries were becoming successful. This only built up their reputation with the higher-ups: Murderers' Row.

The Muskrat Cavern immediately booked her for the next year. The entire place erupted in excitement. It was they who established her fan club. Now the Crow's Nest was another fish to fry all together. You see, one thing about crows, the only music they really like is reggae. Oh, a bit of ska or dancehall here and there, but for the most part the rasta beat is what moves those feathered beasts.

There is something in Rastafarian rhythm that hits musical grooves. Miss Bluesberry took the stage, beginning with a funky jazz bit. They sat up and took notice. Her list of songs traversed blues, jazz, rock, and reggae. She ended the night with a very popular crow song, "Babylon Got Me By Da Dreads, Mon". If it was reggae they wanted, she would deliver. In the end, she was showered with caws for fifteen straight minutes.

She returned to the bungalow after the gig. She took a day off to wander the highlands, unwind, be alone. This is something wise to do. It allows the mind to decompress. After the recent events in her life, it was a necessity. If you endure life altering episodes, it is wise to take time alone to analyze, process, and profit.

It was on this fateful day her life changed forever. She met a male skunk who gave life it's fullest meaning; Mr. Loganberry of the Hills. He was a most handsome skunk, just a shave larger than she.

As the "Darling Doe" was daydreaming under the shade of a spruce tree, she sniffed his scent. Her nose definitely caught the pungent odor of a male. She perked up instantly to see what was what. One thing you simply must know is skunks are possessed with horrible eyesight. The nose is key.

When she was on the precipice and sang to the valley, she could only hear the response of the forest denizens. She could not see that far. She cocked her ears just so to pick up any noise. Sure enough the sound of dried leaves rustling in early autumn carried the unmistakable telltale effect of a skunk.

Once in her purveyor, she hailed to him. He ran as fast as he could. Introductions were made. He was certainly handsome and strong which are things prized by lady skunks desiring a mate. He lived far off, dozens of kilometers away, and had come seeking a mate. Skunks, all animals really, do not have much in the way of nuance when it comes to mating. Courtship is rather brief.

Birds often go through elaborate mating rituals, but then they are feathered, right? There is, after all, that term, "bird brain" for a reason. Man goes through a lot of preamble before all is settled and done. Animals have rituals, but nowhere as complex as the human species. What men go through for a mate and what females go through for the same thing can border on the absurd. He did, however, offer her some flowers and a dashing smile to seal the deal. She ate the flowers.

Love blossomed the way it can only for polecats. She led him to the Beaver Blues Bungalow to introduce him to everyone. It was clear as a bell she was in love and he utterly enamored of her. He was unaware of her impressive talent, only her beauty. She explained her career and what she was obliged to do. It was time to get ready for the journey up to the Pond Gathering at Grand Meadow.

Her entourage got her ribbons, pine cone brushes, and songlist in order. The crows hired by Tutti Frutti and Mon were diligent. The three of them sent word ahead by dragonflies. They were ready to take off once hearing the go-ahead. The others were: Agwe, which means Spirit of the Sea, and Roje, which means Protector. This duo were the offspring of Winston and Beverly from Willapa Bay. They moved north, working with Tutti Frutti and Mon training to become talent scouts.

Agwe always led the way and Roje soared above for security. With Mr. Loganberry of the Hills along, it was easy. Any coyote or bobcat would think hard about attempting to attack two adult polecats. It was a three day journey up the east way by ascending the rapids. They decided to take the swifter southerly course.

The only concern there was a bear. The carnivore would not want to get sprayed, so it was relatively safe. Besides, this time of year all bears did was get fat and go to bed for winter. The route was away from humans. They almost always run skunks over with wheel bearing, metal machines.

Saying good-bye to Beulah and Buford and the band, they headed to Grand Meadow. Miss Bluesberry was as giddy as a baby with a piece of candy in one hand and a rattle in the other. The first night, they found a nice spot in some rocks which provided a good concealment. It began to rain in earnest come sunset. It poured all night. The skunks found love in the cave far from the rest of the world. It was a wonderful thing for this lady. She was more happy now than ever.

The crows sought cover amongst the pines and were quickly asleep. The next day brought them to the highlands and beauty of the Olympics. Late summer flowers blossomed in the strong, alpine sunshine. The crows had double duty to watch out because the skunks only had eyes for each other. Nonetheless, the trail was fairly easy and they kept a good pace.

The second evening's abode was a hole in the wall of rocks. It was crude, not ideal, but they were so in love it hardly mattered. Once again it poured all evening and through the night.

The third day was sunny. They arrived at Grand Meadow. It was abuzz with animals from tiny Olympic chipmunks to massive wapiti gathering around an enormous open-air stage. It was strewn with instruments and creatures who appeared to know what they were doing. This was the last show of the year. Many were now in hibernation.

A Kingfisher was squawking on stage. He seemed to be in charge, or at least thought he was. Agwe and Roje flapped ahead. Miss Bluesberry and her buck stopped at the pond for a drink. As was common in the world of skunks, animals cleared out. They feared being sprayed.

BOOK 2: HIBERNATION

The Kingfisher took one look at her and flapped swiftly. He made a beeline straight for Miss Bluesberry and Mr. Loganberry of the Hills. He had a grin on his beak from ear hole to ear hole, landing on a diminutive branch next to her.

"Oh Miss Bluesberry, I'm Boston Charlie, a good friend to both Mr. Mizithra and Murray. He is going to be here, too. We met the night of Murderers' Row, if you recall. Sadly, Mr. Mizithra is already in dormancy and will miss the entertainment. Did you hear he married? Yes, yes, he wed Miss Saffron. And you found a man. Mr. Loganberry of the Hills, I see, well congratulations and welcome."

"Miss Bluesberry must go now to ready for the show tonight. You, my dear lady, are the featured attraction for the evening's entertainment. And you, Mr. Loganberry of the Hills, come with me and we will find you a home for the next few days. You can get settled, grab a bite, and relax until the drama begins."

"First class, nothing but first class for you, dear lady. If you have any needs, please do not hesitate to ask. My wife, Matilda, wants to meet you very badly. I will arrange it or I am in deep trouble. Ha-ha-ha".

He led her to the stage. Several others directed her backstage to her dressing room where ribbons, brushes, and music were laid out perfectly. Meanwhile, Boston Charlie went to an excellent spot under a stand of firs surrounded by sisal bushes for Mr. Loganberry of the Hills. It was a well established hollow with a nice room for sleep and an area to relax. It was the star's suite, adorned in skunk niceties and a huge bouquet of flowers.

He went right to sleep knowing the show would be later. He wanted to hear her sing. For her part, Miss Bluesberry was jittery and nervous, understandable considering this was her first big event. It did her heart good seeing deer from Hart Den, including Jon and My Deer. Muskrats from the Cavern stood, holding "Darling Doe" fan banners.

A smattering of crows, and some of the regulars from the Bungalow ventured to hear "their" lady. It was a massive assemblage. She had a place backstage to sleep. No sooner had she lay down than she was out for the count, too exhausted from the trip to worry any longer.

"Ma'am, ma'am, ma'am", urged a gentle voice, "it's time to get up and get ready for the show".

It was Horace, of Horace and Boris, two Douglas squirrels who ran things offstage. They got acts ready, on and off stage. They ran security, enforced by Canadian geese. Let me tell you about the Canucks: do not mess with them, ever. They are the bikers of the animal world. If you expect trouble, snag a gaggle of these geese and your worries cease. Ten dozen stood in front of the stage to keep any getting close to the stars.

They were very musically tuned and were there for virtually every event. Of course, Canadians were heavy metal fans, which should come as no surprise. It was a common practice for crows to hire security. Boston Charlie left that up to them. Horace and Boris were go-betweens. All he had to do was get the whole thing going.

Getting everyone to assigned seating was a chore best handled by scolding Stellar jays. They were adroit at squawking everybody to their own sections. They kept wapiti and deer in back as they were the largest. A few mountain goats managed to sneak in around back, too. Then came beavers, otters, muskrats, filling the middle in vast quantities.

Next, came rodents, mice, etc. Finally, on the shore were birds. Out on the water sat hundreds of ducks, teals, mergansers, seagulls, the whole bird world. The rest of the flappers adorned the branches of surrounding trees. When it came to an end of summer festival in the Olympics, the Gathering at Grand Meadow was top of the top in mountain country.

Blue jays escorted, squawked, and scolded various groups into place. Refreshments were carried by an otter service, Robert the Butler, whose members were busy carrying goodies for those in attendance. The noise of anticipation buzzed the air.

Sadly, some species were in hibernation, unable to witness the spectacle. Mr. Mizithra and his colony were over a thousand meters higher in elevation. Snow was blanketed deep in the Valley of the Stones. In fact, due to hibernation, this was Miss Bluesberry's final performance for the season.

This is one reason this festival up in these mountains was so special: soon many performers retired for the year. She would go back to the Bungalow, do a couple of shows, then resign to her cave not far away. She and Mr. Loganberry of the Hills were busy making plans about days alone in her cave, giggling smiles of love. How mushy sweet it was. Yuck.

"Stags and does, bucks and fawns, birds of a feather who have flocked together, critters, welcome to the Pond Gathering at Grand Meadow for the last festival of the year in these mountains. On behalf of all those who make this event possible, we thank you."

"Tonight, you get a treat. First, it will be the comedy of Slick Sammy Seagull. Then, the Beaver Boys bluegrass mountain sound. The next act will be the Santorini tumblers, a family of Eastern gray squirrel acrobats. Finally, you are blessed with a true star: Miss Bluesberry, the "Darling Doe" of the Beaver Blues Bungalow. She is going to sing us a song or two. Let's give a huge Grand Meadow welcome: Slick Sammy Seagull", shouted Boston Charlie to the immense crowd.

A wave of shouts, squawks, hoots, hollers, bellows, and other animal noises arose as Sammy took the stage. He was one of those gulls who got around. He was well known on every part of the Olympic Peninsula, north, south, east or west. He was a famed comic, hilarious, well at least to animals. Seagulls sometimes had talent; he was one of the very best.

"Hey you bunch of animals. I just flew in and let me tell you, I nailed five, count 'em, five humans on the way. Right on the freakin' noggin. Each and every one. A big, white, mushy shite. I swear, I swear".

The crowd exploded in laughter. Pooping on man is the greatest gag animals know. Anything to mess with humans since Thumbs had ruined their lives and habitat for too long. Payback. He held them for a half an hour with jokes about people, bears, coyotes, and cougars. It was a way to attack those none of them could defeat. Humor was their only weapon.

"You know my fabulously furry friends, as a proud seagull I love to fly, aim, and drop", the crowd laughing already, "but you have to know, there are times I sure wish when I was flying, I was a cow", the erupting applause now deafening as he bowed and headed off stage.

Slick Sammy left them in stitches. Waving good-bye, Boston Charlie reemerged and introduced the Beaver Boys, an all beaver bluegrass band. They were a fiddle, banjo, mandolin, and stand up bass. No vocals, they wailed, putting the crowd in a festive mood with sweet mountain sounds. It echoed over the crowd and many began to dance. It is quite a sight to see so many animals doing jigs, especially wapiti.

For their part, Horace and Boris were with Miss Bluesberry, getting her ready. Once the bluegrass boys finished, the tumblers would not take much time. Her band was a collage of players, but were pros, so she was not to worry. They knew her songs. She would have a real chance to belt out her best for a packed house. Mr. Loganberry of the Hills would be up front. She could concentrate on him should she become angst ridden or rattled by the size of the assemblage.

Being nervous in front of a crowd is a common fear. It struck her. She was frightened by such a large mass and new beau seeing her perform for the first time. Within her mind only the worst scenarios played out. She was a victim of self-doubt.

Oh, what a horrible thing. The fear can so envelop you rendering you an emotional basket case. She began to pace, black padded palms sweating, mouth dry. Fear came even stronger. She thought of running, spraying scent, vanishing into the night.

"Now, dear lady", doctored Horace, "Take a few deep breaths and let them out slowly. Yes, yes, there now, calm yourself. When you get to the end of the next breath, close your eyes. Push the last oxygen out of your lungs until they are completely empty. Do it slowly, there, there, now in one more cleansing breath. There, now, relax. See, you feel better don't you, dearest"?

Sure enough, Horace's advise was perfect, easing her glossophobia. She truly felt a bit calmer. He escorted her to the back of the stage. She could see various animals in the water, the ground, and in the trees. It was a full representation of nature's bounty in the mountains. There were literally hundreds of critters awaiting her in expectation of talent they were assured would come.

Of course, there's the rub. She was avoided due to her stink and notoriety as a bad character. Here was a chance to prove all wrong and show her unique talent. Armed with truth, she bolstered her courage and sallied forth towards the stage. She met the band, a gathering of otters, beaver, muskrat, and felt at home. The lead guitarist was a female beaver.

They were very supportive and assured her they were ready to play her songs. To them, she was a highly regarded talent. They were honored to play with her. It was clear with her nearly frozen look she was nervous. They knew what to do to set her right.

BOOK 2: HIBERNATION

"Let's open with "Party Up Them Hills, Boys", shouted out Ishmael, a beaver bass player. It was a strong bluesy, country song which always got the crowd going.

"Yeah, that will wake 'em up. let's rock", Miss Bluesberry heard herself saying.

"And now, now, oh now my forest friends, you are going to hear a voice so pure, so enrapturing, so sweet, you will be telling the tale for the rest of your days. She is the queen of the pond, the songstress of the legendary Beaver Blues Bungalow. She is our special guest star for this evening's entertainment. She is the "Darling Doe". So without further adieu, I give you, Miss Bluesberry!"

Boston Charlie's introduction was magnificent and she, still very shaky, moved to the spotlight of the moon glow to face the crowd. The music began. The festival gang, knew the song, applauded, anticipating a great show. True to their word, the band knew her song and got in the flow of it. She suddenly was overwhelmed by gripping anxiety.

Miss Bluesberry froze. She could not get a note out of her mouth. The band sensed it, jamming an elaborate intro in hopes she would work it out. This gave them the bonus of a chance to gel. The audience was oblivious, merely enjoying the sound. Boston Charlie saw it too, but he was ready. Oh yes, friends, was he ready.

Out of the darkness behind her came a sound. An auditory sensation caused her to immediately turn round shocked in surprise. It was a most welcome interruption. Boston Charlie pulled a fast one. It was Murray and his precious saxophone rising up to the sky. He sidled along side and rubbed her shoulder to his. He turned and strode to the front of the crowd doing his now famous stomp across stage.

He unleashed a most vicious solo, completely different than what the song called for. The band watched in awe his power on the brass. He got more out of a reed than anyone ever. He was in his element. Murray found he loved the spotlight. He proved this little insignificant woodrat was someone to respect.

Unless you try discovering hidden talents, you will never know the joy derived from painting, learning a song, writing a poem, carving wood, whatever. It is the shear pleasure of something you created. That is just plain cool stuff. All of us have some good thing we can do. You may not be the best, but you can enjoy your talent. If you explore the saxophone and find you can't play it worth a darn, try something else.

You may find your talent lies in the harp, tuba, sculpture, origami. Regardless, you have something to do to benefit yourself and society. Try, please, to discover yours. The world reaps the reward of what you sow. Art has real value.

Murray held the crowd spellbound, hanging on each precious note, then faded to the back of the band, almost disappearing. She was forced to stand alone. The musicians recovered to the original song, found the groove and led into the first verse of "Party Up Them Hills, Boys".

"We got us a jug of sweet wine, dandelions plucked right on time. Let's go get Jose` and Marty, go up in them hills and party. Yeah let's go party up them hills, boys.", she unleashed the lyrics with vibrato and fire. Her voice carried above the band, echoed out across the pond, to the shoreline, to the very back of the crowd. The wapiti, deer, and mountain goats began to stomp in approval.

At the break, the lead guitar player, Deirdre, burst forth with the well known riff, carrying sound sweetly across the pond. Then the piano player, a muskrat, Ronaldo, took over the show. He unleashed both hands to play a blues background piece which shook the show. Finally, Murray emerged, ripping a torrent of honey laced notes so bluesy Miss Bluesberry came forth to a huge applause and hugged him as he played.

"Come on up on my mountain, come on up to the hills. We are gonna party, we are cure all your ills. Yeah, let's go party up them hills, boys.", she finished the last verse with a sex appeal that caused many in the audience to blush.

The eventide's amusement tarried to wee hours. The crowd thinned like a middle aged man's hair as the last tune played. It was her signature closing song, "I'm All Black And White", a number written for her by Twigs about life as a skunk. He adored her and loved playing drums behind her. The song was a tribute to his heroine.

The audience were awed and amazed, honoring her with three encores. The next two days were glorious. She received great fanfare from all who heard her sing. Boston Charlie introduced his wife, Matilda. She gushed over Miss

Bluesberry, lauding her with profound praise. It was an uplifting episode. She hugged Murray forever.

Murray was equally extolled. At the end, they made their way together with Mr. Loganberry of the Hills down to the bungalow. It took just two days, but the weather was turning to rain mixed with snow. Once they were down to the pond, the snow ceased, but rain was perpetual.

Murray still had to go to Willapa Bay, but Miss Bluesberry was done for the season. She would do a few nights at the bungalow, then retire with her darling, sweet beau, Mr. Loganberry of the Hills, for a well earned hibernation. They worked on improving their cave for dormancy and comfort. They dug a deeper chamber in the recesses of the rock.

Sharp claws and loving team work soon provided a lovely lover's nest. It would be a warm and happy winter for them. It was a dandy year for both of these close friends. Murray made ready for the flight south, reassuring his mother and grandmother he was fine and all was well. Next, he knew was the big one: the carnival at Willapa Bay. He was secretly excited beyond belief.

Chapter 6 Terry, Chris, and Lee

"Rainbow One to base, Rainbow One to base, over", demolished the silence with a frantic voice over the radio back at the forest office's base.

"Rainbow One, this is base, over", came the immediate retort.

"Chris, this is Terry. I am up at the Olympic marmot colony on hill 423. Chris, this is weird, but they are gone. The entire colony has been abandoned. Over".

"Terry. What? How can that be? There were thirty-six marmots in the colony. How can they all be gone? Are there any dead bodies laying around? Any coyote tracks? Bear? This makes no sense. An entire colony does not just up and disappear. Let me get Lee in on this, hold on. Over".

"No sign of any animals invading. No human tracks, bear, coyote, mountain lion, nada. The colony is just plain empty. Roger, will stand by, over", said a bewildered forest employee.

Terry scanned the entire area. It rained for days off and on causing major changes. Any trail left behind by Mrs. Lip Gloss and the rest of the colony washed out. They traveled single file for the most part save the young ones.

By the time Terry arrived on the scene for their regular inspection of colonies in the Olympics, Mr. Mizithra was busy leading the mass of marmots to the Promised Land.

"Rainbow One, Rainbow One, this is Chris. I have Lee here and we discussed the problem. Take the truck and head east along the creek where that old boat is off the left side of the road. See what is going on in that direction. Lee will go down the south side and scout that area out, too. Do you copy? Over".

"Roger, wilco. This is really weird to have the whole thing deserted, though. Rainbow One, out", stated a disbelieving Terry.

Lee loaded the truck and left the station, heading south. There, at the other colony on hill 402, the entire colony was wiped out, every marmot gone. The whole area looked as if it were demolished by wild animals. Evidence of raccoons and small rodents riddled the area which seemed odd. Nonetheless, the entire colony of 27 Olympic marmots was gone. It was the colony closest to the beaver pond down the way.

The dirt road Terry traversed was the same Murray hid along side when heading home after escorting Mr. Mizithra and his critters to the Valley of the Stones. It was this truck which sped past him in Book One, covering him with dust. Terry raced around the corner and up the hill heading due north. It was on this climb the truck slid to a dusty stop.

Standing in the middle of the road were two wailing coyote pups. In backtracking with them stashed in back of the rig, the soggy body of the mother coyote was found and fished out at the bottom of the rapids. Again, no sign anywhere of the colony. The older ones who hid at the old boat were mysteriously history.

"Rainbow One to base, Rainbow One to base, over".

"Base here, Terry. Lee says the same thing is at hill 402, but there the colony was demolished. You say your colony was just gone, no sign of any trouble, correct? Over".

"10-4, base. No sign of any trouble. Also, the two old marmots who lived in that deserted boat are gone. I found two starving coyote pups on the upper north road. I found their mother drowned at the bottom of the rapids down by the beaver pond. It looks like she fell in and the speed of the water was too much. She hit a rock. Her whole head was bashed in and bloody. Over".

"Roger, well, come on back to base with the pups then, Terry. This is one for the books. Over".

"Roger, Chris, I am going to get some photos for you of the colony, though, so we have a record of it. They are gone. Weird, too weird. Over".

"Roger. Out".

The mystery remained. The trail was glacial by the time the Thumbs caught on to what happened. Two entire viable colonies were abandoned. No real explanation was found. One had been decimated, but marmots always rebuilt near to their old home. Here, complete desertion. They searched and researched, uncovering no evidence. Two colonies simply vanished in thin air.

The last thing marmots wanted was for human beings to know what they were doing or where they were going. Experts were consulted, but no one could think they simply made an exodus and found a place far from pesky intrusions into their lives and culture. Their escape from man was successful and complete. Yes!

Chapter 7 Willapa Bay

Willapa Bay is the second largest estuary on the west coast of the United States of America. Numero uno is San Francisco, but the two have little in common. Millions of people live down south while this northern oasis is dominated by birds. It is a major rest spot on the migration either north or south depending on the time of year. It is massive, over 670 square kilometers (260 square miles).

Most water recedes with low tide, returning to fill the shallow, mucky spots. Small channels are deep enough to be permanent. Crabbers haul up baskets full of ocean's sweet harvest. Oyster boats reap from well placed beds. A full ten percent of the oysters Americans consume come from this bountiful waterlogged farmland. It is a natural paradise. For lovers of razor clams, this is home.

It is here annually, as birds head south for winter, crows throw their big bash: the Carnival at Willapa Bay. Hundreds of thousands of birds from over 300 species come through. It is the safest haven for them to rest and prepare for the arduous journey for warmer digs. This massive party was highly anticipated. Talent brought in was the best. The soiree` lasted three weeks to take advantage of the fun.

Winston and Beverly were pure class. The Ladies of the Lake were the upper crust of the north. Beverly's brother was the revered Lloyd Joseph. They were not like Derrick, Clive, Delroy, Everton, Clifton, Tutti Frutti and Mon. They objected to their sons, Roje and Agwe following those two, but agreed they could elevate the dignity of talent scouts.

Theirs was the compromise many parents make when their offspring decide to follow a path they feel is wrong. If a child is persistent, and we must admit parenting is a bit like herding cats, it is best to find ways to help children succeed. We cannot disown those previously encouraged to be individuals.

A good parent will be careful with judgments which only serve their needs over the passion of their children. It is therefore best to be a parent, not a friend.

Remember, you are not raising a child, you are preparing an adult for the real world. Kids can't help being kids because that is what they are. They know nothing until they are educated.

Murray had flown, but this was something much, much greater. He was going to fly a long way. Some of it would be completely over the Bay which meant 100% water. Were he to fall off the swan, he would die. It was that simple. Fear gripped him as he prepared to leave. He said good-bye to his family and prepared his mind for the journey. Visions of falling off the cliff on his path to White Mountain made him wince.

The birds rose from the pond. The formation swung north to clear the trees, then west, lastly due south. He buried himself in the soft, warm down of the pilot and closed his eyes. It was his humble way of giving himself a meager sense of security.

Swans flapped at tremendous speed along a pathway in the sky past Forks towards the Bay. Since these were the final days of summer and first days of autumn, it was fairly warm. The only concern was the sudden wind gusts and rain disruptions racing off the ocean. The coast is unpredictable. One moment it is clear and calm. In a flash, a whirlwind appears, sending airborne water and torrents of hard, pelting rain.

Just as suddenly, raging weather calms, oceans quiet, rains cease, temperatures rise. Willapa Bay is a force to be reckoned with. No one should take it for granted. Humans die out on those waters or are sucked out to sea by sneaker waves on the coast. People foolishly turn their backs on the power of the Pacific to their own peril.

The trip took hours.

They landed outside Westport, which lies at the far southern end of Grays Harbor. This area filled with humans, avoided by most traveling south. Murray was allowed to stretch his legs and catch a bit of salt air. It was blustery, but nothing out of the ordinary. Life was beautiful and serene. Swans had to refuel and rest. He took a nap under sisal bushes and dreamed of the upcoming shows; he fell asleep with a warm grin.

He was to be there for three weeks. By then, the weather would change in earnest for upcoming winter. The swans would need to go south. Murray would bury himself in the warmth of the Beaver Blues Bungalow and his cozy digs across the hall from his mother and grandmother.

He needed to be home in time for Yom Kipper. His sole stipulation in his contract with the crows was the Sabbath Day to rest. He would be moved to six locations. The first was past the human village of Tokeland. Here was the traditional tribal digs of the Shoalwater peoples. They were an ancient fishing tribe who populated the north portion of Willapa Bay. The party area for birds for the first shows was at the first big waterway, North River.

This is a short, fat, lilliputian river, a mere 50 kilometers in length. Like the bulk of waterways it's genesis is the low Willapa Hills, a modest coastal range whose highest peak is under 1000 meters in elevation. They are soggy mounds, bombarded by rain more often than not. Winds can be harsh, hail murderous, but animals, birds, and water life thrive here and love it for it's absence of man. Most of the flat land was reserved for cows, so animals were left virtually alone.

The native population was never large. When white people moved in, it still remained fairly barren due to weather and harshness of the climate. In contrast, the larger estuary south in San Francisco is packed with millions of Thumbs.

There was a lovely spot beyond the delta. North River empties it's bounty into the bay. Rocks rose a mere five or six meters above the lapping waters, with a large exposed mass. It provided a superior area for performing. Theodor and the formation set him down on the water, then paddled up to the rocks where he hopped ashore.

The band was made up of otters and muskrat. Local beavers did not swim out that far and were not so inclined to try. The group was a more jazzy sounding mob. Murray fit right in with them. One thing about Murray, he was an adaptable little fella.

They had a quiet little alcove out of the wind, where seagulls, terns, grebes, mergansers, wigeons, scaups, pintails, and others gather to listen to the sounds. Most would stay about an hour, then head off for the trip south. The seagulls and some others were around all year. They enjoyed the carnival time, but were more nonchalant than those passing through. The shows were only over in the fall. Flocks heading north, did not stop long enough to make it worthwhile.

For some reason, the hooded mergansers adored Murray as did the California gulls. The cinnamon teals were in love. Mallards were not into jazz. He was not applauded by them. Of course, ducks just quack. If they really loved an act, the noise, called a "Quack Attack", would deafen. Gulls and terns made their sounds. He was a hit, enjoying the day.

It rained at night. There were a few squalls here and there, but they were a hardy bunch and did not worry much. Murray was again a sensation. The swans showed up in time to take him south to his next spot, the backwaters of Willapa River. They led him east by southeast to a lovely area removed from man. Few indigenous people had ever ventured to it. It was very remote.

The area had an abundance of wapiti, called Roosevelt elk by man. There were mule deer, squirrels, muskrat, beavers, weasels, otters, and rodents aplenty. Word reached ahead. A fan club of woodrats rabidly anticipated him. The stage for he and the band sat overlooking a cliff, very near the river.

The view was stunning and was ideal for accentuating the presence of a rodent with a sax. He was on it for the show. The band was tight, right, and out of sight, the Cliffside Blues Boys. Legend had it they were the fifth set of the same animals by species who from generation to generation had filled this region with the best coastal blues known. Each could trace back five full generations to the same instrument.

For example the piano player, a weasel, Plinky had total recall of his grandfather recalling his own grandfather and how his ancient relative began the band. Once the piano was joined with a good guitar player, a beaver, Lightning Jack, the rest was history. It fell in place. The beaver knew a muskrat, who knew an otter, who knew a weasel.

This tight knit bunch of five played forever. They taught their children to play, who in turn continued the band. They did as their fathers and mothers, passing desire to the offspring. Male or female it did not matter, only talent. It showed.

When Murray was introduced, it was in the old club. It was so far off no human ever neared it. It was safe from weather and coyotes, et al. He was shown a nice room to unwind and rest whenever he needed. His two saxophones were already on stage on stands made from twigs and branches, awaiting his lips.

He hit the bed and later awoke to the sweetest sound. Piano notes, bluesy and cloying, filled his fuzzy ears. He rubbed his eyes to fully waken. Then the guitar took over, a lead so solid he quickly made his way to the stage. Once there, he saw the whole group jamming.

The guitarist, five times removed from Lightning Jack, was DeMond, a blisteringly vicious player. At piano Plinky was smiling and staring at DeMond. On bass the eldest member, an otter, Roddy MacNamara. The drummer, fifth generation muskrat on the skins was Zigzag, a blistering tapper with perfect rhythm.

The singer, a weasel, Sweetie Pie, was a powerful lady. She loved to have a strong sound to belt out songs like: "Breakin' Blue Ocean Blues", "Willapa Bay Gonna Be My End", and "The Thumbs Gotta Go". While Murray did not know these tunes, he did by the time he left.

Talk about fitting in, by the last of their practice sessions, the whole band decided to make him an honorary member. They announced it at the end of his gig. After the final show, other musically inclined critters gathered around with everything from jugs to mandolins to autoharps. The jam session lasted to the wee hours.

Murray's sounds were so sweet, they called him, Sugar. He and DeMond were in sync. When he left it was as if they were brothers from another mother. When he had to head south, it was with a tear in his eye. Other than the family back at the Bungalow, this was the closest he ever felt to another group of musicians.

They did a gig down at the Bone River. Here was a group of wild musicians who were into Salsa music. He did not know anything about this type of music. Since the sound lent itself to brass, he found a way to get in on the songs and had a blast. Jacinto, the head of the group, was a muskrat. There were four muskrats and three beavers in this band, the singer was a female Douglas squirrel named, Consuelo whose voice was a journey in sensuality and heat. She flirted with Murray. She exuded femininity in a way causing him to be quite disturbed. For a confirmed bachelor, she was a bit unnerving.

The swans flew him to a spot on the west side, Leadbetter State Park. Imagine looking at the topside of your left hand. This peninsula is shaped like your left pinkie finger. Now, look at your fingernail. This is where the park is. The rest of your hand would constitute the Bay. The left side past your finger would be the Pacific Ocean.

There was a sizable gap in between the park and Tokeland to the northeast. This is marked by breakers, where Bay and Pacific meet. He was taken here for his last stop before the big show on Long Island as the Thumbs call it. To animals it is Carnival Island. It is virtually unfeasible for humans to get there. The autumn weather tends to keep them away. It is a perfect place to have a massive get together.

At Leadbetter, he had to worry about bears because they were plentiful on the peninsula, preparing for winter, eating everything in sight. They had to beware. One thing here is while the Canadian geese were no match for a bear, if one were to appear, they would peck at it in unrelenting waves until it left.

This was tricky. Thankfully, none of the black brutes showed. The band was a good rock group. He enjoyed the time with them. After this much travel he looked forward to his final stop, the huge shows on the big island. It was the ultimate for any animal performer. The legend of the place rocked south, to Young's Bay off the Columbia River, the coast to places like Ecola State

Park down in Oregon, and east to Cathlamet up the mighty Columbia River. Murray heard of it up in lofty Olympic country.

Murray was excited.

Chapter 8 Carnival Island

He rested each Sabbath. He spent the day on Carnival Island in his own room. They had a lovely spot for him and the performers in the hollow of a stump. A hole led down in the ground to a long row of rooms. He found a star on a door with his name on it. He entered, finding his two saxophones on twig and branch stands. A bed of pine boughs and oak leaves, and stand with a wooden bowl of water were awesome. Some fresh water bulbs filled a basket. A chair, too, to sit next to the stand.

He found the bed comfortable and food lavish. It was an excellent place to be alone and reflect. He prepared for a performance tomorrow in front of this massive audience. There could be up to 50,000 birds and animals from Anna and Rufous hummingbirds to the massive wapiti in attendance for this supreme event.

The place was abuzz with sounds unlike any he ever heard. Music flowed from every corner. He decided after his reflection time to venture up to see what he could see. He donned his kippah and climbed up out of the stump. Luckily, most of what was available was less than a Sabbath Day's journey away. He could embrace it to his education and enjoyment.

Since crows were throwing this shindig, there were hundreds of them. You saw a murder here and a murder there. Sometimes they could not quite get to a dozen. The smaller groups were laughingly referred to as, "muggings". Even crows got a kick. These ebony hosts were beaming with delight. The festival was going along swimmingly.

A few paces away from his stump for the stars were groups of four: squirrels, chipmunks, otters, beaver, etc. They sang and harmonized in a different way than he had heard before. It was very appealing. There was absolutely no musical accompaniment, simply voices.

He moseyed on over to lend an ear and heard the sound of four voices, each a small bit above the other in a very melodious and satisfying way. In addition, the quartets would each start by standing together on stage. As they got going, the entire foursome would move slowly around the stage singing in harmony all the way.

"Pardon me, friend. Vat is zis?"

"Oh hey there stranger. This is what is called a "Barbershop Quartet" and it is four muskrats, see there, yes. One is bass, then lead, baritone, and tenor. Various voices range up four octaves singing as one. It comes from the old country in Europe. While men waited to get a hair cut and shave, they would sing.", rambled an informative crow.

"It developed over time to what we have today. Mice from England brought it with them when British ships came. A few mice deserted the stinking ships and stayed here, leaving us this style of singing. It is a very good way to share songs", stated a friendly woodrat standing next to him.

"Thanks. So are zose judges over zere at zat table?"

"Yes, every festival the competition is fierce. The last three years that group of Douglas squirrels over there by the spruce tree has owned the show, but this year there is word of a quartet of muskrats who have a strong reputation. See them over there, yes, those four. They are the ones who are up after this group and the otters just over there. Stay and enjoy, you will love this stuff."

"Danke, friend, I vill", passed an interested Murray.

He made his way closer, pushing various critters out of the way until he slithered up front to fully enjoy the experience. Murray was moved, simply moved by the effect of four voices united. It was so pure, so sweet, and without any musical accompaniment.

A capella was his new buzz word. He had to take this sound back to the Bungalow. Three voices in harmony, the lead carrying along with a slight variation on the tune just got to his soul.

Group after group came, sang, and went. Soon half a dozen quartets performed. The final groups emerged. An earlier quartet of chipmunks held sway so far. First, was this select group of muskrats, the Furry Four. Their song was a truly old one, "In The Good Old Summertime". Needless to say, Murray was struck at the perfection of their harmonies and how precise they were with each note.

When they finished, the crowd howled in delight and approval. They bowed then moved off stage. The emcee was an animal Murray had only heard of, but whose species did not live in the mountain country. He was intrigued. The animal was a rabbit.

Now some of these are native to Willapa Bay. Some are castoffs from humans who buy little fuzz balls for the observance of Easter, then throw them away out in the cold, vicious, unforgiving world. Humans buy these cute little bunnies for kids to celebrate the holiday. Children quickly tire of pets. Parents often dump the coney in the woods. How cruel can this be? You buy a living being your child does not want. They do not know how to care for it, so you cast it away to survive in the wild.

By so doing, people show complete disregard for animals around them. How dare they misuse an innocent living creature in such a way, simply discarding it once the shine has worn. Sad, sad, sad. So, the little guys and gals are tossed in the wilderness to survive. The Thumbs figure if it is an animal, it is able to survive outdoors. Most are killed and eaten within the first 24 hours after they are cast away as living refuse.

He saw this rabbit, a throwaway named Juan Rabbit, on the stage spouting away and became interested. He was three times larger than Murray. He was black save a spot of white on his nose and chest. The nose was fuzzy and, well, cute. He had large, soft brown eyes and long, slender ears laying against his head and a fuzzy nub of a tail. The hindquarters were immense, showing tremendous leaping ability.

He was having the time of his life and encouraged the crowds to shout for the quartets and their choice of the best. The other rabbits loved the guy. He was a live wire. The shouts for the muskrats, the Furry Four, were the loudest. Next up was the famed Douglas squirrel group, the Bare Branch Brethren. They won many championships. It was almost impossible to beat them. The muskrats had a slim chance.

"Well, here I am, your funny bunny, Juan Rabbit. It gives me the greatest of pleasure to introduce to you the fearsome foursome. The best in the west. The barbershop quartet they set the bar by: the fabulous Bare Branch Brethren."

With ultra consummate timing, they moved forward and in unison hit note after note in perfect harmony. The song, "Singing For My Life", was a most moving piece. The words were simple and sweet, affecting the audience. No one could doubt their talent. It did seem they were the best, even better than the up and coming muskrats.

However, as they hit the height of the number, Oliver Wendell Douglas squirrel, a tenor, made one critical error: he coughed. He expectorated. The audience gasped. The mistake was unavoidable to miss. He had a bug fly in his gob as he inhaled to hit the note. It caused him to gag and cough. This was the opening no one expected and won victory for the new champs: the Furry Four.

"Well, folks, we can see the sadness from these noble squirrels. An errant bug flew right down Oliver's windpipe, but those are the breaks in the rough and tumble world of barbershop quartets. It is with great delight we now crown a new champion. Without further adieu, the Furry Four", shouted Juan Rabbit, his white nose twitching across his smiling black face exposing big bunny chompers, his ears perked hard upright.

Murray found himself swept up in excitement. The crowd clapped and clapped for them. The defeated group came to shake paws in praise for a job well done and earned. They were noble in defeat as we should strive to be under similar circumstances, yes?

A frothy swell of fur emerged as a number of other muskrats hoisted the four champs up on their shoulders, toting them around the crowd in revelry for the victory for their species. The squirrel's reign was a long one. Now, muskrats ruled the marvelous day.

Tiring of this, Murray spied a large area with many spots upon which to alight. The trees were filled with perching birds of a feather from the North American coast up to Alaska. This place of gathering was perfect for them to rest for the trip south. This festival given by the crows on Carnival Island was something every traveling birds and local animal anticipated with baited breath.

The areas on the ground, fallen trees and the like, were filled with groups of varieties of rodents, rabbits, beaver, otters, etc. It was an amazing gathering. The only rule was: no one eats anyone else. This was a hunting free zone enforced by a massive flock of 500 Canadian geese. This sturdy 500 were charged with protection and were always successful. In fact, they were patrolling the skies in large formations to scare off any raptor who might try to find a quick meal. Raptors do not care about or honor bird rules.

In front of the crowd were an immense amount of musicians and instruments, some devices Murray had never seen. They were tuning weapons of pleasurable sounds, chattering amongst themselves in the various sections. Woodwinds sat here, brass there, percussion in back, and strings, my oh my so many strings. They ate up the entire front row on both sides. In front of the semi-circle they formed stood a stump with a large beaver standing on it, baton in hand.

Behind the conductor, up a meter or so, was an immense stage which opened at the back to one of the broad inlets on the island. This made a perfect place for swans to land. You see on Carnival Island for the end of the year festival swans were going to perform Swan Lake. Now look folks, people have performed this ballet since Tchaikovsky wrote it over a century ago. It is changed a bit here and there to fit the times.

Nothing, I mean nothing can prepare you for the sight of real swans performing Swan Lake. There was only one change they were going to have this year according to the emcee up front, a proud, bold white egret named, Ivan Trobosky.

"Forest and feathered friends, one and all, please hear my announcement. Tonight, due to the illness of the tundra swan, Torhild (the fight of Thor) is being replaced at the last moment due to a tear on her left foot. As all know, she plays Odile, Baron Von Rothbart's daughter. By great fortune we have with us a newcomer to Carnival Island."

"She is a Mute Swan, who got caught up in a jet stream and ended up here from Russia. Miss Svetlana is a blessing. She performed the ballet back home and brings her expertise to the role for us to appreciate and enjoy".

At this point a swan came forth with a few differences. Sure she was white, most are. Her head's shape was a bit larger, neck a slightly different curve. She had yellow on her beak. The two American versions' beaks were entirely black. She was lovely, bowing gently before all to behold her. The curve of the nape of her neck glistened in the sunshine just so. Other male swans were ogling her as well. Ah, yes, the sex appeal of the foreigner with an accent made her the center of swan attention.

He made his way to a place called, "Rodent's Row". Mice, woodrats, chipmunks, squirrels, and the like gathered in one area. Beavers went up and down the row tossing bags of seeds and nuts for any who raised a hand. Murray was pleased to be able to rest and hear new music. Behind him were the larger animals including deer and wapiti.

The manner the stage area set provided those perched in trees to enjoy the aerial view. He sat, viewing a perfect panorama. Those on water had a spectacular space. Ducks, mergansers, grebes, gadwalls, coots, cormorants, pelicans, snow geese, et al, and seagulls of several different species sat floating on water, bobbing up and down, loving the show.

He saw instruments up front and by gabbing with a nearby field mouse got some interesting facts. One, instruments as well as other parts of the festival were in plain sight. How could they avoid the prying eyes of man? Easy, at the end of the entire three weeks, beavers take the lot under ground to store in lodges until next year. Whole groups of otters, muskrats, and beaver helped clean everything to ready it for next year.

Two, he found the variety of instruments for this show: Woodwinds: piccolo, 2 flutes, 2 oboes, 2 clarinets, and 2 bassoons. He especially liked the oboe for it's sound. Brass: 4 French horns, 2 cornets, 2 trumpets, 3 trombones, and a tuba. He fell in love with the French horn. Percussion: timpani, cymbals, snare and bass drums, triangle, tambourine, castanets, tam-tam, chimes, and the most interesting, glockenspiel played by a deer who hung it from his antlers and struck it with his front left hoof.

Then up front, Strings: 14 first violins, 16 second violins, 12 violas, 10 cellos, and 8 double basses. He found the sound of the cello his favorite. The symphony began to play under direction of the conductor, the famous beaver from across on the peninsula, Lady Regina P. Fremont. Her legend on the Bay was immense. It was considered to be a real boon for Winston and Beverly to land her this year. She missed the past two with previous prestigious engagements elsewhere.

Instruments were handled by various creatures. One such critter he only heard of in the mountains. This species did not favor heights: a porcupine. They were on the heavier string instruments, rabbits in a long line and in rows played first, second violins, and violas. Seeing forty-two Belgian leporides sitting together playing strings in unison was a sight to behold and a sound which stirred his innards.

BOOK 2: HIBERNATION

There were beavers, otters, muskrats, and the lone deer tending to various devices. The triangle was taken by a sweet little chipmunk. The tam-tam was in the paws of a short tailed weasel. His brother handled chimes. It was hilarious, watching the weasel play the chimes, he was so nervous.

Act 1: the story began. It was a love story between prince Siegfried and a lady, Odette. She became a swan because of a bad guy named, Baron Von Rothbart. In addition, were the queen, Benno, a friend of the prince, Wolfgang, his tutor, Odile, the daughter of the abominable dude, Baron Von Stein and his wife, Freiherr Von Schwarfels and a herald, a footman, and various court type characters.

As stated by Ivan, the players were swans, both trumpeter, largest of all birds on the North American continent, and the tundra or whistler swan. It was so lovely to see individual actors fly in, land on the water, then take the stage. Murray was intrigued.

Act 2 came after a brief pause, then a longer break. Act 3, another long break. Finally, Act 4 where the lovers both end up dead and on the way to heaven. The story was one of love, deception, purity. He adored it. Murray dabbed his eyes at the end with the back of his fuzzy paw.

Svetlana did an exquisite job, given the new crew to work with on this ballet. All applauded the performers with great zeal. It was an eye opening experience for Murray. His entire musical knowledge was of mountain music, the blues, bluegrass, rock, jazz, and salsa. Nothing he heard came close to this union of a hundred different instruments joined in telling a tragic story of love and purity. It left him reflective.

Sabbath day's end called him home inside the stump to finish the day in prayers of thanksgiving. Murray was very happy to have seen two such incredible, moving forms of music in one day. All he had to do was walk around a bit. His mind was swimming in sounds, sights, and smells. Carnival Island was a magical place.

Night fell hard as did his eyelids.

Murray arose early, stretched, yawned, and found his way to the food tent. It was warm and sunny in early Pacific Northwest autumn. Anything and everything he could ever want was at his beck and call. A fat juicy fresh water bulb, a couple of lovely water bugs, and some nice green clover filled him to the brim.

Next, he sought Ivan Trobosky to see about his musical performance at Carnival Island. He wanted to get ready, find his place in the line-up, and who in the wild world he was going to play with today.

"Oh, my friend, Murray the saxophonist. Yes, yes, yes you are to be appearing this afternoon on the main stage with the Willapa Bay Blues Band. They are the very best. I will take you over there myself personally and introduce you. Your saxes will be up on stage. The stagehands here are the very best. The crows would stand for nothing less than the highest quality for every little thing. Beverly is a real stickler, so we cater to her each and every needs. Lovely these crows, simply lovely".

"We have a number of acts going on starting with a weasel magician in a little over an hour. The crowds will begin to gather once everyone has eaten. Did you get to eat? Oh good, well, you follow Prickly Paula the Pugnacious Porcupine. She is a comic, the funniest lady on the Bay. Once she gets the crowd pumped up, all you have to do is blow your tiny lungs out and gel with the band. If you can achieve that little feat, you will be living in the tall corn", stated Igor with certainty.

He performed a few more chores, sending orders via barn swallows. Dragonflies were done due to weather. Once dragonflies went in hiding, barn swallows, less reliable for sure, took over as the messengers of the wilderness. Birds were larger, but had one glaring flaw. Barn swallows weakness: at the drop of a feather they would desert their delivery to follow a flying bug. They could not resist the temptation to chase and eat.

Ivan led Murray to the huge backstage area. It was packed with animals and birds who had either performed or were in preparation for a show. The place was alive with electricity. After weaving through a myriad of performers, they found a quiet spot where four animals sat with guitars and drums. This must be them, thought Murray. A sullen bunch, and typical of musicians around the globe, they looked like they had been awake for six days without sleep.

"Furry ones, this is Murray, saxophone. Murray, this lady beaver on your left is drummer: Miss Bang, Boom, Bang. Next, this fine otter: Dylan

O'Tanner, is lead guitar. He is regarded as the best of the best on this Bay or any other. Next, he bites so watch him, too late, sorry, is this short tailed weasel: Giovanni plays piano with more gusto than any around. Finally, the beaver banging bass like a meat cleaver: The Sir Franklin, Esquire."

"Murray, this band has been rated the best for a very long time. Get acquainted and do some jamming, line up the songs. You are on in about four or five hours depending on how things go with the other acts".

"Oh, wow, I totally forgot. The singer, the singer is out. Poor Pegeen. She got shot by a hunter and became dinner. Murray, she was the best singing Pintail we ever had. She got tricked by a floating duck dummy. A Thumb got her with a shotgun blast. We had to find a substitute".

The band members hung their heads in sadness and shock.

"Weird as it is, we found a bird who is not from around here. She ate some fermented berries, got blind drunk, fell asleep on a load of logs headed for Longview, and woke up at the mill. She was disoriented and flew due west until she got here. We cared for her. She is ready to earn her keep as she said".

"Her name is Corinna, a Red Northern Cardinal, you have to see her. She is a dusky soft brown with red hues and a thick, husky snub beak. Her shape is much like a Stellar jay. We have nothing around here like her. Anyway, she came from Louisiana and is stuck here for good. We found she sings blues with this incredible southern drawl. I think you will find her impressive. Well, gotta go", Ivan vanished, leaving Murray to fend for himself.

"We will do best, Murray, to go away from the rest of the mess under the stage. It is by the water's edge. Then we can jam and see what we shall see. All we know is crows came back from the Olympic mountains saying you were something extra special. So, let's go have some fun and see where you fit in our tight little group. We have a new singer to break in, so this will be fun", shot the strong voice of Miss Bang, Boom, Bang.

They moved the instruments. A few stagehands helped with piano and drums. Murray carried his own sax. He had not played for a few days. He was ready to wet his lips and whet his appetite with some music. Sure these furry four were the best, but he trusted his talent, finding himself happy not nervous.

It was a lovely, private spot under some trees, next to a small shoreline beach just a few meters wide; perfect. Once the instruments were in place, stagehands split the scene. Murray grabbed his sax and chomped down on the reed a bit to moisten it. As he did, a beautiful red and soft, dusky brown bird fluttered in, landing next to Murray. Settling down on a handy stump, she ruffled her feathers. She introduced herself to her new band-mates.

"Hi, y'all. Please, call me, Corinna. I want to get to know each and every-one of you. After I got the vapors from eatin' them there berries back home outside of Baton Rouge, Louisiana, why, I woke up on the way to little old Longview. I was flustered. I flew and flew west, followin' the river like back home on the Mississippi, only there we fly south."

"Thanks to the kindness of strangers like Ivan and the others here on Carnival Island, I have a chance to sing and shine. So, y'all, here I am", said this bird with a curtsey and sugary, sweet voice dripping with honey.

The males, drooling, fixated on her natural beauty. She was quite a bird. This was one extremely good looking example of aviary delight. Miss Bang, Boom, Bang was not so impressed and motioned her forward. She handed her song sheet music. Corinna the Cardinal selected a strong blues tune with a big back beat, "Take Me Out Of The Blues Game".

It would demand the best from The Sir Franklin, Esquire, too. The bass had to take the lead over everyone. The drums were to be handled by Miss Bang, Boom, Bang who acted a bit stunned by Corinna's bold choice.

Dylan O'Tanner went over, strapped on his guitar, and strummed it a few times. It was in tune, ready for use and abuse. Giovanni sat down on his stool to his keys and hit a few to gain some instant rhythm. Murray got his reed moist, gave it a love bite, and a blow or two. All things considered, his alto sax was ready.

The six looked at each other. At the count of four, Giovanni hit the initial keys in a beautiful entry. Not knowing the song, Murray looked at the sheet music on the piano. He found his part in time to chime in with his stellar introduction to this new super group. He was in perfect time. He played along with ease, sensing the same from the band. He melded to the background in time for Dylan's sliding guitar riff to bring the whole thing around to the first verse. The players were on the same page in an instant, bass and drum in full command. The only thing not yet tried was the new voice. It was her time to rise and shine.

"When we met my love, I thought you were so far above, all the others I met before, now all I want is to show you the door", and now to the chorus, "Take me out of the blues game, it's a cryin' shame, got no one but me to blame, so baby take me out of the blues game", belted out the syrupy Southern belle.

By the time she had ended the first verse and chorus, the whole group were in awe of her pipes. The members of the Willapa Bay Blues Band forgot the late Pegeen the Pintail in favor of this foreigner from the south. Not only did she master the song, but was in perfect timing with Miss Bang, Boom, Bang and The Sir Franklin, Esquire.

"I done told you love is not a game, but it's always gotta be the same. I hoped you would be different, better for me, but you was just a name, and worse cain't you see. Take me out of the blues game, it's a cryin' shame, I got no one but me to blame, so baby take me out of the blues game".

As she hit the final sultry, powerful note, Dylan O'Tanner's six string ax came to life. He exploded with such power and control all who heard it in the vicinity gathered around, listening to the sound of this virtuoso. He wove a serenade around Corinna with his mystical notes. Still, he was a pro and soon bowed over to Giovanni and his eighty-eights.

He let loose a torrent of notes, jamming hard and fast with both hands in a display of weasel talent unleashed on unsuspecting crowds. It ended. He nodded to the new guy, waiting for Murray to let 'er rip.

He did. Boy, did he. Murray was with a band whose reputation was worth every accolade ever given and more. So, with all the power he could muster, Murray strode forth in the little group and let go a series of notes melting any opposition in an instant. It was so melodic, so sweet, he could hardly believe his good fortune.

Murray felt as if he were above his pay grade, but soon discovered by their response he was at home. It is interesting with any given ability to find people who are of equal or superior talent. Once you get to this sweet spot, your own qualities rise to the occasion and you become a better performer. Sit down with your guitar, for example, in a group of really talented players. You will become tighter, more in tune, and confident in what you can do. They wrapped it up, nodding in agreement; the newbies were just fine.

After a good practice session and going over the line up of songs, they felt confident enough to go watch a couple of acts before they had to go onstage. They had enough time to play eight songs. They were the final act of the day. If they were really popular maybe an encore. It was uncommon, but not entirely unknown.

They meandered as a solid mass. As a musical group, they needed to stick together. It was just under some tall fir trees. They had to watch for birds above. Birds do not care where they go, nor whom they hit below.

Up on stage was Slick Sammy Seagull. Murray caught his eye while he was looking the crowd over and nodded to his friend. The comedy flowed out of this feathery fellow. Sammy traded comedy like he was dealing winning hands of poker. He wound up his show with the same joke about being a flying cow. As usual, the gag brought down the house and he flapped off stage to Murray.

"Murray, you old woodrat you, I am glad you made it. What do you think of this place? Hey, who is that wild lady bird with you? When do you go on, last? Dang, the finale. You must be the hot one tonight. I always knew you had the chops, you wily old rodent, you. Very proud indeed. You are doing the mountain folks proud."

"My next gig is back there after the Crescent Cedar. I will give a report up at the Blues Bungalow. You had better be on the top of your game", he laughed out loud as he waddled away to go drink some water and eat a cracker or two. Sammy loved to eat crackers.

Next, was a rock group, the Sir Douglas Squirrel Quintet and their big hit, "She's A Gopher Mover". They played to adoring young critters. It was light, airy, and fluff, perfect fodder for those who knew little about music. What he heard yesterday and what he just finished practicing, was the real stuff.

Still, you had to give it to the five nut gatherers. They had a good sound. It was catchy. While youngsters crowded the front of the stage and had to be held back by angry Canadian geese, the next act got ready. It was a lady porcupine, an animal whose very appearance stunned Murray.

He knew she was a rodent, chubby and adorable. She had spikes to defend her very soft underbelly. Were any assailant or human being to get a hold of her, they would shriek away with a handful of quills embedded in either hand or paw. She bore dark fur and two soft umber eyes. She had massive, sharp, strong, long black claws at the end of each foot for climbing trees which is where they sleep.

The other thing he noticed was how bawdy she was. Unlike his own timid nature, she embraced life with an overwhelming zeal. This Paula was cuter than a bug in a rug. All sorts of animals swarmed around her and she ate it up, giggling and laughing to the stage.

Once at the backstage stairs though, her complete countenance changed. A few handy Canadian geese blocked the way so no one could claw at her; this was her time to shine. She needed to ready her mind. Paula gathered herself, shook pinions in place, took a long, deep breath, let it out slowly, then emerged from darkness into the spotlight.

She was home now and lovingly embraced it. Her ritual was to stroll from one end of the stage to the other. Next, she made her way to the middle of the stage, up front and waved to the crowd. They were in trees, water, and covering the ground. Critters were already laughing and she had not so much as spoken a syllable.

"Folks, folks, folks. So I woke up in a tree this morning, climbed down, rubbed my eyes, and here I was. It wasn't a dream. Oh no, oh no, oh no, I'm your worst nightmare. That's right, folks, it's Prickly Paula the Pugnacious Porcupine scaring kids with my quills, giving the boys some thrills, and showing how sweet the laughter spills. So sit back, relax, but don't ever back into me, you will never forget it".

"Oh, is that quill mine? Well, pardon me for gettin' sticky", Paula spouting her most famous tag line.

The crowd exploded with hoots, caws, antlers shaking, and every other way to make noise they could produce. That sow had them in the palm of her paw, toying with them for forty-five minutes. Several juncos and finches fell out of trees laughing so hard they could not breath.

Mice, gophers, deer, otters, it did not matter, the audience were either holding their sides or rolling on the ground. She ripped out stories about trying to date with a back full of pinions. Human beings, oh she tore into them with brutal, humorous savagery. She told a story about a near-sighted wapiti who tried to woo a brown horse.

"You know what I love? I simply love to drop out of a tree right in front of a family coming through my territory to have a picnic. Hey, you Thumbs over there, I'll give you a picnic. Hey you dopey, invasive Thumbs, who wants the first quill"?

She tore into their preservation efforts, imbecility in the wild, their smell, doing an entire bit about the garbage they produce and how they call animals dirty. Joke after joke, gag after gag, stories stuffed with power packed funny closing lines filled her period on stage. She held the audience spellbound the entire time.

The reason, Murray surmised, they loved her so much was she was genuine. She really knew what it was like to survive in the animal world and bore a few well hidden scars as testament. She was one of them, with the talent to tell the truth with more humor than anyone else in the Bay. They adored her to pieces. The applause for her was the best of the entire day. The Willapa Bay Blues Band had to shine were they to supersede her presentation.

"Folks, folks, folks. Thank you so very much. I gotta go climb a tree and get some rest. You wore me out. The fabulous Willapa Bay Blues Band, personal friends of mine, are up next to rock your world. They have a new singer since Pegeen became Peking duck last week, rest her soul. They have an additional saxophone player, a talented woodrat from way up in the Olympics come down to share his lungs with us".

"The one thing I wish you could do for me is crowd surf, but you would have barbs in your paws for the effort. Before I go though, always remember and never forget, if you get in a tight situation, quills up"!

They exploded in laughter, love, and idolization for this lovely lady. She was so funny, the break between acts had to be extended to twenty minutes. They set up instruments for the band and let animals and birds recoup. She wowed them with laughter and hope, a rare combination. She descended the stairs and instantly became a diva again, the life of the party. She was up the stairs fifty minutes; for her it was the whole day.

Murray was moved by how professional she was. It must have taken so much for her to do both things at once, a meager three stairs to attain them. It made him realize how tough she must be. How truly talented she was to stand up there alone, controlling the entourage and mob who followed her like leeches. He was glad for his mountain retreat and the safety of jamming in the Beaver Blues Bungalow.

They ascended the stage, got behind their instruments, and waited for the introduction. They were going to play eight steaming, hot songs. As for Dylan, the only thing any creature who flew, walked, slithered, or swam was going to remember was them. Backed by the other permanent members of the band, he enforced the need for a superior performance.

He did not doubt either newbie, but one screw up and he would tear into the perpetrator personally. That rule went for the whole band. This was their show. They were the best. It was up to them to prove it to everyone else. Once with their devices, each member felt positive vibes. It is amazing what thinking as one does for a group. Corinna the Red Northern Cardinal, exotic songstress from the south, ruffled her feathers, then shook them in place. She was ready.

Ivan Trobosky stood to the front of the stage, waving to the crowd to quiet down and let him speak. It took a bit. A few security Canadian geese had to break up a scuffle by some Eastern gray squirrels who were picking on much smaller chipmunks over a better seat. The squirrels were booted. No one can argue long with a gaggle of angry Canadian geese. Ivan waved wings high overhead, finally gaining control of the crowd.

"Oh forest folk, oh now, now it is time for the star attraction. Winston and Beverly personally went up to yon, far and distant misty Olympic Mountains to retrieve for your entertainment and enjoyment, a woodrat with an amazing talent. Sugar Murray is joining your favorite band on sax. Voted number one three years running, the incredible Willapa Bay Blues Band", the crowd exploding in raucous approval.

"Now forest friends, we have a sad announcement. As you may have heard, Pegeen the Pintail, whose soft voice sang many a sweet song, was gunned down by a hunter last week and had to be replaced. We would ask you to bow beaks and muzzles for a moment to remember our friend and all she gave to us", reverent silence followed.

"So, my friends, we have for her replacement a lady bird from the south. She is from Baton Rouge, Louisiana, a place many, many kilometers away. It is too far for most of us to ever travel. Yet she ended up here. We discovered she has an amazing singing voice. She worked with the band. They found each other mutually worthy".

"Now, for the final act of the day, I, Ivan Trobosky, with generous sponsors Winston and Beverly Crow, plus the management of the Carnival Island festival, and in conjunction and coordination with the greater South Willapa Bay Animal Society for the Arts, it is my great and humble pleasure to introduce the best group anywhere: the Willapa Bay Blues Band". Igor Trobosky's introduction was superb.

Dylan exploded, playing notes with a professional purpose, as the emcee vaporized into the dark background. No porcupine was going to get the best of him any day or any time. Miss Bang, Boom, Bang was on top of it. The Sir Franklin, Esquire thumped bass in superior fashion. Giovanni let loose an explosion of keys. Murray instantly found himself in front, strutting and stomping like back home.

The fear which gripped so many left this little woodrat behind. He gave it everything, introducing himself to literally thousands on the island. The largest, most popular festival, it attracted myriads of birds of over 300 species from tiny Anna hummingbirds to immense Trumpeter swans who flew Murray over the Bay and back. In addition, a plethora of rodents, otters, weasels, muskrat, deer, wapiti, gathered.

Dylan was so good, he could control the audience on his own. With the addition of his three other regular chums, the sound was simply the best Murray ever heard. After an excellent musical introduction, Dylan slid up against Murray's back. They jammed as brothers on stage, bonding musically.

Then, from her position center stage, Corinna fluffed her feathers. She stepped forward towards the crowd in as sultry a way a cardinal ever could or would dare, for that matter. She was so open, so comfortable, those closest to the stage began to crowd up against the Canadians to get in and hear every note she oozed. She had not so much as uttered a word.

The song was a blues classic, "The Bird Call Blues", and it was one of the very best the band performed. Dylan and Murray were swimming in notes of music unified in a beautiful, powerful display. Miss Bang, Boom, Bang was banging and booming away at her kit with excellence. She was so taut, it hurt. Her brother in musical crime, The Sir Franklin, Esquire, was right up on her side, thumping out a bass line described by listeners as so groovy, it was sick. Giovanni found spots here and there to fill in perfectly. The sound was right, tight, and outta sight.

Then, Corinna began to sing.

"Well, woof, woof, woof, caw, caw, caw, and ha-ha-ha. Hello babies, I'm Corinna the Red Northern Cardinal", she dripped over the melody, bird hips swaying softly side to side, "I'm all southern belle, baby. So deep in the south, I come from the delta's mouth. Oh yes, y'all, yes indeed. We are here to toy with the Carnival Festival crowd. Will you let me to entertain you?"

"Now y'all just set back, sugar, and enjoy what we got to offer each and everyone of y'all", she seeped the subtlety of a train wreck. Murray swore he saw a rabbit faint and fall to the ground dead away at her sexy voice and wicked strut.

They wailed away on eight songs:"Take Me Out Of The Blues Game" the finale. After the "Bird Calls Blues", "Dead Skunk Stinkin' Up The Road", "Sweet Seagull Blues", "Flyin' South On A Swan", "Ridin' The Zephyr", "Gettin' Quilly Wid It (the porcupine song)", and "The Burrowin' Blues".

Each number provided individuals in the band time to shine. They worked as a sextet, no egos bigger than the other. Each member knew their spot and took it, did not horde, did not denigrate. They formed a combo, as Dylan envisioned with the addition of Corinna. It was a real band, solid. Murray was just a dynamic bonus from up on high in sweet mountain country.

The audience responded with more and more vigorous applause. The trees were alive with every bird of a feather getting together in the glorious weather to add to the excitement. Critters on the lawn and birds in water were reveling in the sound. Ivan sprung to the stage, waving his wings overhead, nodding his long, yellow bill up and down. He could not compel them to stop. He simply shrugged his shoulders, turned to the band, succumbing to the gathering's desire: encore time.

Moved with pride and delight, Dylan did not hesitate. He nodded to Miss Bang, Boom, Bang who let rip a drum line so sick, so clenched, so strong it instantly slaked the song thirsty mob. The Sir Franklin, Esquire added a bass rhythm on top of her drums deep and funky, sweet as dark chocolate. Next, Murray, feeling his oats, tore out a blast with his sax so power driven, Dylan looked at his new chum with an open mouthed smile. Giovanni, not to be outdone, threw out a rock-a-billy riff which tore up the place.

Dylan tossed in his semi-hollow body adding to Giovanni. The tune was "Misty Muskrat Hop", a raucous, long, hard jam. Even the bass got a solo. The lyrics were belted out by the Red Cardinal Queen: Corinna was a real hit. She sang the whole song without so much as a stumble in any part, though she had never heard it. The number finished, on a long, hard wrap up featuring Dylan at his finest.

They bowed and stood back, looking at each other in utter amazement. Giovanni burst from his stool, strutting back and forth in an energy driven explosion of joy and humble pride. Dang, dang, dang this was epic. This was the best performance for any of them, ever.

Dang, dang, dang, it was so perfect. He stared hard at Dylan and Miss Bang, Boom, Bang. His stunned face said: what did we just do? The two of them bore the same expression. Only The Sir Franklin, Esquire donned a massive, knowledgeable grin. The big dude was happy. True to his desires, Dylan had a star in Corinna. She was added as a permanent member right there and then in agreement. Murray was praised by the entire crowd, given his own standing ovation at the end. The rest of the band stood back while he garnered well deserved praise.

Then Corinna was forced to stand alone as well as she absorbed the honest adoration for her supreme talent. The crowd was won over early on and her new reputation had a perfect genesis. She had a home. Finally the whole band came on stage, each receiving appropriate applause for their individual efforts. At last they bowed together as one, Dylan and Miss Bang, Boom, Bang at the center and the rest fanning out on either side. A party was held for the performers.

At the end of that long, long gala affair, Murray found his way to the stump, climbed down, found his door, and fell in bed with a thud. He was asleep with a huge grin on his mug in a matter of seconds. What an incredible day, the bestest danged best day of his whole furry life.

BOOK 2: HIBERNATION

Chapter 9 The Carnival Continues

Murray arose in the morning, satisfied with the prior day's effort in his new band, but restless. He stirred and climbed out of the stump. Early rising had the desired effect, solitude and time to reflect on recent events in his modest life. He wandered alone on this huge, precious island in a quiet forest. Few mammals or birds were up, so he made his way north.

He spotted the edge of land as the tide lapped up on shore. All was still. Haunting, dense fog lay easy on the morning water of Willapa Bay. Above the pea soup, was the low ride of Bear River Ridge, as if islands bouncing in a sea of clouds, south of the Naselle River. While not lofty, it's modest peaks peeked out in early rays and glisten with morning dew. The sight was moving and he considered life golden.

The silence of the moment was broken by an indelible, yet beautiful sound. He was on the far northern end of the island. The music was a lone wood flute with a moving hollowness. At the same time, it fervidly engaged him. Such music, he mused, as it enraptured his spirit.

He came closer to the tones, spotting a lone figure sitting cross legged on a large curved branch broken off eons ago. The dead stalk arose in an arc with a perfect spot upon which to perch, an excellent place for solitude. The image was quietly moving. Murray's feet rustled the leaves causing the music to cease. The lone, shadowy figure turned around. He could not make out who it was, save a woodrat.

"Hello friend," softly came Murray's words, "I vas vandering on zis quiet morning und stumbled across ze music you vere making. It is so schtilling und serene. What is ze instrument you are playing?"

"Ah, yes", stated the dark one, "It is my Shakuhachi, a Japanese flute. My name is, Momo, and I, like you fair one, am a woodrat. I just happen to by a lady one."

"Vell, Momo, if it is alright mit you, I vould simply love to sit und listen to you play. The effect of your flute is amazing. It tickles my ears mit exhilaration. I play ze saxophone und am here to perform for ze festival. Perhaps you could come und listen to us zis afternoon? Ve play late in ze day. I vould really like you to come und give us a listen."

"My day is planned by my family. We are performing acrobats on another stage. If my mother and father give me permission, I would enjoy it. Now please just sit and listen to the music. We will enjoy it together."

Something in her words stirred him. Murray felt honored to be asked to appreciate such sweet, sensitive notes. The flute combined with the quiet of the morning produced a sound which was full and at the same time quite lonely. He sat for a long time at her feet, looking up and loving the soulful sounds she produced.

After a time, they walked back together to where everyone was milling about. They went over to her parents, Fumio and Fumi. Her mother, Fumi, looked very suspiciously at Murray. Fumio knew him from the concert the day before and did his best to welcome him. From his way of thinking, Murray was a fellow musician and might want to wed their available, young daughter. Come to think of it, Murray's mother and grandmother would be on Fumio's side. They would love to see him marry. Then they could toss the moldy potato skins and have a litter of grandkids to dote over.

These four went to breakfast together. The rest of the band was up and came to sit with them, reveling in good company. Murray promised to come see them do their act. Momo agreed, with her parent's blessing, to come listen to him. The band started to lovingly pick, asking him when he was going to kiss her and if they should get the wedding cake ready.

Murray, a confirmed bachelor, coughed and wheezed at the very thought of matrimony. It may be for his good buddy, Mr. Mizithra, but no, no, no, not for him. Momo may be a very cute little woodrat, but he was not interested in anything of that nature. He simply appreciated her musical talents.

Since they did not have to perform on stage until late in the day, he decided to go and explore what else the festival had to offer. There was so much to see. The Highway 101 view of the island sees a stand of trees and a short, broad grassy plain in front upon which the humans trod. Rarely did any of them cross over to the island.

If they did, few ventured off the beaten path they had erected so they did not tromp through the undergrowth. Trees are dense, forming a perfect rampart against the Thumbs prying eyes. Behind this wall of green, safety reigned supreme. They had areas set up for the smaller acts. This way, the audience was up close and personal. This was the stage Momo was going to be using with her folks.

Soon, he found a good sized crowd gathered around one of these smaller platforms. Up on the dais were three male fishers and a female mink. They had a jazz combo running hot for early in the morning. The mink was dishing out scat, a style of jazz vocalization, with great aplomb. It seemed the whole crowd was hypnotized by this enchantress. She let go with a torrent of unspoken words strung in the most sexy way he ever heard a song sung. They were made up jumbles of rhythmic noise.

She made Miss Bluesberry seem almost tame in contrast. The fishers were wailing away on drums, bass, and flute. They had a sax, but he was happy to hide in back, not perform. He wanted to enjoy others with clear talent do their thing and wow an assemblage.

Fishers are an example of man's intrusion. Hunted to extinction for fur, fishers were re-introduced and quickly repopulated the region. This is why man should keep his distance from critters.

"Say, who are zese folks?", queried a curious Murray to a nut chewing Townsend chipmunk standing next to him.

"Scheherazade and the Fisher Three", stated the rodent between chomps on his meal, "They play here all the time. They live on the island and do this all year. They could jam for hours. It is really fun to see them".

"We take the floating Tundra swan ferry service across from the mainland and stay the whole three weeks. The wife and I are here without the pups. Her mother is watching them, so we could have a second honeymoon. She is still asleep, but I love this quartet. Music is our thing, my wife and I".

"Yeah, me, too, thanks".

"Oh, sorry, my name is Fitzhugh, Fitzhugh Townsend. You want to know anything, I'm your chipmunk. Everybody knows me and my wife, Guinevere. We come every year. Both our families have been attending this event for over ten generations. This must be your first time?".

"Yes, I flew in from up in ze Olympic Mountains on Trumpeter Swan Airlines. I am mit ze Willapa Bay Blues Band. My instrument is ze saxophone. Zey brought me here to join ze group for zis show. I go back up north und into ze mountains in a veek. Zis is an amazing place for me. My whole life has been up in zose hills".

"Wow, you must be the famous Sugar Murray. We are going to come see you today for the headliner. What an honor. Your show yesterday is the talk of the island. Everybody spread the word. You are going to have a packed house today. My wife is going to be very jealous when I tell her".

"Tell you vat, I vill leave word mit ze Canadian geese. You und Guinevere can come backstage und listen to ze show. How vould you like to surprise her mit good news versus making her green mit envy?".

"Seriously", said Fitzhugh, wiping nut crumbs from his mug, "It would be an honor, a privilege. Wow, she will flip out completely. This is too much. She loves music and thinks your band is the best. To stand backstage to listen will mean so much to her. Oh thank you, thank you".

"No, no, truly, danke. Anyone who appreciates music needs to have access to it. Ve benefit from you who keep our songs alive. You clap und hoot und holler. Zis is vat we derive our own sense of worth from. You confirm our talents are redeemable. Please, bring Guinevere und you can come to ze party afterward to meet ze other members of ze band".

After a few more words with Fitzhugh, he moved along to sample something else. The place was a veritable smorgasbord of acts. The Fisher Three with Scheherazade were a great group. She was something else. It would be a real gas to jam with them. He could see himself immersing deep into this kind of jazz. It was earthy and pure.

He reveled in this performance in particular. Hearing other forms of music built on his to create more, new tones. He saw muskrat jugglers, a rabbit mime, and a magic show with a beaver so feeble he could only bring gawkers and mockers. Murray pushed on as he heard a commotion ahead in an open area. Curiosity got his interest. He followed the mob.

Nearing the noise, he donned a broad, knowing grin. In the midst of the crowd gathered around a long rectangle, he spotted two hoops, one at each end. A bunch of blue herons, green herons, and white egrets were running up and down. The blue and green herons were his old chums, the Blue Heron Globetrotters. The opponent were a squad of white egrets, the Washington State Generalissimos.

BOOK 2: HIBERNATION

As he approached, the score was 50 to 20. The Globetrotters were running up and down, performing wild tricks as they passed, dribbled, and dazzled the opposition. Those fortunate to watch on the sidelines were entertained to the utmost, laughing the whole way through.

When the game was over with a predictable Globetrotters lopsided victory, he waited a bit for stragglers who simply wanted to fawn over the team. He moved sneakily along side of the green egret, Clyde the Glide. He instantly recognized the beat up kippah on Murray's tete. Clyde nudged Pee Wee, who nudged Larry, who slapped Charles up side the head with his wing. The team got in close and surrounded the old chum.

"Well, if it ain't good old mountain boy, Murray. That hat of yours has seen better days. You birds ever see a brim like this? Your kippah is worn out, friend. What is a hillbilly woodrat doin' down here amongst the Bay set?", Charles asked with his typical endearing, snobbish attitude.

"I am here mit ze Willapa Bay Blues Band. I play ze sax for zem. I flew mit Trumpeter Swan Airlines. Zey vill take me back home after zis. Hey, you must hear ze news. Mr. Mizithra got married. Yes, he did. Ve got all ze way up to ze Valley of ze Schtones, und he met a very nice lady marmot named, Miss Saffron. She is a Misses now, hee-hee. Zey are very happy. My mama und grandmother are safe at ze Beaver Blues Bungalow. Oh, it is so nice to see you fellas again. I loved ze game".

After a bit of friendly cajoling from the feathered athletes, he said it was time to move. They assured their pal they would flap over and give his group a listen. The Blue Heron Globetrotters liked blues, but were more into reggae. They hoped maybe he would muster the bravado to play some funky, rasta sounds. Murray made his way along the row where acts on both sides of it were doing their level best to draw crowds. Some were clearly better than others. He counted himself blessed to be a featured performer with the best band on the island.

A large group was gathered on both sides of a watery inlet which ended like a long, wide yam at the shore. It weaved back a bit. This gave a clear shot for any duck type bird to fly in and coast to a lovely stop. And no fear of raptors diving upon them at a moment of true vulnerability.

Throngs of critters lined water's edge, shouting at intervals. Since he could see nothing over the heads focused on the water, he forged a way to the front. The larger animals were happy to allow a little fellow to get a good look. A sight met his eyes. It was something he had never seen nor imagined possible.

In the water, a dozen buffleheads, all males, wand shaped head gear and heads, faced straight ahead. All the ducks were in a row, swimming in synchronized fashion. In an instant, as one they dove down, flipped over, and showed a long line of duck butts, wriggling tail feathers along the shoreline. The crowd howled with delight.

They flipped back up and turned right as a single unit, then back front, then left. This was done in unison. No one got out of line; they dare not. They went under twice more, wiggling rumps with vigor. On the final go, the lead quacker flapped, followed by the eleven, one by one. He rose above the crowd, circled left, exposing soft, white under feathers.

He turned again right showing his dark upper side. Each of them then turned heads left while still in flight. It gave a dramatic appearance. They lit on the water and, without so much as a paddle, ended at shore's edge. Upon halting, they turned to the left and bowed. They did the same thing to the right, marching out of the water to an awaiting gathering of other buffleheads who surrounded them with quacks of praise.

"Vat in vide world is zis?", he asked a muskrat standing next to him.

"Why darlin', this is Duck Synchronized Swimmin'. Why it's just about the most popular show on the island except the main stage. I am here for the third year in a row. I love this sport. I live with my family over in one of the little estuaries back over by the Bear River Ridge. We come here and have oodles of fun all day long."

"They are so coordinated and majestic. They are good. The Coots won last year, Gadwalls the year prior. It is time for a new champion. You should stay here and watch. My name is, Miss Tabitha, tee-hee. My folks are somewhere. I hope y'all have a fun time, Murray. You sure is cute".

They had to flirt with him. He stood watching this amazing sport. Each group was a dozen, male or female. They had to be the same species. In the end, a bevy of Barrow's Golden eyes, stole the show. Handsome brown heads with striking golden glowing eyes, in a straight line, blew the crowd away from the moment they appeared in the air in four groups of three from four points of the compass.

They merged seamlessly to a perfect row of twelve ducks as they landed, each holding their left wing in the air for effect. This took an extreme amount of muscle control. By the time they were done, it was unanimous. This was an upset. The first case any Barrow's were so highly rewarded.

Murray enjoyed it, then made his way to the big stage. He would practice with the band so they could tighten up the sound and figure out the song choice for the show.

The best new musical was, "Fur". It was about years ago and the "hippie" movement which hit the peninsula generations of critters prior. One thing about animals, they do not vary very often. Any talk of change was met with terse resistance. It is the same with human beings.

Most like things to stay relatively equal. An even keel is the goal. Very few animals were like he and Mr. Mizithra in quest of new vistas, creatures, and adventure. The show was controversial. Murray decided to sit for a bit and watch. The practice was not for an hour and he figured the show might be interesting.

He hailed a passing otter server and got a nice water bulb and some clover. The audience was in front of him. Many were return listeners and thought it the best musical ever. An entirely different viewpoint was held by others. They felt it rubbish, of no artistic value. The decision was left to each individual. Actors took their places. The cast was six thespians and a narrator, a chattering Douglas squirrel who talked rapidly.

It was a bit confusing. It had to do with decision of a young otter and two friends asking whether or not to leave to safety of their little cay in the Bay or leave, defying parents. They could seek new land across the huge, vast Willapa. Were they to stay, they would have to go and defend the lands against coyotes. They were not sure they wanted to fight.

It opened with a song, "Almost Cut My Fur", a tune which asked the brutal question of whether or not it was right to look like any other old otter or make an impact and shave off fur with a sharp rock. The look would disrupt the elders completely. The play had a confrontation with the son, Alphonse, and his mother, Tootsie. This song was, "Mama, Kin Ya Let Me Go?". You could see the hole in his mother's heart and soul when he shared his intention to dishonor the family and swim with his two pals, Ignatius and Aloysius.

The story follows in quick succession. They swim off, singing the song, "Fog On The Water". Alphonse is trapped by man. He is imprisoned and sings, "Under My Thumb's Thumb". He struggles to adapt to be a pet. He strains for freedom. Life away from home is not what he thought.

Alphonse lives in a cage. He wants to be able to do what he wants to do; eat what he wants to eat. They put him in a collar and, tethered, drag him from place to place. The humans wanted to show off their prize.

To this he sings, "Unleash My Leash, And Gimme Some Quiche". It was a humorous number. The audience responded in kind.

Finally, he escapes finding his companions' pelts on a trapper's wall, causing him to lament their deaths. This is the emotionally wrenching tune, "Up Against The Wall Willapa Bay Mother". It was a sad country song which evoked the terror animals dread. It is man who hunts with the goal of species obliteration. Alphonse makes it home a changed otter and sings the last tunes: "Virgo/Let The Moon Shine In" and "Fur".

These were the two best songs in the whole musical. The first dealt with his arrival back home, looking to the future towards the end. The entire cast joins in with a great crescendo. "Fur" was really a wrap up of the whole experience.

His somber conclusion: while new things and places were great, adventures exciting, there was really no place like home. Alphonse resolved to live in his family's world, trying to change and improve it versus running away.

Murray enjoyed Fur and made his way backstage. The rest of the band had arrived and went back to the same spot to practice. They would play most of the same songs, but add two more to keep the blues going strong. They were: "Johnny G. Beaver", and "Moonlight Flight".

The former was an old classic about a guitar slinging beaver from up in them thar hills. The latter was a chance for Dylan to shine. He needed a truly raucous solo. He said if he could not explode once in a while, he would go batty. They took out "Bird Calls Blues" and "Burrowin' Blues".

BOOK 2: HIBERNATION

It was good to shake up songs, so those who returned did not feel they were being cheated. New tunes always worked well. They decided to add a jam session so other musicians could join them. They had to take out one more song, "Sweet Seagull Blues". The line up was tight and came off with perfection. They were even better in this show. The jam session went on for fifteen minutes and was wild, whacky, and wonderful.

Fitzhugh and Guinevere, came backstage and to the after party. He saw Momo in the crowd. She waved and sat through the whole show. He came looking for her afterward, but she was no where to be found. He gathered her folks wanted her to come home right after to practice for their huge acrobat competition the next day.

The prize was valuable. Over a dozen teams partook the challenge. The whole day had been a fantastic, unforgettable one. He again went to sleep in an instant. His mind raced with the excitement of the day. He felt himself a truly fortunate little, old mountain woodrat.

Chapter 10 Reality

Torrid sounds ripped Murray from sleep with shouts and shrieks of terror, pain, and death. With suddenness, his slumber was interrupted by loud noises outside his door. It was the voice of many different species of animals. Where he stayed were other performers. Beaver, otters, and muskrats went to a lodge swimming under water to a hole and slithering in to rest and be safe. The birds were on water or in a tree.

For the rest, it was these rooms inside the stump or any of the other dozens of holes in the ground one could avail themselves of spread out over the island. Murray stumbled out of his room and the stump to see what was happening. No sooner had he come out than he was grabbed by the nape of his furry little neck by some unknown, much larger animal. He was off the ground and led to the water's edge. It was still dark and he was frightened to the core. What was going on and why?

He felt himself being tossed in the air. He landed in a warm bed of down. It was Theodor. The one who nabbed him was Torbjorn. He gave Murray a nod. They took off as one strong bevy, becoming a wedge in flight. It was black as midnight. Nighttime clouds hung low. The air still. He dug his face in

Theodor's feathers. They flew due east. They passed six posts in the water along the highway, next to the Naselle River Bridge.

The red and green lights of the structure shone brightly in the dark of night. They flew under the bridge straight up the Naselle River. Once 500 meters past the bridge, the swans took a right, along a swampland which lay below with a few minor little streamlets meandering.

They ascended Bear River Ridge. The low valley receded rather far, unassailable by man or large animals. The swamplands made sure of this fact. They flew straight up in the sky over 500 meters in elevation and landed on a small pond high above the destruction below.

At the shore of this pool, mice and woodrats awaited their arrival. Murray was escorted ashore. The swans were exhausted. They were not anticipating such an emergency flight in the middle of the night. The emergency rescue squad floated on water, nibbling in the pond. They had no intention of going anywhere. Their sole task was protection. Some thing sinister and disturbing had taken place.

He wanted to know what it was. They went to a safe haven, a small cave directly behind a tiny, gurgling brook babbling from the rocks. It was a large space inside. He was invited to sit and rest, then given a cup of water and offered bugs and clover in a bowl.

Murray found the repast to be just the thing. He felt revived. He was completely puzzled by the events and wanted clarification to solve this mystery. The warm sun was beginning to rise above the haze of morning. It was going to be a lovely warm day, weather wise.

What just happened?

"Well, Murray", spoke Shadow. He was an aged Pacific Jumping Mouse, a mountain sage, "It appears from early reports flown here by cormorants the Carnival Island Festival was the target of a gaze of ugly, angry raccoons. They wanted a certain comedic performer to kill. She is a porcupine named, Paula. It is her jokes which they found so offensive. She had to die".

Murray was struck to his core. She was a friend. He was bewildered. Up in the hills, they accepted the fact animals had the right to speak their mind. You may disagree with their opinion, but they found it fair and open to let all speak. If someone's words were deemed offensive by others, rejection would suffice. It was unheard of in his land to kill any critter for saying things not favorable towards another species or even an animal from the same one.

When denizens put down woodrats, he absorbed their opinions and moved along. No sense fighting a bigot. They cannot be turned from evil. They only find others as evil to with whom to socialize. Only others who are rotten want to be with lousy people or animals.

He was encouraged to rest. Dawn approached. There was nothing to do. Crows would let everyone know when the all clear sounded. Until then, he was going to be there for a day or two. He was to simply stay and take it easy. The fact he was alive and well was enough. Shadow showed him to a room with a nice bed of mountain moss.

He immediately fell asleep. He was safe. He hoped Paula was. He hoped the rest of the band was, too. Then he was reminded of Momo and her folks. Were they okay? It was a very upsetting episode. He knew news would come later, so he bedded down and slept hard. After a long rest Murray arose and made his way out to the common room.

A lady mouse, Sunrise, offered food which he gladly accepted. While he ate, she explained the rest were out, getting things ready for him. She refused to say what they had in mind, but assured him it would be a wonderful experience.

It was a nice day, so she encouraged him to go outside and welcome the glory. He spotted Shadow and a few other rodents. He motioned Murray to come down the hill and follow. He went along a windy trail by the water's edge, then up another knoll. Just on top of it was a small pool of water. He instantly noticed something different about this liquid; it was steaming and had an odd fragrance. He stood looking at it and wondering what this was.

"Vat is zis?".

"It is a natural hot spring. We sit in this for relaxation and to help our tired old bones from hurting so much. It is wonderful, therapeutic, and pleasurable. We insist you give it a try. Everybody loves to sit in here and wile away the hours. You will want to stay in here all day long. Join us and reap the benefits of healing waters".

He waded in the water and sat on the opposite shore. He had a small spot lined lengthwise with a branch stripped of bark. Arms rested on the twig and leaned back against the side. To Murray, it looked very inviting. He got in the water. It was 41 degrees Celsius (105 Fahrenheit) and very nice on his skin. He worked over where Shadow was sprawled out and assumed the same position along the branch to his left.

The most important thing to happen in this balmy fluid was the relief of pain in his left leg. The sun hit him square in the face, basking Murray in golden rays of friendly warmth. Others joined them and sat for a time talking and relaxing. It was a very healing moment.

The respite in the pond gave him time to escape the consternation of the previous evening. It's horrors were vivid in his ears, echoing again and again with screams of injured and dying animals. Thanks only to the swans did he evade death. Once again, he counted his blessings.

When he got back, they rested. He drank a large amount of fresh water and got some more sleep. Later in the day, he arose and came out to see if he could get something to eat. Sunrise tossed him a plateful of this and that and he stuffed himself full. It seems fear mixed with a hot sauna made one both hungry and thirsty.

He stuffed his face and slurped down some mountain water, the first since leaving home. It tasted fresher than the stuff down on the island. As he sat there, a few barn swallows flew up to the cave home and spoke with Shadow. He and another Pacific Jumping Mouse named, Marama (Moonlight), the largest, most stoic mouse Murray ever met, conferred and headed towards him, grim looks crackling their faces.

It seems a gaze of raccoons, perhaps a dozen, swam across the narrows, made their way in secret to the southwest end of the Festival's grounds. Scaling the high cliffs, they slew a scurry of Douglas squirrels standing guard at that end of the area. Once past them, it was simply a matter of stealth and taking time to spring their attack. The goal, as stated by the only one they captured alive, was to kill Paula for her harsh jabs at these bandits.

Of the dozen who attacked, four lay dead and one was captive. The other seven got away, but were damaged in one way or another. A prickle of Paula's pals defended her admirably, killing three and cornering the other with a solid wall of quills. She was unharmed, but the same could not be said for the rest of the place.

Apart from the death of squirrel guards, a handful of woodrats were slain, sleeping in a small cavern. The raccoons tore it asunder, killing all inhabiting therein. They slew a couple of the crows, though not Beverly or Winston, who were devastated by the ordeal. Two Northern Pintail ducks were beheaded.

A couple of Keen's mice were killed by simply getting in the way of the marauders. His entire blues band was safe. They were in the lodge of The Sir Lincoln, Esquire; Corinna was safe, nesting in a hemlock tree. The news was to stay where he was for the next two days.

The six posts he whizzed past before passing under the bridge were attended by a sextet of either Brandt's or Double-crested cormorants. The place was in fact called, "Cormorant Corner". These sleek elusive winged devils were the best spies fish could buy. They could swim submerged for long distances and easily hide from searching eyes.

Since they were so dark, when they arose, it was difficult to make them out in murky water. These cocky ladies and gents would sit on these six posts all day, coming and going here and there at odd times. They sat on the watery beams, gabbing to each other, spreading wings proudly to absorb the sun's warmth.

In creating such a laid back facade, they were able to fool others. Hired as spies, they provided excellent results paid in fish. So, to protect Murray, the barn swallows would return to their regular duties. The cormorants would take over the job of ferrying messages back and forth from and to Carnival island.

Marama and Shadow invited him to a tree protected clearing, two meters across. There, a small lump served as a stage. The sunlight shone to provide excellent illumination. They had some instruments lying about in front. They knew him to be a musician, but had no saxophones to play. He was offered a harmonica. He began to practice with it ardently. He had seen others jam on one. He had blown into one once or twice backstage to see what it sounded like.

After a few minutes, he found a few key notes, slid his lips along the sides of the mouth harp, and nodded towards the mice. He would give it a go. He had no idea what to expect, so planned on simply listening until he felt comfortable joining in the mix. The last thing he wanted to do was embarrass himself. Playing a new way was a challenge. At least jamming with these fellas would be a welcomed distraction from the evil below.

A woodrat, Pixel, sat at the piano. He was the singer. Shadow had a nephew who played drums as did Marama, so they had two drummers. The bass player was another, but unrelated, woodrat named, Hayes. He was a somber fellow, who had little to say. He played, he ate, he slept. That was about it for his life. Good old mister dull, that was Hayes. Sunshine played tambourine and sang harmonies.

She was Shadow's sister. Shadow and Marama played guitar. Marama played his slide style. He found an old, narrow medicine bottle left behind the Thumbs. By placing it on his finger could slide it along the neck and make an amazing sound. Murray had never heard anything like in his life. He was transfixed. He had so much to bring back to the Bungalow to improve the band it was overwhelming.

Once they got underway, the double drums steered a wicked course followed instantly by bass renderings from Hayes. He was the largest woodrat Murray had ever seen. His nickname, Buzz. He barely moved, thumping strings in deep rhythmic time to the melody. Sunshine threw in her banging tambourine. The piano began to assault keys as if they were enemies to be crushed. At this exact moment, the Pacific Jumping Mice erupted with dueling axes in a mind blowing tune with speed and style moving Murray to his little toes.

He did not know what to say or do. He was frozen in admiration at the talent. This was a type of rocking blues, but unlike any he ever heard or imagined possible. The double drums and vapid bass provided a back wall of sound from which to launch their dual threat. The keyboards and tambourine filled in pockets of emptiness nicely. Once the guitars were both playing, it was up to Murray to add his two cents worth.

He had no idea what to expect. This was an entirely new thing. He took a deep breath, shrugged his shoulders, and wailed in the square holes of the mouth organ. The sound came sweet and vicious at the same time. He sucked in, giving it another totally different effect.

He exploded forth in a steady stream of air. He knew not to blast into it, but give it an even flow to balance the sound. He moved here and there along the thing, finding the right places, making mistakes. For a first go, he did okay. He did enough damage and nodded to the boys.

They took off on a musical rant which was so good it sounded as if the instruments were singing instead of being played. The piano had a totally bluesy boogie woogie solo in the middle of the jam. This was simply them loosening up and getting ready for the real music to follow. After about ten minutes, they wrapped it up, explaining to Murray this was Swamp Music. Way back up in these lows hills, the rest of the world seemed too distant and hectic for them.

They jam all day, soak in the hot tub, eat, and sleep the night away. Very few predators ever got up to this area. It was very steep on the south side, which dropped to the massive Columbia River. The swampland was over 1000 meters in length and 400 wide. It was lined with high hills on either side, impassable to all but the lightest of killers. The only fear they had was from above. It was virtually impossible for a hawk or falcon to get through the brush.

With such ease of living, they had only to concentrate on family and music. Food was plentiful, as was water. Homes were safe. They had an excellent warning system in place, so nothing to do but jam. He was at once homesick and yearned for the safe confines of the Bungalow and his friends. His mother and grandmother were on his mind. He found a tear in his eyes. Soon he would end this unbelievable adventure and be flown home. He was awestruck at the entire affair. He had learned so much in his travels, but this was the ultimate.

They played and played for hours. His lips were worn out by the time it was over, but felt himself an able harmonica blower. Two more days passed quickly. He was happy for the retreat. It was an excellent way to rebuild strength. Shock and fear can drain to the bone. He was grateful for the hospitality of the mountain mice and friends on Bear River Ridge.

A couple of Double-crested cormorants flew in, landing with a giggle on the pond and paddled up to the shoreline. Shadow nodded to Murray who turned and thanked the gang for the great time. He hugged Shadow and Sunshine. The swans arrived in a few moments and soon he was wending his way back to Carnival Island and learn the fate of all concerned. These events surely put a damper on the whole festival. One thing for sure, he would not take life for granted.

Chapter 11 Aftermath

They arrived. He waved good-bye to the swan escort who saved his meager little life. He headed for the main stage to learn what happened. Finding a rumpled Igor Trobosky, Murray was updated. A dozen raccoons with an ax to grind hit in the wee hours and dispatched the squirrel guards posted merely as a precaution.

No one expected to be attacked. This sort of thing had not occurred in generations. The retaliation was immediate and swift. Not only were the raccoons done in, but the raven who led them through the barrier, an evil creature named, Nevermore. He was tried, found guilty, and pecked to death by Canadian geese who lost one of their own in the struggle.

On a sad note, Ivan informed him Momo's folks were dead. She was unharmed. Her folks gave up their lives to protect her tossing Momo in a tree. She scurried aloft to safety. They were buried with others slain in an open grass glade deep on the island. The Sir Douglas Squirrel Quintet was now a quartet. Ducks were buried at sea, per their requests. It was a sad episode and would never be forgotten. They even got the rabbit mime.

For her part, Beverly's health suffered. She was made to rest and would not partake of the closing celebration. She could not bring herself to face the fair goers. Winston, old and sad, was master of ceremonies, a task Beverly always rejoiced. Now, she had not the heart to attend.

He made his way backstage and found Miss Bang Boom Bang. She was sitting on the edge of the stairs, head in paws. She lost a friend in one of the chipmunks murdered. Her mourning was deep. She said Dylan was getting together with the powers that be. What were they supposed to do with the party crashed and demolished? The skinny was they were going to play, then tomorrow a wrap up jam ending a day early.

Murray found he was going to go home. He would be there in two short days. While upset about the whole disaster, he was elated at the prospect of returning to familiar digs. They were playing in two hours.

BOOK 2: HIBERNATION

He acknowledged the time and headed off to find Momo. She was back at the water's edge, alone, sadly playing her Shakuhachi. Pouring emotions out through music was her way of coping with the loss of the two beings around whom she lived a happy life. She was a young woodrat alone on the island where her parents were slaughtered by vengeful demons with black, masked faces. Murray approached her. She ran into his arms and held on for dear life.

He was really the only other woodrat she knew. Most of the acrobats were other species. The competition for tumblers was called off due to the death and injury of so many participants. He held her for a long time, stroking the back of her head, rocking her gently from side to side.

After time, he asked her to come to the backstage area and stay in his room that evening. He asked one of the staff to make her a bed in his dressing room. His request was filled. He took her to rest before they had to perform.

Murray then went back to the practice area where they jammed for a bit and selected a half dozen tunes. They were going to cut it short as it was still a bit of shock. They played, were awesome, were applauded, and went to bed early. No party tonight, the pain was too fresh.

The final day was a lot of pomp and circumstance. Winston wound up the show with a memorial to the fallen. He had praise for defenders by giving a medal to the head of the Canadian squad. He thanked the artists. In spite of it, they loved the music and ballet. The comedy and juggling were awesome.

Murray and the band played one last time. He incorporated the mouth harp for the first time on "Ridin' the Zephyr". It played better than sax. Instincts were correct. The band recognized and appreciated his growth. They were now close friends. He expressed his love for each. The next day, he took Momo off and spoke to her at length.

He had an idea. He wanted her to come back to the Beaver Blues Bungalow and be his wife. She agreed with a gush of joy. He could not think of life apart from her and felt responsible. He adored Momo and the way she played the flute. From the day he heard the lilting music and watched her play, Murray's stoic position as a confirmed bachelor wilted like a bunch of broccoli.

They checked with Theodor. He said his wife, Terese would be happy to take the lady. In the confusion of the night of evil, he lost his old kippah, making due without one. The staff of the show, in recognition of his contribution to make the shows successful, presented him with a shiny new one as they were about to board the swans for departure. Chipmunks, Fitzhugh and Guinevere found some cloth and sewed it together. His music soothed them and made life better.

The band, Fitzhugh and Guinevere, the Blue Heron Globetrotters, Ivan, and everybody came to shore and wished them a safe trip. They were asked to come back next year. Murray agreed before he got a chance to think it through. He had learned much, played more, mastered new types of music and had a new instrument to play.

He made tremendous friends. Murray met and fell in love with a lovely woodrat. In the same breath, he saw death and destruction. It was a trip which deserved a time of rumination, intent on having once they arrived home.

The flight was in several stages. Wind played havoc with them at one point breaking the formation, but they arrived safely. The swans landed and bid them a fond farewell; south for winter they went, singing The Ride of the Valkyries. Murray and Momo were now husband and wife in a quick ceremony performed by Ivan Trobosky, whom they discovered was the fellow. It was common to find love at the Carnival Island Festival.

He led her to the entrance to the club. Being a day early, he was not expected. Having a young, lovely lady with him took everyone aback. His mom and grandma were overjoyed, welcoming Momo as family instantly. They made a big deal over his new kippah. He was happy to be home.

He spent some time talking to Bette, Twigs, and Tobias. He had a lot of new music and styles to share. He yanked out a juice harp and smiled. They were surprised and joyous. A new instrument meant more songs to play. He gave Bette a human bottle for her finger to learn slide. The gang met Momo. Bette began to call her, "Lil Sis".

Murray was home, married, and very tired. He went to sleep soon after arriving. He and Momo did not get up until late the next day. The rest wanted to hear about his time away and the famous Carnival Island Festival. By the time he was done telling the whole sordid tale, they were shocked he got back in one piece.

Murray was, too.

BOOK 2: HIBERNATION

Chapter 12 Mr. Mizithra Arises

After six weeks of solid sleep, Mr. Mizithra and Mrs. Saffron awoke from slumber. Living off body fat, they lost weight and were anxious to have fun. During this time, young ones arose for lessons. It was happily time to tell his tales. This was his biggest contribution to the colony.

Each day stories were shared. They were done for the betterment of the youth. Armed with these fables, they could learn how to safely defend the colony and themselves. They covered a wide range of topics. He took his task seriously. The pups did, too, reveling in his way of teaching.

There was time for his new wife and marriage. Having been in wedded bliss before, he knew what made things work. His sole purpose was to ensure Mrs. Saffron was happy in her home and with him. The way to have a successful marriage is to automatically think of your spouse before yourself. Honesty holds the key.

If both people involved in this institution were of this mindset, the notion of divorce would dissipate instantly. This attribute Mr. Mizithra possessed in abundance. His only thought was her comfort. He sought her out to make their new home better. He dug a bit here and there to oblige her requests.

In due course, she became pregnant. In the Olympic marmot world, a litter was up to six young ones, but this was rare. Two or three was much more common. Females generally only had a litter every two years. He was very pleased and doted on her with joy and happiness. All he could think of was becoming a dad again and having another opportunity to take on this noble task. Losses of the past lurked in the back of his mind, black clouds on a sunny sky.

For a human to become a parent is a daunting responsibility. You do not simply make a baby and let it fend for itself as does a turtle. You take the child, adorn it with love, wisdom, care, and preparation for life as an adult. Such was the case with him, though he was a rodent.

He felt the same sense of responsibility as a human being. Shuffling about in the safe warmth of their colony, Mr. Mizithra was able to hold daily classes with young ones. Adults who were awake and not busy sat in. Those of Mr. Milk Chocolate's colony would come to hear him as they had never enjoyed the experience. His lessons were worth hearing over and over again. Each time someone heard the same tale, they learned something new. It was a way to add a layer, if you will, to a twice told tale.

In the retelling, they were able to garner more from each story, thus adding to their personal knowledge. This then could easily be passed on to others and applied to one's own life. The deep underlying message of each was the reason he was sought after by the colony. It was plain, no one could spin a yarn the same way as Mr. Mizithra. Talent simply rose to the top over time. He was the cream of the crop.

What follows are some of the tales. When he was done with a nugget of truth, he would always go back to his beloved to see how she was coming along. Mrs. Saffron was of major concern. She was bearing her first litter. She needed personal attention lavished upon her. Secondly, he lived with pain of the past, his first wife and children. The image of their death haunted him eternally and never completely left his thoughts.

When you witness trauma, it is embedded in your memory banks. How you deal with the horrific pictures is a personal thing. However, in handling them, it is a bad idea to do so in a destructive fashion. Human beings are the only living creatures who slowly kill themselves with vices. If you witness something which shakes you to the core, there are positive, successful ways to help overcome.

Chapter 13 Kai the Kestrel

On one of his very earliest travels away from the colony, just after his wife and young ones were killed, Mr. Mizithra went on a journey alone. This was before he had become acquainted with Murray. As a young three year old Olympic marmot, he was frisky, strong, and quizzical about the world. His parents taught him about mountain life in proper marmot fashion.

He knew his four whistles and how to seek cover. He learned early on about danger from above seeing a cousin get snagged by a gold eagle. He was never seen again. He did not heed warnings from others. The same was true of coyotes and bobcats. Anything which hunted him was well known, caution his middle name. The only solace gathered watching a chum die was his stupidity improved the whole. His mistakes could cost the others. Sadly, his death helped the survivors.

It was a sunny, late spring day and Mr. Mizithra climbed to a good high spot amongst an outcropping of large rocks. He wanted to see what he could see and adjudicate which direction to head. This was a time to wander and think. He was still in mourning and quite disillusioned with the world in general.

After the loss of his young family, nothing seemed to have much purpose. The feeling of the colony and family was he needed to get off alone to reflect and let his emotions go. Were he to remain at home he would be less prone to releasing inner pain. The solution was not only for his own good, but relieved them of the responsibility for his healing. They had done all they could. It was up to him to reason within himself. This was an action of love they knew was a bit hard.

The sun shone brightly and he was soon caught up in thought. He forgot about looking around and chose to simply enjoy the scenery. Not too often had he ever been so alone. There was not another marmot in sight. He found rather than being frightened being alone, he enjoyed the solace. He discovered he could clear his mind and think things through. Trauma does a lot to shake up the mind.

You never know how you will deal with something dramatic until it happens. A solid foundation and a good support net serve to make these times more able to be dealt with positively. In the long run, finally it is up to the individual to react to sadness confronted by the realities of life.

Wind whistled about him, tossing his outer fur about. He liked this activity and found his eyes beginning to shut. A nap was just the thing. He turned to find a good spot to rest. If he were lucky a nice hole would be there, abandoned, and ready. He saw a niche` in the biggest rocks. Sure enough, he spied a nice hole in which he could lay and get a few winks.

This bit of good fortune had him wearing a silly, lazy grin when out of nowhere a blur of brown plumes slashed in front of his face. He pulled back in instant reaction at the intrusion. He looked about to see what in the world it was.

In a flash, the brown blur of feathers assailed him. It was a bird, perhaps 25-30 centimeters long with a 55 centimeter wingspan. It was a small raptor, not able to do much to a mass like him. The entire weight of the little fella was 120 grams. The tail was not long. Mr. Mizithra was eight kilos. The mathematical differences were astronomical. What this mini monster was attempting was beyond reason.

All he wanted to do was get rid of this pest and nab a few zees. The miniscule assailant made one last pass. Mr. Mizithra snagged him out of midair by the tail. He hoisted him upright and held his other paw against the neck of the bird. It twisted and wriggled, then relaxed, resigned to the fate coming his way.

"What in the world are you doing? Unless you are altogether insane you must know a marmot is not the target of whatever you are little man."

"I am a kestrel and you are my prey. I must find food."

"Well, kestrel, I am too large for you. Please cease this foolishness. There is no way you were taught to hunt marmots for supper. You need a nice small mouse or one of those hideous bats which flap about in the evening sky. There is no way you could ever fly hard enough to get off the ground with me, much less any way you could kill me. Again I ask you why you are attacking? Be quick with the answer you wear my patience in hard fashion".

"My name is Kai, strength, and I am on a quest. The others, they have all killed and been successful. I alone am mocked. I alone am shunned. I have only been able to slay the easiest victims. Other kestrels say I am a poor excuse for a raptor and chase me away".

"So, were I to be able to bring in the biggest beast of all time, they would praise me and give respect. Thus, in spying you from above, with your plump rump, I thought it would make a perfect target. You are much larger up close. I apologize. Please let me go. This is painful".

Not wishing to injure this curious bird, in a moment of compassion, he let go. At the exact second, Kai chomped down on his front right paw and tore off a digit. Mr. Mizithra screamed out in pain and dropped his hold on the raptor. He grabbed his mitt with his left one and did his best to squeeze out the pain.

Blood spurted. He winced at this sudden turn of events. The front finger was gone, the kestrel carrying it off in his beak, squawking to beat the band with delight. He made a bead for a crest over the hill. In a moment's time, the sound of others of the same species screeching away at Kai's return with Mr. Mizithra's digit.

The beast was small, but quite a good looking fellow. When it comes to size, the kestrel is puny. Peregrine falcons are much larger. Once you get to the bald or golden eagle, the size is immense. One of those big boys could scoop up a marmot and whisk it off for dinner with no problem. While majestic, a bald eagle is not as good looking compared to the shear handsomeness of a male kestrel.

The head is a steely blue with stark, black eyes rimmed in bright yellow feathers. The beak is saffron at the nostrils, then black and razor sharp at the front. Polka dots of black adorn the front of this beauty whilst rusty brown feather dress the back. The head and neck are streaked in black. Feet are bright yellow, talons short, sharp, and vicious. Compact, picturesque, deadly spelled this little guy's credentials. A lady kestrel is sedate in appearance, but no less attractive.

Mr. Mizithra raised his right arm showing proof of his tale. There was indeed a digit gone. It did appear it was torn off by something very sharp. The scar left behind was violent in appearance. The adolescent ones huddled in fear. Their young minds soared with terrified imaginations. They vividly envisaged the scene painted. Each could almost feel the pain when a finger was ripped from his paw. This thought took hold of them. They shuddered in fear.

"While still in agony, I managed to find a way to enter the tiny hovel carved out from whomever was there before me. It was important to get the blood to stop, to keep it from infection. If you get an infection in your blood, it is very hard to halt. This is one of the fastest ways to die. If you are ever hurt and caused to bleed, get it stopped and cleaned. If not, then you will be pushing up lupines".

"After a bit, things calmed down. I got the paw washed in a nearby stream. Once back in the hole in the ground, sleep overcame me. The only problem with the place was I was bigger than whomever was there before me. My feet stuck out. The rest of me was able to curl up just fine. I made the best of it and got some sleep. Due to my injury, the thought of digging was not present. There was no way for me to dig a tunnel for the pain would be too great. Rest was best, so I slumbered away the evening".

"When the dawn's early light shone through, the first thing I felt was pain, only this time it was from my left foot. It was sharp and dire. It was debilitating. Reaching for my foot, I saw indeed a toe was missing. Again I heard the screech of that little so and so as he made off with another part of me."

"Once was bad enough, but twice was simply too much. He was surely an evil little man and no sense of honor. I made up my mind there and then to get even with this tiny monster. If I did not, he was going to eat me piece by piece".

"For several days, I hid in this cubby hole, eventually digging it out enough for a decent place to rest for one as big as me. This gave time to mend my wounds and recover. Food was handy as was water, so it was simply a matter of forbearance. Patience is an excellent attribute to add to your repertoire, my young charges. It is essential in life's pursuits. It is the long-suffering who are victorious over life's lengthy course".

"The kestrel flew past daily several times. Sometimes he would taunt me to come out. He was hungry for more marmot meat. A few times he brought other kestrels and shared his exploits loudly. A boastful fellow, it was a weakness I intended to exploit to my advantage. An arrogant adversary is easily defeated. Over confidence is a hindrance in battle."

"This enemy does not respect his opponent thus stripping himself of the vital weapon. Respect is not fear, it is wisdom. It is knowing your foe is seething with hate as are you. This demands an honest weighing of strengths and weaknesses on both sides. You must accept they will do anything to win. Controlling emotions allows one to see more clearly the challenges before them. This weapon I used to my advantage on Kai".

"Employing some imagination, I got into position. The water by the stream had a slow spot, a small pool good for getting a drink of water. I could see he was going to try and get my tail. It was one of the taunts he tossed at me repeatedly while circling overhead."

"He and his pals would try to poop on me as well and buzz me while I was above ground. While humiliating, it was a way to set the enemy at ease. A relaxed antagonist is easier to spring a surprise on than one who is vigilant. He was above me, two of his compadres with him."

"My back was turned, but I could see the pool's reflection. As Kai came full speed towards me, I stood still as if oblivious to what action was occurring behind. Precise timing was required and was performed. As he came to the end, his talons shone forth, beak open to chomp off my twig, my precious, stubby tail."

"Just as he prepared to pounce, I spun around, took a hold of his legs, and did a perfect marmot wrestling toss, shoving him under water. I put the full force of my body against his and won the position battle. It is only because we are such accomplished wrestlers I was able to make such a deft move."

"Though he struggled mightily, I did not relinquish my grasp. Soon, he was gone. I finally released my hold on his talons. He lay limp in the water, smug look drowned off his dead face. I held up my right paw, cursing him for his actions against me. Then, I turned my gaze towards the sky. Hoisting his lifeless corpse aloft, I shouted a warning for kestrels to beware Olympic marmots. I said this was the fate for any who crossed paths again. We would remember Kai and kill every kestrel seen."

"Young ones, the lesson is simple, but most important to learn. An enemy, no matter the size or attractive outer shell cannot be taken lightly. While Kai was much smaller than me, he inflicted lifelong pain. I am no longer a whole marmot because I did not respect his ability. Instead I disregarded him because of his size."

"I am permanently debilitated for my own arrogance and pride. Learn from this young ones, heed the lesson I teach. All about you above ground are they who wish to eat you. Birds, beasts, and man are out there looking for a tasty, fat, young marmot to roast and eat. Mind my lesson, learn the message, and beware the harm. Do this and you may just live to a ripe old age."

Chapter 14 The Four Whistles

This lesson began in the very oldest of times. Mr. Mizithra passed it along to his charges as any caring teacher would. This education was essential. It dealt with safety. After food and drink, refuge was the most important thing for the colony's needs. Each animal species has their own way. A bird can fly off, a fish can disappear in the depths of the water, and a rodent can hide in it's hole. Any time an animal, bird, fish, or insect is out and about in the wild world, a host of predators abound. Finding effective ways to be safe were taught for all. Such was the case with the four whistles.

The ancient tale was told as follows: in the earliest days, a strong Olympic marmot rose above his peers to lead a large, robust colony to the heights we enjoy today. Other existing families are a result of branches from the original tree planted by one named, Mr. Mountain. Even as male marmots go he was massive. He was in his fifth year and held a quick mind and strong will.

He guided his following away as native tribes began to hunt them in the lowlands. Mr. Mountain led a large fellowship of over fifty males and females. They took no elders. This was a clean break from the past. Mr. Mountain had approval from the five families from which he selected his new settlement group. They headed up Elwha River, past Windy Arm, Whiskey Bend, up river east of Mount Meany.

Each step took them higher in the mountains, further from their past. Crystal Peak rose to the west. They were in the neighborhood of Buckinghorse Creek. Up there the river is more to the liking of marmots. They do not swim. Here they found fallen trees across the narrows, so need not fear crossing water.

They had free range. It was remote. No man, few predators, and a hard place for a raptor to strike made this place ideal. They found a nice spot near the river, under a large stand of fir trees with thick bushes surrounding them in a clinging, ground hugging fashion. A good sized heap of large rocks to dig tunnels helped. It was a very nice point to find.

They agreed to this particular location. Digging began in earnest. Due to the immense size of this body, they had to have a larger layout underground. They dug several large halls with rooms off of each for individual families. They dug a series of entrances and escape tunnels. False passageways were created to fool predators. If they came in, they simply went a long, wrong way. Fleeing marmots could escape with ease.

They had an underground water source as well off the back of one of the immense rocks found above. It was like an iceberg where you can only see the top ten percent of it. Below, the ninety percent had water dripping constantly, forming a little pool.

BOOK 2: HIBERNATION

Come winter, the setting was excellent for their needs. Hibernation was very important. Food was plentiful, as well. It was agreed this was a consummate spot to establish a colony. Mr. Mountain was an excellent leader. The joy of the vibrant group was overflowing. They were very happy indeed.

Trouble began slowly, building rapidly. With a colony this size it was easy to miss the genesis. A young female disappeared one cool, early morning. She went to find some new greenery and heard of a lovely patch around the bend. She went alone to explore, with no need for fear. No one had been taken by a coyote or cougar. It was more than a day later and she never returned. Her mate was a nice fellow. He was heart wrenched, and went to seek her. He never came back. A search party found no trace once he cleared the corner.

Mr. Mountain laid his scent in the bushes on the corner to mark their territory. A few other males did so to show how many there were. Even if it were a large hunter, they felt the mass of them could ensure some form of safety for wives and children. It was a sense of honor and unity the marmots felt for this new, emerging effort.

They wanted to succeed. They felt Mr. Mountain was the best choice to lead. The loss of even one young lady was too much. The fact her husband was overcome with grief was not his fault. Theirs was deep love. They thought of ways to ensure safety of the whole.

The lost were a new couple, well trained by their parents. Not long after this, a young pair simply vanished. They were out near nightfall, romantically catching the last warm rays of the day, embracing. All who last saw them placed the duo at the tunnel's entrance, hugging and eyeing the evening sky.

After an extensive search they found no evidence. Four were gone. This concerned them. A meeting was called immediately. Another leading male spoke up, stating the loss of four with no evidence was too much. He wanted to go down the mountain. It seems a few others were with him. He wanted permission to go back to the old home with the quitters. Mr. Mountain was prepared, producing a letter from the elders given him as they were leaving. It stated clearly once these pilgrims departed up the Elwha River, they were no longer part of the old colony.

They surrendered any hope of reinstatement to their former family. Once they left up river, the choice was made. Their fate was their own. Love was expressed, but with firm conviction. The stand was eternal, no coming back would be allowed.

Some immediately recoiled from their choice and asked forgiveness. Rebellion against authority was not a marmot trait. Mr. Mountain understood the fear and desire to escape. He said, 'Here we are and here we will stay'. They focused solely on safety. Then another sad incident transpired. It was near evening time once again.

A few ladies were gathering violets and clover for supper. A swoosh came in the air. One lady was lifted up and out of sight. It was an owl. The tufts above evil, yellow eyes, massive size, all clearly markings of a Great Horned Owl. He flapped away with with his meal. They flashed down the tunnel to Mr. Mountain, and spilled what they had seen.

This was the evidence to confirm what most had suspected, a silent, deadly killer from above. It solved the immediate problem. They would keep from going out close to eventide. However, it did not cure the larger and scarier one, the safety of a colony. They took time and suggestions from members.

As a new settlement, everyone needed to be a part of decisions. They had Mr. Mountain as leader, but no elders. It was a colony of young couples. Experience was in short supply. Inspiration can come from anywhere if you open your eyes and look about.

The idea came from a young lady, Mrs. Violet and her husband, Mr. Rocky. In the large main hall, the whole body gathered for a huge party on some enchanted evening. They became separated and were soon at the far ends. She could see him, but could not attract his attention. They were an especially close, affectionate couple. Due to this, they had invented a special whistle of their own, low pitched, which descended lower still.

It was so unique, when she blew the sound he immediately turned, spotted her across the crowded room, and ran to her awaiting arms. As he did, Mr. Mountain's mouth dropped open in stunned amazement. Others came over and pulled at his shocked body.

Finally Miss Rainbow, got his attention. She urged them to assemble. Incidentally, Miss Rainbow later married Mr. Mountain forming the most powerful family in the hills. The counsel sat for the rest of the evening. The party was canceled, the rest sent to bed. They labored all night. By morning, the solution came which Olympic marmots use to this day.

Four whistles were the reply. This vocal quartet solved the problem. Since it's inception centuries ago, no changes have been made. All marmots subscribe to the identical form of safety. Any who try something different fail miserably. Due to the duo who inspired this safety system it was called: Violet Calls Rocky or VCR. Once implemented by the colony, no owl ever got them again. No coyote or bobcat, no bear or cougar was able to get more than one victim.

They knew complete safety was not feasible, but they could stop the mass deaths incurred from the talons of an owl. They recognized it could just as easily be a raccoon or eagle. It could have been a fox. Air was not the sole solution. They needed a complete safety net for the whole gang.

This outcome is still in practice in Mr. Mizithra's colony as well. Now, when one thinks of Olympic marmots, you must always keep in your mind they like to make racket. This is not some mute mob who sit around silent and solemn. On the contrary, they make the mountains ring with their shouting. They are noisemakers.

When wrestling, they nuzzle up to each other, hug, kiss, and frolic. They are a humorous, loving crew. As a result, vocally they growl, yip, and even chatter teeth. Humans who watch these rascals at play are always adorned with grins. With such a noisy throng, it was only natural they harness this talent and make it work for them. There was even fun in the solution. The results are as follows.

First, the FLAT call. This is a general whistle designed to let the rest of the colony know one's location. This is not a warning, merely letting the family know you are alright, at such and such a safe location. It is a flat, single note. It is low, softer than the rest. With a colony of more than fifty it was vital to have everybody accounted for at all times. This single, flat note is heard constantly. They tend to be a chatty species and enjoy staying in touch via voice. Remember, they do not have cell phones.

Second, the ASCENDING call. This was a distress call lasting about one half of a second. It begins with a "yip" and ends with a "chip". As the warning goes, the call rises in volume and scale. By the time it is done the voice is up the soprano chart. The nice thing about this one was it's ability to travel in the areas in these mountain peaks and valleys. The sender of the call would yell it in each direction, then pass it along. Ten or more could then warn the rest of the colony to seek safety.

This cut down on deaths greatly. It worked to perfection once they got the system in place. Whistles did the job early on by warning the appearance of a wily coyote. No one died. He tried to get to one fellow, but was effectively eluded via a well hidden escape tunnel under a nest of sisal bushes. He made a call, sent all over the place, then scurried down the burrow, out back, and off to safety, while the coyote dug a dead end.

Third, the DESCENDING call. This is a distress call, but goes down instead of up. This is mainly to warn against raptors. It was this call which Mrs. Violet made to Mr. Rocky. No one used a low tone which then went down the scale. Due to it's unique call, it was assigned to bird attackers. With it, they warned others without sound reaching descending raptors.

It cut effectiveness. The losses by birds went to an astoundingly low number. This whistle alone made the whole method worthwhile. It made Mrs. Violet and Mr. Rocky proud of their contribution. This twosome founded their own family later. Others eventually selected them as elders. Their offspring were strong, helping members of the family. It was from these two Mr. Mizithra's mother's family had it's roots. As stated in the very beginning, he came from a long, distinguished line.

Fourth, the TRILL call. This is a general warning, used on humans to scare them. The thought was the immense size combined with a dozen of fuzzy, yellow fanged creatures trilling like mad might scare off city folks with sunglasses, sunscreen, cameras, and Bermuda shorts. It's purpose was to seek safety. It was used by the young, who were not yet trained in more subtle and difficult warning whistles.

It was very effective when a gaze of raccoons hit. They only killed a single marmot, one Mr. Fog. The gaze came on him quite by surprise. Unable to get off any other sound, he blurted out a loud trill before they tore him to pieces. Due to his sacrifice, no others were slain. This call was known as a Fog. Thus, marmots could protect the whole.

Less were taken and harmed. Mr. Mountain and the rest were praised for solving the problem. It made him the first legend of the Olympics. As the colony expanded, some went further in the mountains and set up their own tribes. Each took with them the Four Whistles.

Mr. Mizithra taught them to the young ones daily, over and over for no other method of safety was found to be superior. This story would be retold by them as they became parents. The story would pass along the generations, benefiting them as it had their parents. The parents of Mr. Mizithra's students would reinforce his lesson. They took the young ones up after hibernation and practiced with them until it was second nature. This way the colony survived and thrived.

Chapter 15 On Meeting Murray

One of the stories Mr. Mizithra loved to tell was when he met his best friend, Murray the woodrat. On his way home after the bad incident with Kai the kestrel, he chanced upon a scene which was to have long lasting consequences. One of life's oddities is how individuals or episodes can come full circle in our lives and how those same circles of ours can intersect with the circles of others. Dizzying, isn't it?

This can have either positive or negative results. It proves no one lives only unto themselves. As we have seen, life is not something which lasts forever. One is here one moment, gone the next. That reality must be accepted to understand the true meaning of life. Mr. Mizithra faced facts without fear. To do otherwise would cause his demise.

Coming home after his sojourn of grief in the high mountains and wounding by Kai, he waddled back towards the colony. He knew he was going to have to continue life without his beloved wife, Mrs. Christmas Special and daughters, Miss Cake and Miss Cookie. He found himself daydreaming along a dirt road. It fed eventually to his colony. He did not like to follow man made roads. It was too easy to be run over by one of their monster machine thingies.

Still, he could see a shortcut. He had to follow the roadway for a small period, then turn to go downhill. In the tiny vale below he had to cross a rocky place to descend to his family. He knew were he to press hard, by nightfall the familiar landmarks of home would welcome him back from this injurious misadventure.

As he crested the hill and was about to turn from the road, he saw a shack off to the right. Curious, he crept ever closer. The place was clearly empty. No sign of man met his eyes or filled his nostrils. As he got a bit closer, a scene of terror unfolded. Below him was the unmistakable form of a raccoon in some vicious motion. He looked harder. Mr. Mizithra made out the shape of an old friend, Ernie the Pacific Northwest banana slug. The raccoon was attempting to get a hold of him, to make a meal of the slippery yellow fellow.

Ernie was slimed up to shun advances of the determined killer. He was a smallish raccoon, betraying his age. He was only about a kilo more than Mr. Mizithra, who was full grown. A raccoon can grow to over twenty-five kilos and kill a marmot with little problem. This was more like a teeny bopper, a juvenile delinquent attempting to be a big boar. He had Ernie in a hold with one hand and was trying in vain to get his mouth on some part of the slug. Due to slime, he could not close his teeth.

Mr. Mizithra spotted his friend and hastily waddled. He could get in between, allowing Ernie time to escape. He sped with all his marmot body could totter. A small brown figure emerged from over by the shack, darting at the much larger raccoon. Coming up from behind, the animal became visible as a male woodrat.

He flew full speed at the raccoon's back, launching at his head. He chomped the left ear of the beast who screeched in pain, releasing his slimy grasp on Ernie. To his slothful credit, the slug began an immediate slither for his hole in the side of the little knoll he called home. He left a slime trail behind as he wriggled for safety.

The woodrat leaped off and made for the shack. The raccoon turned, spotted the rodent in full retreat, and ran at full gallop for his new target. Woodrat a` la king would make a fine meal any time. The heck with the slug. He had a lead, running as fast as his little legs could carry him.

It was obvious the woodrat was going to be caught and killed before he could reach the safety of his hole under the left side of the shack. The raccoon's pace was quickly overtaking the new victim of his hunger. It was only a matter of a second or two and he would nab the interloper to his slug dinner.

"Not so fast, jerk", shouted a meddlesome Mr. Mizithra.

He flew in midair off a well placed large rock on the back of the raccoon. They tumbled. Mr. Mizithra came about first. He launched into the banded bandit with vigor, tossing him several times over and over with taut marmot moves. His colony's wrestling paid dividends more times than one could mention. At once, the woodrat was on the monster.

Suddenly, two against one. The raccoon screeched and rushed them. They made an aisle, chasing him up around the side of the shack onto the road. He ran pellmell with a wild scream of deformed anger being denied not once, but twice a meal. He did not handle defeat well, no, not at all.

It was at this precise moment, a tourist jeep rounded the corner going a bit too fast and careened over the side. The feeble driver jerked the wheel right, attempting to get back on the route instead of driving ahead and slowly merging onto it.

The result: the vehicle leaped in the air, slamming on the road in a violent and noisy ring of destruction. The driver turned to a straight line down the alley, slowing. The tire on the driver's side rear hit the raccoon in the side of the head. Stunned and bleeding, it went over the side, disappearing downhill. After the jeep went on by, they could hear the howls of the intruder echo through the canyons as it fled the scene.

Once the dust settled, a woodrat stood facing an Olympic marmot. Both were dirty, bloody, and battered. They stood frozen in time, staring at each other. Ernie was safely in his hole in the wall, recovering from the ordeal. The whole episode took only moments, but it saved a friend, got rid of a nasty murderer, and left these two to sort it out. They stood, stoic, checking out each other.

One wanted to size up the other. Bravery is not common. Coming to the defense of someone of another species was not ordinary. Sure, the vegetarians got along and covered each other with warnings if a predator showed an ugly mug. Otherwise, everyone pretty much watched out for their own kind. It was the way it was. For Murray to come to Ernie's aid was impressive to Mr. Mizithra and vice versa.

"My name is Mr. Mizithra. I am from the colony over the rocky outcropping past the vale below. If you do not know, I am an Olympic marmot. Thank you for your help with that vile raccoon. Ernie may be a Pacific Northwest banana slug, but he is a good friend."

"No vorries, ally. My name is Murray. My mother, grandmother und I live under ze ranger's shack. Ernie is also my friend. He loves ze music my family makes. I am happy to attack zat young nasty beast. He is called, Ralston. He is an upschtart, zat rancid raccoon."

"Even ze senior ones of his species do not care for him. He is bad news. You are ze first marmot I ever met. My, but you are rather galactic. It vould be my sincere hope ve could become friends. It is evident a large fellow such as you vould be of great help to vone as schmall as me."

With that, the two shook paws, remaining friends all of their days. The bonds of friendship were strong and built by mutual respect. They had fine adventures together. Murray had never married. Mr. Mizithra was a widower, so they had time on their hands. After meeting each other again several times the notion of going around Bear Town and the pond below became plans for the future. Next spring, he and Murray went their first exploration of the world.

As he finished, Mr. Mizithra had the exact smile he always had when he told it. The thoughts of those days gave him a warmth inside no one else could possibly understand. Friends are precious and should never be taken for granted. Friends are family not of the same blood. They can be more trustworthy and reliable than kin. These two held a strong tie which grew stronger with time. Mr. Mizithra met his family and Murray visited the colony. Their exploits were told and retold by Mr. Mizithra with utter delight.

Chapter 16 The Day Of Terror And Pain

The most difficult story to recant, with a tear in his eye and somber tone, concerned the death of his wife and daughters. It was an important lesson. It proved the need for a constant vigil and dire consequences evoked by the actions of others. Someone you do not even know can alter your life forever. This fateful possibility was such an instance. It came on the day of terror and pain.

Spring came early in his third year. A frisky, madcap Mr. Mizithra was ready to get out, see old friends, make new ones, and meet a female. The urge to procreate was strong. He felt a real need to meet a lovely young lady, woo her, marry, and start a family. It was the marmot way. He was no exception.

The day was a sun filled spectacular. Warm breezes wafted upward from lowlands where heat abounded. It melted the snow about them. It was still very early in the season above ground. Heat rays radiating off the white stuff gave wicked waves which distorted vision. It was fun and Mr. Mizithra was on full display.

First, along came Mr. Guar Gum, his dear, old chum, approaching in a menacing fashion. He surged at Mr. Mizithra with glaring, yellow teeth. He raised his arms overhead and leaped at his friend with glee. The two wrestled over the grass, down the snow, then to the bottom of the hill. Mr. Guar Gum let go his grip.

His pal rolled over, grabbed him, tossing him over to the snow. He slipped even further down the hill. Mr. Mizithra flew down the slope on top of his buddy. They wrestled around for a few more minutes, then lay beside each other completely spent and out of breath. Both laughed with complete abandon at the shear joy of the moment.

"Well, Mr. Guar Gum, you old so and so, how was Sleepy Time? I did not see you at any of the lessons. You think you know everything, hmm?", jabbed the ultimate student.

Mr. Mizithra was a bookworm, his parents teachers. He excelled in learning everything he could about his people and their ways. He was the star of the Four Whistle class. The history of their people was his favorite. Classes on safety were difficult, but he attended every one, so he might improve any weak spot in his life.

He knew the value of being a necessity to his colony. This fact was drilled in his head by other marmots all around: parent, teacher, leader, friend. It was important for each individual to take a part in society. No one could be a part of the group and not participate. Each member had a purpose. He took his seriously.

Mr. Guar Gum, by contrast did not share his zeal for learning. He was happy to learn enough with little commitment. He did not see the reason to learn as much as his childhood chum. The two were born in the same place. Their parents were close, so naturally became friends. Sadly, laxness showed. He was slovenly, never serious, and had few prospects. He was one it seemed destined to a life of selected mediocrity.

This was his choice. Mr. Mizithra loved his buddy, but did not share his lack of drive. Mr. Guar Gum gave him the nickname, "Big Galoot". Mr. Mizithra wanted to do his best to help the family he loved up on the hill overlooking Port Angeles. It gave him a sense of pride.

Mr. Guar Gum urged him to follow back up the hill to a spot where a bunch of other young marmots hung around. Like human teenagers, they tended to be awkward and shy. Males would try to stand out. Girls would giggle. Since Olympic marmot males are so much larger than females, it was common for boys to show off with feats of strength. For these fellows, as you already well know, it always came down to wrestling.

As the boys approached the teenagers, a pile of males were paired off, flipping each other, tossing to the side, or leaping in the air to fight. Girls were off to the side, enjoying the fracas. Hither and thither, a body would fly in air, tussled about by a hearty foe. It was dirty and hilarious.

Mr. Guar Gum saw a spot, nodded good-bye to his amigo, and dove in the fray. He happily vanished in the din. Mr. Mizithra looked for a likely foe. He saw a lone arm sticking out of the dusty atmosphere. He snagged it as a way of initiating a wrestling match.

Per the code, it did not matter, it was a big, old pile of fun. This was their way. Humans do not grapple as a way of friendship. No one goes out and wrestles their date. All I can say is: go with the flow and enjoy their culture. He flipped his unknown victim to the ground and dove at him. It was not a him. It was a her. Glint filled eyes hit him like a thunderbolt. He froze for an instant which was all she needed to flip him and to leap on his chest. She was smaller, but a powerful lady. He grabbed her by both feet and twisted her to the ground.

They stood, facing each other in mock anger, arms lifted to wrestle. Others were engaged in battle. The whole scene was alive with fun and childhood madness. It was a time for young ones to get to know each other. Soon most would find their special one.

Courtship for marmots is brief, weddings a time of sweetness. You are human, so a marmot is a cute, fuzzy little pile of fur. You cannot fully appreciate what happened to Mr. Mizithra when he got a full look at the young lady. He froze. As far as young lady Olympic marmots go she was a devastating beauty, a real knockout. Surprisingly, she did not leap at him, rather she stood as a statue, gazing at him with full open eyes.

As far as male Olympic marmots go, Mr. Mizithra was a handsome fellow. He had as of yet no scars to mar his mug. They were enraptured. While chaos raged, they stood alone in the crowd. It was sunny, peaceful, the air full of love. Without knowing her name, he was in love. Without so much as knowing his, so was she.

The moment was broken by none other than Mr. Guar Gum, who knew both, wrangling this get together so they might chance to entangle. They never met, but he knew both quite well and felt their personalities would mesh nicely. He was more right than he could have hoped.

As a matchmaker, he knocked it out of the park for a grand slam. He emerged from the dusty mess grinning from ear to ear at his apparent success. He grabbed each with by a big, meaty paw.

"Well, well, well, so you two finally get together. The whole colony has been waiting for this day. It's like everybody knew but the two of you. Mr. Mizithra, son of Mr. Gingersnap and Mrs. Cupcake, teachers at school, meet Miss Christmas Special, daughter of elder Mrs. Lip Gloss and late Mr. Gin and Tonic, whose memory honors the entire colony."

"She has been raised in a special area of the settlement and has not been around most of us. My father is her uncle, so we are family. Mr. Mizithra is my oldest friend and closest neighbor. Our folks have been close all our lives. Both of our mothers were raised together".

He stood smiling, turning his head back and forth over and over until he had to laugh. It was clear he was the third wheel. They needed to be alone. He nodded good-bye and bowed out. They did not notice him leave. Their eyes were locked in complete adoration. He led her to a promontory far from the others who disappeared to them all together.

For hours they gazed into each other's eyes, held paws, speaking in low, soft tones. Tossing safety measures aside, they sat alone falling deeply in love. By the end of the day, he went to her mother to ask for her hand. She was overwhelmed with joy as she knew his folks. He was an excellent student with great prospects. Approval was instant. He went home and told his parents who immediately got a hold of Mr. Moo Goo Gai Pan to perform the ceremony, which he did in splendid fashion.

With the advent of autumn, they hibernated together in a special room he dug off the main chamber of his parent's hole. In spring they would dig their own. For now, this worked. Besides, the new Mrs. Christmas Special and mother-in-law Mrs. Cupcake got along famously and wanted to be together as much as possible. Soon after the long nap during which they lay curled up as one in a warm, safe fur ball just for two, she found herself pregnant.

The joy the young couple experienced was contagious. They quickly shared it with his parents and her mother. The thought of becoming a grandmother was a moment of pure pride from Mrs. Lip Gloss. She lost her husband and son to a human being driving a truck without lights on, going too fast to stop. Both were run over, the driver not bothering to stop. Her hatred of the Thumbs was real and had an honest reason.

Come spring, two lovely daughters emerged with beaming new parents. They were, Miss Cake and Miss Cookie. Miss Cake was a few minutes older than her "younger" sister. They were a pair of nuts, to say the least. They had the vivacity and good looks of their mother, the curiosity and strength of their father. Both were healthy and inquisitive. They wanted to explore the world and find out absolutely everything.

All three grandparents quite naturally bragged and doted on the girls. No other young ladies were as sweet, smart, or demure as these two darlings. And woe betide any who argued the point with Grandma Mrs. Lip Gloss. She did not get to be a female elder, something not too common in a male dominated society by nature's design, being timid. She loved the girls, pouring her pent up love on them.

In the early days coming out of Sleepy Time, a lot of snow was still on the ground laying around them on the hills. They were up around 1800 meters, so snow melted slowly. It was always so much cooler than those living down on the waters of the Strait of Juan De Fuca. Snow in the lowlands did not stick around long. Marmots could not survive down there as was proven by history.

When Mr. Mountain took pilgrims up in the high places, he saved his species. Those left down below were soon decimated by predators and humans.

They disappeared, the remnants surviving by hiding in the lofty peaks of the Olympics.

Mr. Mizithra and Mrs. Christmas Special got together a picnic and led the girls to a little emerald isle in the midst of an ocean of snow. They could frolic about in a warm, grassy oasis and play in the white stuff. The day passed in lovely fashion. The sun was crisp and warm, food delicious, the family time memorable.

After a rest, they took to riding on their parent's stomachs as either mom or dad propped a pup up and slid down the embankment on their backs. They would run back up and do it again. After a bit, the girls wanted to rest, so Mr. Mizithra said he would take one more slide down, then they could head back to the colony. The fun was so pleasurable, he did not wish it to end. A bit of the kid remained in the adult marmot.

He slid down the slope with great delight, ending up with his head pointing downward, feet in the air. It was then he saw a sight which never left him. At the top of the hill was a blur of black fur. It was a female black bear with two cubs. Too late he caught the telltale scent of the monster. At once he heard shrieks of terror and pain from his wife and daughters.

He leaped to his paws, starting up the incline, slipping and sliding in wet snow. He finally reached the top. A scene of blood and carnage met his gaze. All three were dead, being consumed by an ebony trio of murder and death. With no thought of ever living another moment, he dove at the nearest bear cub, chomping down on it's nose with all the force his four teeth could muster.

The miniature monster squealed in pain, dropping the remnants of Miss Cake, grabbing at it's wounded black beak. The mother bear reacted with fear and surprise. She got hold of Mr. Mizithra, scratching his face deeply. The swipe took off a portion his left ear and sliced a gash above his left eye. She tossed him over the side down the snowy slope.

Blood trailed behind his limp body. He was unconscious. What happened after remained unknown. It was Mr. Guar Gum who found the bloody scene. The bears ate three females. Only bits of fur and blood lingered, staining white snow red. Mr. Mizithra was no where to be found. Later, he was discovered at the bottom of the hill by his father.

Four males retrieved his limp body. He was cold, his outer layer of fur blood soaked, clinging to him. His head and face were baptized in blood. The chunk of missing ear was seen to. The bleeding stopped once the area was cleaned. The gash over the left eye was another matter.

It took the help of some new friends who lived on the next stream over and had heard of the ordeal to mend his eye. An otter doctor, Mrs. Molly O'Shea O'Grady came, mending his wounds. She was well known, staying until he recovered.

A few patients came over to the colony. She tended to them. Any spot worked. Physically, he recovered, but inside was demolished. His parents saw to him as did Mrs. Lip Gloss, herself a total mess. Together, the three elder ones helped the young widower. His family gone, he had no motivation to continue. It was evident he was in a state of flux and could take dramatic measures even to the point of leaping from Miss Lily's Peak.

It was here a young, distraught Olympic marmot threw herself off to her death because the young male she wanted spurned her advances. Mr. Mizithra was down, but that did not cross his mind. He thought deeply of what he might have done differently. He realized there was nothing and this was simply life. It was the pain of death. It was the bloody sight of lifeless bodies being eaten before his eyes which haunted his mind.

Time passed. He found a way to get through an hour, a day, then a week, without breaking down. He went on his grief trip, meeting with the evil Kai, then his new and best friend Murray. Ernie, he already knew. Saving him from an attack by Ralston the raccoon made them better friends. He made an enemy out of the critter whacked in the head by a jeep tire.

Ever in his mind, always present for the rest of his days was this scene which went from heaven to hell. His physical scars healed, but the hole in his heart never did. It could not. If you lose someone close, especially if due to some form of violence you witness, the image stains your mind permanently. For someone to expect you to "get over" something so traumatic is unrealistic.

People do this because they do not know how to speak to one who has suffered a horrific event. It takes one endowed with compassion, experience, love, and an open heart to help. Mr. Mizithra did not find a special someone who could help him until Miss Saffron. She saw warmth buried deep within.

The pain of life placed a sort of shield over his heart. She was able to remove the impediment with a healthy assault of true love. He melted in her loving arms and impetuous ways.

This story always ended with a smile and a tear. The smile was for the love of his life and his daughters. It was also for his new love who bridged the past to the land of the living. The tear was for the pain and past terror and joy of what he now had. By not surrendering to pain, he overcame, found new friends, love, and a reason to live.

All in all he considered himself a very fortunate rodent. Those who heard this heartfelt story silently left him alone after for his voice always tapered off, simply muttering a few unintelligible words and weeping. Stories such as these are the best, dear friends.

Chapter 17 Ollie

The most enjoyable yarn spun: Ollie the Obtuse Osprey. Oh, what joy he had telling this whopper. Now, some was true. There was an osprey and he was not very successful at catching fish. The rest is either suspect or outright horse hockey. Old Mizzy would take out his palette of adverbs and adjectives with this snake oil saga and paint it as gaudy as he could. The story grew wilder with each telling.

It seems Mr. Mizithra met this osprey on his first foray in the wild with Murray, the year after his wife and children were slaughtered. Life is brutal for animals, so any form of humor was greatly appreciated, lauded for it's inherent gain. Everybody likes a good laugh. When it is at someone else's folly, so much the better.

The osprey is a raptor who only eats fish. Other than the peregrine falcon it's the most widespread species worldwide. They are magnificent birds, with devastating yellow eyes, sharp, hard curved, black beak, and a large body of 60 centimeters and 180 centimeter wingspan. Osprey are adorned in brown and off-white feathers.

They live in large aeries near water: a river, pond, or lake, returning annually to the same nest, which they build higher each year. One simply must improve one's home, don't you know. Offspring inherit it. Ancient civilizations wrote of and admired them from Romans to Buddhists. Their prowess at hunting fish was enough to impress dozens of various peoples around the planet.

Mr. Mizithra and Murray took their first trip, along side the pond, upstream to see Dr. Molly O'Shea O'Grady. He wanted to thank her and meet her husband. Though they had not met the fellow, everybody said he was a riot. It was impossible not to adore the otter. He was just one of those lovable chaps you are fortuitous to meet. They were lost and needed to find their way to the delta to the targeted stream.

Once they saw it empty into the tarn, they could make the way up to the home of the otters. With little more than vague directions, they were on shore, water lapping up on their toes. No other animals were found. It was useless to try and talk to a gnat, so they sat down and waited for someone to chance by.

Soon, a shadow came hovering above them, circling the water. The fellas ducked under some cattails and peered up in fear at the menace in the sky. Since the time with Kai, he learned to respect anything with a sharp beak. Their ability to strike from above at such breakneck speeds was daunting. Both feared the wrath from above.

This was especially true for little Murray. He made a snack for any raptor. Since it was up high, the sunshine made it nearly impossible to spy the predator. It took a while to spot. Once they did, it was clear he was only looking in the pond water. Nothing on shore seemed to be of interest. Now a bald eagle will go along doing the same thing, but will fly past any water and continue to hunt marmots, woodrats, carrion, or anything else it could shove in it's hungry beak.

Suddenly, the shadowy figure shot for water at lightning speed. As it neared liquid, twin talons shot forth, razor sharp, black as sin. The bird bolted back as it hit water. In this way, an osprey hits water, but not too deep. If they went down too far, death was very possible. Whatever was swimming lazily below this beast was doomed. Speed was a raptor's best weapon. Combined amazing eyesight, they present deadly consequences for anything in their purveyor.

A large splash and the bird hit water with deft velocity, hoisting immediately back out the water, wings flapping desperately to gain air. Sadly, no fish wriggled in it's talons. Nothing but water dripped off these deadly claws.

BOOK 2: HIBERNATION

It flapped skyward, then lit on a branch not far from the trailblazers. The boys were equally stunned.

"Say there, we would like to ask you a question if it is not too much trouble. Sorry you missed your fish. I am a marmot, this is my traveling companion, a modest woodrat. We were hoping you might be able to help us find a stream which leads up to some otter friends of ours. Do you know of any such a small delta which leads due south leaving this pond?"

"Thank you ever so much for noticing I missed the freaking fish. Go ahead and rub it in why don't you? It happens to me more than anyone else and is rather embarrassing. It is what you might call a "touchy" subject with me. I may well be an osprey and live on a steady diet of fish, but it does not mean I can't clobber a sassy, fat marmot."

"Get my drift? Good. My name is, Ollie. Now, as to the delta, it is just over there where the pond makes that little curve. Do you see the clump of trees there, yes, and the small mound beyond? Well, there the stream empties into the pond. It does form a delta. No salmon can ascend, but small fishes get up there all the time. Why is it you wish to go up stream? This seems an odd thing for rodents to do. What is the mystery?"

They noticed something rather unique, obtuse, and odd about Ollie. When looking straight at his face, his right eye pointed in a completely different direction than the left. It was clear to Mr. Mizithra why he missed, Ollie the Osprey was walleyed. With his vision so adversely effected, it was no wonder he was a fishing failure. If he could not get anything to eat, he would soon perish. They decided to see if they could help him and offered what they felt might be a solution.

They would go together up the watercourse, Mr. Mizithra to thank the doctors, Murray to have his injured leg examined, and perhaps the doctors to do something with Ollie's whacky eyeball. Three was better than two. He agreed to travel with them in hopes of being able to help his plight. It took several more attempts before he finally got a fat lake trout. Once he ate the trio headed uphill, due south.

He flew above, providing an aerial escort hitherto unknown. Mr. Mizithra and Murray feasted on greens and lilies around the pond. It was not until the ends for life. If you need an osprey, call for Ollie the Obtuse Osprey and I'll come flapping."

To this day, Ollie comes around the pond as it is a permanent part of his territory. Whenever they see each other it is always a time of fun and love. The trio was odd, but experience tied them together. Friendships are forged for a reason. It is a wise person who values associates family and friend. Not many will reach out to help. Anyone who does is someone upon whom you ought to heap great amounts of respect. As you live, certain people will rise to make life better for you. Wisdom comes when you learn to appreciate the rarity of such souls. Enjoy the good days, for they come by too seldom.

They learned a bird with a marmot with a woodrat being helped by an otter did not mean anything bad. It is amazing how narrow-minded people are. If you do not peer outside of your personal blinders you are limited to what you can learn. It is the one who is willing to read or be taught about a subject from a different angle who benefits most.

Learning from alternate sources can reinforce your knowledge, solidifying it as true. It can prove you need to learn more. Perhaps what you know is not the full truth, merely a portion of the whole. If you can learn everything about a topic and gain knowledge you will be a better person, more educated, and share with those who seek the same wisdom.

Learning from one source is merely indoctrination. The trio was better individually and as a whole due to the journey upstream. Each grew as a result and were better for mutual effort. This should teach the benefit of teamwork and friendship. Apply this and profit, but never forget the vision of Ollie the Obtuse Osprey splashing in water only to return fish-less. Humor is one thing which makes life worth living.

Chapter 18 The Folks

We all have kinfolk. In the case of Mr. Mizithra, his were simply the best. He always appreciated them. A loving mother and father made his life one of prosperity. He grew up learning the ways of his kind from two marmots who knew and taught the rest of the colony. The family on the ridge functioned together in certain key areas.

next time they came down this way they made it around to th
the pond and discovered the Beaver Blues Bungalow thanks to l

It took two days, but they finally located the otter lodge o
of the stream. For Murray it was a brief, swift, swim across of
or eight meters. For Mr. Mizithra, it was an effort in terror. It to
minutes to find a place he could ford the foaming, white water a1
The doctor emerged, introducing them to Graedy.

He was a hearty, boisterous, earnest fellow. They instantly 1
ment. After the initial greetings, they turned to the patient at har
fellow with a wall eye. Graedy had an extra minnow on hand and
air to Ollie who swallowed it down with zeal and a shower of tha1

Molly's advice: cover the good eye with a patch, forcing the
to look forward with exercise. It would take time, but it was po
could keep him in minnows with no worry. They were plentiful. I1
their home until the bad times. Mr. Mizithra and Murray had no
and wanted to visit the otters.

It gave time to look after Murray's bad left leg. An encou1
crafty fox left a severe injury. Graedy was able to do some work. Afte
pain, he straightened it. In time, it was better. The pain which he e1
daily abated, fully recovered. He found on stage it could hinder his
and would often ache after a performance.

After her prescribed time period, the patch was removed.
eyes lined up perfectly. For the first time he was able to see with 20/
Initially, it was hard to fly. He was used to the weird way he had alw
He flapped into an alder, dropping to the ground on the first try. Mr.
Murray, and Graedy had to hide to keep from Ollie seeing them.

They were dying of laughter. It was a humorous moment. The)
wish to disparage their chum, but it was funny. The second time, he
straight, flapped over to the small pond above the otter's place, and do
water. He came up with a nice sized trout. He flew back down to the ot
laid the rainbow at their feet.

"There is payment for services rendered. Thank you, Molly. I c
fish with the best of them. Thank you, Mr. Mizithra and Murray.

One of great value, were the classes by Mr. Gingersnap and Mrs. Cupcake. His mother's family established the settlement a hundred generations before with Mrs. Violet and Mr. Rocky. His father's line came from the legendary Mr. Pilsner, a raucous madcap fellow who lit up the Olympics with his chums. Their stories were epic.

When his father's family came to the colony, Mr. Gingersnap was recovering from destruction of their home by bears. The refugees made for Mrs. Cupcake's and were immediately accepted. It was not long after they met, fell in love, producing two young ones. Mr. Mizithra and his ill-fated sister, Miss Maple Syrup.

That winter, the snowfall was not enough to keep the colony safe from damage. Nature is unforgiving, coarse, and pure. The Olympic marmots in particular need a certain amount of snow or they face peril. It is the way they endure the difficulties of high elevation living. For us, snow is an impediment (or a great way to have a day home from school), but for the rodents up in the mountains, it is a necessity.

The snowfall was inadequate. The temperature did not sink enough to stop melting, causing a flood in the colony. Miss Maple Syrup was sleeping in her first hibernation when the walls collapsed. Only by shear luck did the rest of her family get out. The loss of this young one was a tragedy, but they knew it could have been worse.

Marmots learned what occurred when not taking every safeguard. Safety is more than looking out for the enemy and whistling. They discovered the need for education and set about to teach the others from their deadly error. They took pain and loss and turned it positive. It was named: The Miss Maple Syrup Academy in her honor. Other marmots sent their young to receive their philosophy. It was chocked full of common sense wisdom passed on over the generations.

They had another litter of young two years after Mr. Mizithra, then one more two years after when aged. Mr. Mizithra had six other siblings. He had little to do with them due to the events which formed his life. He was more of a loner, though bore no ill will towards them.

He loved the young, but was too absorbed in his own life. They had his folks which was more than sufficient to amass a wealth of knowledge. His travels gathered wisdom. He was smarter in many ways than any of the rest because of his leaving and eagerly learning as he explored.

He became the yarn spinner because of an innate talent he was born with, nurtured by his folks. They were marvelous marmots. They built him up, later showering their other six youngsters the same measure of love. All of their offspring were stalwart members of the family. Their pups saw Uncle Mr. Mizithra as special.

His stories and tales made their colony unique. No one else had a storyteller of such renown. What his parents knew poured into his empty head, filling it with knowledge. He churned it to wisdom by application. Few mistakes, quick learning, and wise choices helped him become a marmot of distinction.

Good parents produce good adults. This is true for birds, beasts, and human beings. You only have a brief period to impress your young charges before they go out in the cold, cruel world. Taking as much time as is necessary, disregarding your own needs, is what parenting is. You made the thing, now love it and take the best care of it you possibly can. Raising a child from birth to walking, to talking, to the terrible twos and beyond is pretty danged cool stuff.

Children left to their own devices will almost always find trouble as they did not know better. The best parents, like Mr. Gingersnap and Mrs. Cupcake, see duties in rearing offspring as a privilege, not a chore. It is an honorable responsibility. Their children were not a burden to be shed, satisfying only their own wants, but were seen as a blessing. The loss of their first daughter taught them this harsh lesson. The winter before he and Murray went up the White Mountain, his parents passed away in their sleep.

They were found in spring curled up in a furry ball of love. He was wrapped around his wife, caressing her, his face resting in between her ears. A look of serenity painted both faces. Their chamber was sealed and everyone said their good-byes. They loved each other, loved their pups, served the colony. Their recollection in generations to come evoked love. In life, we can aim for little better than to be remembered by those who knew us the best as individuals of character and love.

Chapter 19 Murray And The Fox

Before the journey up the White Mountain, Murray had an incident which left him marred for life. It was the source of both embarrassment and pain. He went out for a stroll which became a hike. Soon he was lost. The curse of daydreaming nailed him and death was very close, too close actually. Not paying attention to one's surroundings is very foolish.

It is like people whose eyes and attention focus on a tiny, glowing screen while they walk. Time after time we see people injure themselves because they simply walk off a cliff, into a lake, or traffic, phone in hand. Stupid is stupid, but this is ridiculous.

Murray lost himself in thought. This set in motion a near death episode. When he returned to find Mr. Mizithra, the story frightened him. Taking safety for granted is a foolish thing, deadly for a woodrat. There are no police or other safety measures available. If you are out in the woods, you are alone and have no backup to ensure safety.

This is one of the major reasons the colony of marmots was such a good idea. There is safety in numbers. Forty pairs of eyes are better than one. So, when an animal ventures forth as did our duo, worry about being secure was real. One false move, one error in judgment in the woods, and you were some predator's meal.

He went out one day, meandering about. Frema fretted about it, but her grandson said not to worry over him. He came across good, old Ernie. They chatted a while, then Murray headed down the road to the woods over the top of the hill. He wandered over the terrain in an effort to push up the bluff, take a rest, then watch the world pass by. He wanted to be alone to contemplate his navel and other vital matters.

It was simply a day of leisure, something no sensible creature would do. Still, he forged ahead, pausing here and there for a sniff of the air or a nibble on some crumble of food on the forest floor. He would stand up on his hind legs, too, and have a peek at where he was.

Sadly, his imagination got the best of him and, engrossed in thought, he was soon totally lost. He could not trace his steps. He had foolishly crossed a small creek several times, his scent gone. In a state of immediate panic, too late he realized the unmistakable scent of a fox filling his little rodent nostrils and heart with fear. He would make an instant and satisfying repast for the beastie.

A red fox can smell mice from 100 meters away. It was obvious he was in deep trouble. He knew his scent would carry. Due to the acute hearing ability of these attractive hunters he would be found if he were to so much as rustle a leaf. He froze in place for a moment. He had to figure a way out of this deadly circumstance.

Let's face it, coyotes are ugly. A bear is a monster. Cougars are simply terrifying. A raccoon looks like a bandit on the prowl. But when it comes to good looks for a bad dude, a fox is, well, foxy. It is no wonder in Russia they are attempting to domesticate them. They look sort of like a dog. Their size is nice, so it would be a real feat to make them tame animals. Humans have not domesticated a new species in thousands of years.

Fox possess incredible, envious, hearing, sight, and olfactory senses. They are compact, highly adaptable, and successful. It is true they are very attractive. The hides are prized amongst fashionistas of dead animal skins. Vixen are hunted for pelts. Wearing their dead carcasses is a sad remnant of the past whose time has quite thankfully been gone the way of the dodo bird, dinosaur, and the mullet.

There he stood, frozen in place like a mannequin, not permitting so much as a hair to move. Murray took his bearings. The stream bed was the closest thing to safety as far as he could think in this panicked moment. It was a windy spot where the waterway swished back and forth, fighting a difficult course. It was the only thing which provided a glimmer of hope.

He slowly moseyed to the brook's brink, looking for a particular thing. Spotting what he wanted, he moved very slowly, but purposefully to the spot. The musk of the fox, a male he now whiffed, came ever closer. He was being hunted and fear almost stopped him from functioning.

Murray inched ever closer to the water's edge. The sound of rushing water blurred the noise of his movement enough he was able to get in position. It took more time than he felt necessary to ensure completion of this goal. Nothing is worse than feeling the ticking of a clock as each second betides doom. He crawled deep, low across the creek to avoid detection. Fully doused, he finally made it to the far shore, slowly raising himself to the rocky shore.

The stones, rounded by the passage of water, grit, and time, were built into a small, cylindrical mound. The lump was the result of winter thaw. When ice and snow melt each spring, water eagerly seeks the easiest way to fulfill it's obligation to gravity. Once an adequate amount of the wet stuff has done the same, they form a trickle. When enough trickles get together in a cascading watery festival, they form a creek.

When you have a sufficient amount of these, you get a river, which leads to the ocean, evaporates and turns into clouds with rain and snow. This is a process which has gone on for more than a few weeks, too. It is what we call the weather.

He could smell the fox quite strongly now. It was only a matter of time before they would meet and he would die a horrible death at the paws of this proficient hunter. Let's face it, folks, no one wants to die a violent death and be eaten. This does not sound like a whole bunch of fun, yet is the fate of many animals in the wild. Murray knew of the skill of these rodent seekers.

Others succumbed to them in the past. He resigned himself to the inevitable scourge to come. The only chance was do precisely what he did. He could not outrun the thing. He could not defeat it in battle. A fox was over twenty times larger than he, so no chance. He could neither run or fight. He could attempt to hide, to so disguise himself the fox would be unable to locate, kill, and eat him.

Desperate times call for desperate measures and he took one. Once at the heap of rocks, he pushed his pointy nose in a small niche` between a few of them. There were hundreds of rocks piled on top and a number slipped off as he pushed ahead. The wetness and smooth shape of the stones made them deadly weighted demons. Were he to be hit by one, death was a distinct possibility. Dauntless, he forged ahead. He was struck by some tiny pebbles, but managed to push into a little hovel.

Slowly he turned, eventually facing front. Not a moment after he quieted down did an ominous shadow cast itself across the water onto the bank on the other shore. He was a massive adult male red fox. In these hills, they were eliminated years ago, but man in his infinite wisdom brought them back. Now the critters of the forest had yet another killer to face. Thanks a lot, Thumbs, way to go.

Vixen are excellent rodent hunters. It is their major source of food. As far as the reynard was concerned, this was the next meal in a series of past similar meals and ones he would have in the future. While to Murray this little woodrat was pretty danged special, to the fox: simply supper. After finishing a Murray meal, he would belch and say, "Next". It was no big deal for the hunter, but the hunted resented the heck out of it.

Murray did not dare breathe. The fox leaped down to the bank of the stream. He slowly scanned the horizon with his golden orange, narrow, sharp eyes. Possessing excellent vision, the slightest movement would meet his gaze and pursuit could begin. Due to the swim across the creek, the scent Murray emitted was missing.

The fox sniffed and smelled the air, to no avail. He knew a woodrat was close. The aroma of the potential dinner filled his nostrils. He raced over to the spot. He froze in place, head slowly turning. The fox paused, looked quickly around behind, locking his stare on the wet pile of rocks. Something attracted his attention. Murray's little heart beat like a drum in terror. Death was but moments away. He dare not so much as blink.

Just then, a loose stone slid off the top where the fox nudged it. The rock proved Newton's theory rather well, landing on the ground right in front of the predator. A dozen smaller peebles fell with the large one, causing a small landslide. Annoyed, the fox turned back to the scene, leaving the mound of rocks alone. He plopped down on the pile to rest his rump. Patience is a wise hunter's tool. He had a whole bunch of the stuff. He sat in the same spot for a long time.

Frustrated, after 15 minutes, he decided his quarry had escaped. He went back the way he came. Soon, the sound and smell of the fox was history, Quiet once again filled the forest. Only the babbling brook made noise. After a very long wait, Murray emerged from his rock of salvation. By burrowing in the stones, he found a way to survive being fodder for a fox. Animals do incredible things in order to endure. He had done what was successful. Given the unique circumstances, Murray had proven wise.

Hiding in this odd, rocky place, swimming in the icy alpine waters to shield his scent, and remaining entirely still for a long time did the trick. He was grateful and shaken. The ordeal set him back. He was a fool to lose his way, an utter and complete fool. A residual disaster lingered in the aftermath.

When the fox came and sat, he did so on a rather large, flat rock. The stone in question was nestled up firmly against Murray's upper left leg. The pressure point lay directly on his hip. The pain was so intense, he had to bite his lip to keep from screaming.

The weight of the fox, angle of the stone, and Murray's own position made the perfect storm which befell his limb. He could not budge. He could not let out a painful moan. He could not breath. With no idea what was going on underneath his own butt, the fox was injuring Murray for life. The woodrat was unable to move it at first.

He dragged his body over to a small tree. With all the effort he could assemble, Murray pulled up to a place to hide on the limb of a slender tree and rest. He was in profound trouble, far from home, lost, gravely injured. He could not walk. Were he to stay in this place long, any number of hunters would gather and find a way to get him.

A raptor from above or a bobcat from below could get him in no time. He had witnessed a zipping goshawk peel a mouse off a tree in it's talons as it whipped around the trunk. With no forewarning, it came, it saw, it conquered the mouse. The entire episode was an unforgettable memory. He knew this fate was a distinct possibility. He had to rest. Exhaustion got the better of him and soon was fast asleep.

He awoke. It was nighttime. He was hungry and chewed on a bit of wet moss. The pain was pounding. He could not mend it. There was no hole. No blood flowed. His hip was crushed under the weight of fox and stone. An injured woodrat would not last long. If he was not able to get home he would soon perish.

The only thing he could think was follow the morning sun east. He went back to sleep. In dawn's early light, he shinnied down the slender trunk for the creek. He got a long sip of water, then swam across, dragging his useless left leg. He rested against the pile of rocks for a bit, then over to the ledge above. It was both arduous and painful.

He found some food along the way, even snagging a bug in midair. The protein was appreciated. It took hours, but he was finally able to get to a recognizable place. The sights and smells were familiar. No sign was found of the fox. Progress was slow. After eight hours dragging his dead leg, he spotted the road. Not far from it, he soon got far enough to spy the home shack.

A truck, later a car zoomed by. He was able to avoid them. Human beings were a nuisance. It took over an hour after he spotted the cabin to be able to call to his mother for help. Murray was haggard, exhausted, tired, thirsty, and deeply wounded. Frema heard his frantic calls. She and Nediva emerged from their hole in the ground, helping him in the tunnel.

Due to his deteriorated condition, they dragged Murray to his bed. Nediva got him comfortable and provided some broth. He begged for food and was instantly given a water bug. It went down quickly as did the broth. More food was brought, finding a happy home in his empty belly.

The left leg was tended by the ladies, but they did not possess much in the way of medical skills. The best they could do was see to his dietary needs. As far as recovery, all they could do was help him hobble about inside the chambers under the shack. It was not safe to go up. He spent weeks down there in the shadows, trying to heal.

"Okay, dearie, you can go outside today, but I vill be right by your side. Lean on me, son, mama vill always help her baby boy."

It took several weeks, eventually he walked on his own. Murray had to focus. There was no room for lollygagging or laying about feeling sorry for himself. By the next spring when he and Mr. Mizithra went up the White Mountain Murray was in fine shape.

The day his marmot buddy showed for the next adventure, Mr. Mizithra found his pal stretched out on a pine cone. This stretching exercise, while it appeared to be a spot of leisure, was in reality a tool used to get his left hip loose. He had learned how to make it work and was now able to run. While he had to rest more often than before, it was better than the alternative: dinner for a fox.

Interestingly, Mr. Mizithra learned an engrossing fact when he met Momo. Murray's new bride was an expert in acupuncture. Over time his left hip was better by her shoving pine needles in exact places around the injury. By sitting on him in this precise position, the fox crushed blood vessels smashing muscle tissue around his hip bone.

He would lay with these needles sticking out of his hide. She would lovingly place each one where it was needed and stroke his sweet head. Traveling days were gone. All he had to do was walk a bit and strut on stage. This his hip could handle. Murray was a happy woodrat.

He outfoxed the fox. Ad-libbing out of immediate necessity, he found solace amongst the most primitive of sanctuaries, rock. By staying perfectly still and doing all he could to stave off death, he managed to survive. By shear determination and will he made it home to the safe bosom of family. Thanks to mother and grandma, he recovered. Thanks to his wife, he was able to improve to the point where pain was merely a twinge. He could live with it.

What lingered with Murray and the lesson Mr. Mizithra taught, it was practical. He had not taken proper precautions. Were it any other rat, no one would have heard the story. He was fortunate in location, blessed with the result, and resourceful in how he survived. He was, however, foolish for getting lost, forgetting the danger to dream. Dreams are great, but not to the point of discarding reality.

Safety rules lives. It does yours. You lock doors and windows, make your home, hearth, work, and car safe. It makes sense. You put on warm clothes when it is cold outside. When you do not protect yourself, you are open to harm. You can be killed for lacking awareness.

Not turning your back on the ocean because of sneaker waves, not hiding under the only oak tree during a lightning storm, you know common sense, are necessary bits of wisdom. Murray forgot and nearly paid with his life. Keep vigilant in your own life and be happier for it.

Chapter 20 Rayen And The Night Ninjas

This saga is one which Rayen the American Red Robin observed and shared with Mr. Mizithra one morning when he was out for a stroll. Rayen is a North American tribal name meaning Blossom, because her soft blue egg hatched in a thick mess of spring blossoms. Her mother doted on her and felt her a wonderful young bird. Rayen was as true to her species as any other red robin in history.

It was she who informed Mr. Mizithra the saying, 'the early bird catches the worm', referred to them. It was not from a sparrow or a flicker, a finch or a chickadee. No, it came from robins due to their habit of being the first to rise and hunt fat, succulent worms for breakfast. They are pure protein and ever so filling. A slurp down the beak, down the gullet, emptying in the gut. Yummy, she said, simply delicious.

The very thought of eating a slimy worm made Mr. Mizithra gag. It held absolutely no appeal. If there was a worm handy, she could have it to herself. He would not bother her about it, no way no how. Robins are a very common thrush. Their song is lovely and contagious. You can listen to them sing all day.

It is impossible not to love the little dark heads, white stripe around the eye, and telltale reddish-orange chest giving unequaled distinction. People welcome the sight of a redbreast. Robins are the harbingers of spring. They arrive both before and after a storm, which is how Mr. Mizithra met Rayen one rather blustery day.

The summer squall suddenly hit, swiftly, as they tend to do up in the Olympics. Our favorite rotund rodent was toddling along, seeking new stuff to discover when a churning mass of black, fast moving rain clouds hit. Precipitation beat down, no warning. The wind whipped, turning a previously lovely day into a living terror.

He spotted a small ledge with some thick Bracken ferns growing over it's edge, dry ground below. This was a perfect spot to hide out until the tempest passed. Finding shelter in a quick outburst was a boon, in more ways than as Mr. Mizithra soon discovered. He found the recess larger than it seemed. He spied an animal under the safety. It was a female robin, or hen, masked deep in the shadows sitting very, very still.

"Do not fear, Miss, I am no threat to you. I am Mr. Mizithra, an Olympic marmot out on a jaunt around the area. My home is up on the far ridge. The storm caught me unawares. Please, do not fear me. Relax and let us enjoy this respite together. I do not recall knowing or even meeting a robin, but your song is both familiar and very lovely."

"Thank you, kind sir. This is a quiet place of refuge. The storm was sudden, but the second it is over I must leave. The worms, you see, the worms come out and it is feeding time. While we wait, I must tell you about a most recent event which occurred not far from this very spot."

"It was first thing this spring when we robins ventured back up the mountain after returning from migration south. On a rather early, chilly morning, earlier than usual and I decided to be the very first bird out to snag a nice, fat worm for breakfast. The only problem was it was a bit too early. Trouble found me in a major way. I am lucky to be alive. It was terrifying."

When Rayen got up it was not quite yet light and visibility limited. She landed on the ground at the base of a large spruce tree to await the rising of the sun. It was in this early time the evening shift went to sleep for the day and the day shift emerged from safety to find sustenance and a new day in the wild wilderness.

She wanted to get a huge meal and prove to her round of robins her value to the group. Rayen was a sweet young hen, and Mr. Mizithra was deeply intrigued by the story. Within a few minutes she realized the error of getting too early a start. The lesson was not to test the barriers which separate animal life. Nasties come out at night and can easily make a meal out of a foolish robin.

As she sat under the relative safety of the tree trunk, with blinding speed five dark figures suddenly swooped down at her sides, surrounding Rayen. One of them clung to the bark of the tree directly above her. They were modest sized, covered in fur, but something was extremely peculiar about them. To Rayen, they looked kinda sorta like squirrels maybe? She was not entirely certain as her knowledge was not extensive.

"They had dark fur, a small head with large eyes and ears. The belly was lighter in color. But what was of puzzlement to me was from their front paw to their foot along each side was a fold of loose skin. It was attached at the wrist and ankle. Otherwise, I would say they looked just like some type of a squirrel, but unlike any I had ever seen before."

"We are the Gliding Ghosts, the Night Ninjas of this forest. We are small both in number and size, but make up for it in stealth and power. Proudly, we are known as Humboldt's Flying Squirrels. My name is, Nicolo, the Victory of the People. It is my task to not only form and train this group, but to find and slay the enemies of our kind.'

"Next to me is my sister, Allessandra. Her name means, Defend. When it comes to defense, with our gliding and fighting techniques, she is the best. She can soar a full fifty meters. Most of us can only muster about forty or forty-five, so she it incredible."

"You see, we maneuver in the air when we spread out gliding from tree to tree. Flying squirrels form a winged square, steering with our highly maneuverable wrists. The nice, fluffy tail is to slow down. Next, is my dear friend, Valentino. He is the Strong and Healthy one. He forces us to workout, to fly daily. He makes us test our wrists and tails to ensure proper attack abilities."

"Next, is good old Edoardo. The eldest of our quintet, his name means, Rich Guard. He has the best eyes, has the most stamina, and is our wealthy benefactor. As far as emotions and needs go, he is the hub of our wheel. Everybody goes to Edoardo for consultation."

"Last and definitely not least in anybody's mind, is the love of my life who refuses to become my wife, the darling, lovely, and most wise of us all, Sofia. Her name means Wisdom. Sofia is too wise to marry me. She solves our problems which are legion as we seek out trouble as a rule. They are not many Humboldt Flying Squirrels, but we are feared."

"The bobcat is a foe as is the owl, but our biggest source of concern is the raccoon. One in particular has been lurking around here for years. You may know him. His vile name is Ralston. I see the very mention gets a rise out of you. Well, he killed my father and mother. He and his minions are the scourge of the area".

With her words, Mr. Mizithra's jaw fell open in disbelief. After some time, he was able to extract where she met them and what time of day. She told him it was easier to do earlier in the morning, an hour before daylight. Rayen identified the tree and to wait for them to come to him. Simply being there the Night Ninjas would sense his presence and appear.

The rain ceased and she bid him good day. Off she went in a blur to the small open field a flap or two away. Within a moment her head jabbed the soft earth, parting the blades of wet, green grass extracting a big, old, red worm. It was consumed in an instant. She nodded back at her new marmot buddy and hopped over a few meters to see what else may be available as a meal for one hungry red robin.

He did as told. Sure enough he found the spruce and huddled up underneath it. The weather was chilly and moist. Dew clung to the soft grass causing moonlight to shimmer down, offering it's odd silvery gleam. He did not have a long wait. Nicolo glided up front, Edoardo plopped in position above, while the others took up defensive positions around him.

Mr. Mizithra explained who he was and of meeting Rayen. He told them of his knowledge of Ralston. They were flabbergasted to discover Mr. Mizithra was partially responsible for the injury to the raccoon's head. The blow apparently sent him to insanity. He swore to kill anything he could. They told him the raccoon specified a certain marmot and woodrat as targets in his ranting.

"Come to this tree at this hour, any time you need us and we will come to your aid. Trust this, my friend. The Gliding Ghosts soar about this forest. We are unafraid of those larger than us, for we have found with the Ninja arts a way to protect our numbers. We will keep an eye out for you. As you can see, our eyes are rather large", at this Nicolo gave a little laugh, "so it won't be difficult to see a fellow as large as you. Know this, sir, we are allies. Call us, we will come to your aid. You have our word on this from all of the Gliding Ghosts".

They hugged. The quintet scurried up the trunk and in a flash were gone. This new knowledge was fantastic. Having met Rayen, he was able to learn more about Ralston, who was becoming a real menace. He was a full grown male raccoon and could kill Mr. Mizithra with no trouble.

He accumulated some valuable warriors if needed. These vicious gliders were sincere. It was their goal to add more Humboldt Flying Squirrels to their unique home in the Olympics. As with the marmots, they had nowhere else to go. Imagine an army of these deadly rodents soaring into battle. Mr. Mizithra placed this new wisdom in his memory banks. One day, he knew, these allies would be of great assistance.

Once again humbled by the random events of life, in seeking shelter from a storm, he found a new friend. She led him to some more clues to his enemies. It provided two new sources of help. Rayen was always around in the hills. She would sing out and he could whistle back. This they did on several occasions, becoming fast friends.

As to the flying squirrels, well he was extremely impressed. Their silent stealth, unity in motion, their power gave him a real sense of awe. He had never seen something so small able to kick butt, but there the five of them stood. Allessandra was the one who frightened him the most. It was something in the glint in her eyes which shook him to the core. He was glad they were on his side.

The young ones loved this story and knew of the Night Ninjas who flew through the skies. None of them was allowed out at night because of owls and coyotes, so there was no way to meet up with them. Since they were a mystery, the tales about them grew taller and taller with time. Of course, our Mr. Mizithra was more than happy to embellish just a bit here and there about his encounter with them.

Hey, it's his deal, what?

Chapter 21 A Nice Slice Of Marmot Anyone?

The season before the trek to the White Mountain, our dynamic duo were off south on their journey to the Queets River country. It was a fine early summer's day. They were having the time of their lives. Away from family and the colony our two best buds were romping along through a huge open field brimming with abundance. The warming sun brought the world to life. Anything with a beating heart came out to meet the day.

The air was electrified. Butterflies fluttered by, colors providing the most spectacular panoply regalia against the green of trees and meadow. Barn swallows swooped, snagging fat, lazy gnats with gusto. All sorts of other birds, juncos, finches, jays, robins, etc. swarmed about filling the air with sound. Good times abounded.

They were on the tenth day away, drifting west of the stream where Graedy and Molly dwelt. Their lodge was a few kilometers away, but they were on the way to Queets and were in too much of a hurry. They would swing by to visit them on the way home. With only hundred days to accomplish goals, priorities were important lest they come up short. Mr. Mizithra had to fuel up for the impending hibernation.

He shoved fresh greens in his gob while waddling along. This trek did not have the urgency which cropped up the next year. The charge from Mrs. Lip Gloss was clear and pressing. They simply had to get away from the imposition of the Thumbs. This current trip was for exploration. It was a fantastic period high in the Olympics. Summer always brought beautiful green and sights to fill one's eyesight all the glorious day long.

They traversed a small hillock beyond the immense field. Once at it's edge, they peered down. It dropped a deep, sharp grade. There were gaps here or there where the earth between top and bottom eroded, leaving nothing but raw dirt behind. It was very difficult to get to the bottom. A small but swift creek blocked the path. Murray had little trouble making his way down, then swimming the stream, though he did drift quite a ways. The current was fleet, but he got out safely, snagging a bite to eat while awaiting his chum.

Mr. Mizithra had to find an alternate way. Murray's path was not something he could perform. He began to scan the horizon to his left and right, selecting the left as his best choice. The mound sloped many dozens of meters downstream. He could work the way across. There, water was deeper and barely moving. A sort of pool plopped itself comfortably on the spot. He felt even a fat marmot could make the other shore. A nice, little stony beach provided an easy entry to wet stuff for an unwilling swimmer.

Murray was busy chewing, glancing over the immediate area, while Mr. Mizithra made it downhill. His pace was more a sped up waddle. It looked odd and did not get him going much faster. It just made his marmot rump wiggle vigorously.

As this knoll descended, the path to water became clear. He would have to hit the rocks, then into water by the pool. It was a bit of a swim, not too much. The current was almost non-existent. He felt comfortable with his choice. He did not enjoy water sports, but when on the road exploring, one cannot be choosy. You must go with the flow. How droll.

With no forewarning, Mr. Mizithra was suddenly hit hard across his back. The unmistakable sensation of a slicing motion across the left side of his belly stung. Fur flew, blood flowed. He rolled over twice and began running as fast as he could without any idea of what was behind. He was not about to turn around either. The pain from the bleeding wound was as if his side were afire. He dare not turn to sneak a peek.

The unmistakable odor of coyote struck his nostrils. When it comes to enemies, coyotes are number one. They kill more marmots than any other critter, including man. This one was hard on his heels. Only with quick thinking he would be able to make it out alive. He shouted a trill or, as we know it, a Fog. The whistle reached his resting companion who instantly made for the edge of the stream to ford it and help. It was a good thing Mr. Mizithra taught him the four whistles as well.

The only thing in sight was a tree up the other side of the hill from where he was moving down by the water. It was a long shot for a coyote is

BOOK 2: HIBERNATION

much faster than a marmot. The difference became more apparent when Mr. Mizithra had to go upward. His short legs and chubby body did not bode well in a foot race with a coyote. It's long, lanky legs with a body built for speed easily overtakes prey. He could feel the striking of the paw pads smacking the ground with force as the hunter closed. It was going to have to be so madcap, even Mr. Mizithra could not visualize it.

On the fly, with no other thoughts or options, Mr. Mizithra did a complete head fake. He jerked his head right, simultaneously moving the body left. The noggin stayed aiming starboard, toward the water's edge: his goal. The body took a completely different line port, darting for the tree up top of the knoll. His brain holder turned back front and sped uphill. The surprising move caught the closing predator off guard. He fell into the stream.

Murray was on shore by now. As the coyote hit water, he launched a rock in the left eye of the bad guy. It did not hurt much, but the beast turned attention to the rat. Murray ran as he threw, diving in an old, small burrow just above the water's edge. He was entirely undetectable to the temporarily blinded coyote, who went back to his intended target.

Mr. Mizithra scurried without delay. He had to get to the tree. He spied a black hole in a space in the base of the tree. He quickly made out what appeared to be an escape tunnel on the far side. From what he could surmise, it was an old, abandoned marmot home. It was a small one, so perhaps a dozen or so dwelt here until something caused the colony to disband. Smaller settlements were common.

Not every marmot wanted to be in a large group. The small ones get by with less effort. This abandoned dwelling set up under an alder tree on the rim of a small ravine, a stream meandering on by. The brook swelled in spring, fizzling to a trickle by late autumn. Rain returns, replenishing the river bed. It was now easily crossed. Murray did what little he could. It bought the time allowing his buddy to squeeze in the old tunnel and wriggle in the dusky chamber to safety.

With vision in one eye blurry, faked into falling in the stony stream bed, the hunter was blood thirsty, full of pure, unadulterated rage. He was going to not only catch this chubby meal, but tear it to shreds for what he was put through. His pride hurt as well as his body. He could see the marmot's big, furry rump rambling uphill towards the tree.

Then he spied a hole in the base of the trunk and knew it was, at best, a place of temporary salvation. The coyote's nostrils and eyes filled with the smell and sight of blood, a good sign. Mr. Mizithra was bleeding out. If unable to get dinner one way, he could wait until the rodent died and get it another. It mattered little to the wily coyote. He waited for prey to croak before. He was content. It was easier than tromping in the underbrush trying to chase down a fleeing meal.

The coyote was impressed by the head fake. No other marmot had tried that move. He slowed to a trot, assessing area and target. It would be a matter of a couple of hours, time enough for the meat to tenderize. He could nab a nap. He had but to wait by the tunnel's entrance, an eye on the escape route. All would be well in his tummy come supper time. The slice to the side was deep enough to cause perpetual hemorrhaging.

He got one razor sharp nail of his left front paw in marmot flesh. He got a chomp of fur with his mouth before it rolled away. Thoughts turned to his aching eye. He could not be certain what happened. He hit wet pebbles after slipping due to Mr. Mizithra's madcap move. All the coyote could recall was sliding. As he stopped, a stone hit his left eye. The pain was atrocious. He ached from landing on hard rocks in the riverbed.

With no further thought of injuries sustained, the hunter made for his prey with renewed vigor. As a full grown male, he could bound up the incline with blinding speed. The pace set by the marmot would be easily overtaken, but the danged rodent got in the old tunnel before the kill succeeded. Seeing no other viable option, he laid down to wait, tease, and ask how the wounded warrior was doing. Food was at hand, all he had to do was bide his time. Supper would be fresh marmot tonight. Delicious.

"Marmot meat, marmot meat, oh how sweet to eat my tasty marmot meat," taunted the knavish coyote, "You are going to taste so good. While you lay there bleeding to death know it is I, Borya (the warrior, the fighter), king of the coyotes who dines on your flesh this evening. You will die, oh scrumptious, plump one. When you do, I will get you and feast on your flesh. Lay there. Chew on that while you bleed the precious fluid of life".

Mr. Mizithra knew his horrible speech was true. Were he to remain in this place for long, the gash in his side would bleed out. He would die. If he were able to escape, Murray could help him down stream to the doctor's lodge. Perhaps he could make it and they could save his life. It was his only hope. He had to find a way to make escape feasible.

He now moved about very slowly so as not to arouse the coyote's attention. He knew better than to respond to the jerk. It was clear the old escape tunnel was not going to work. Part of it was collapsed. A root had grown it's way down through the middle of the hole. The place had been empty for years, now dilapidated.

Then fortune smiled. He spied a glimmer of sunlight glistening off the shiny water of the stream below. The course of time was evident. The reason for evacuation was twofold. One, the root system was overtaking the mound as the tree grew in size and length. Two, erosion, over time along this side of the creek, cut a deep gash in the riverbank.

It left a good sized hole below the colony, roots dangling freely. They were not quite to the ground. Still, this was why the place was vacated. It was unlivable. The moss cloaked ridge would shrink enough with erosion the tree would topple in the stream as the circle of life churns along.

"Vell, vell, vell, how ze heck are you doin', Mizzy?", whispered a familiar voice.

Shocked, a fuzzy face poked itself out from the sunny basement. Murray shinnied up a big root and found his chum in a bad way. He came over quietly. He took a good look at the wound. Mr. Mizithra was in dire agony. Hemorrhaging lessened, yet unable to fully stop without attention from Dr. Molly's knowing hands.

He had water to offer, a kippah full. He had a few violets to munch. These eased some pain. Having Murray by his side gave great comfort to the injured rodent. The avenues of escape were minimal. Run, no. Were he to try, the coyote would eat in a flash. It came down to brain power. They had to out think a highly successful hunter. Together they hatched a wild, diabolical plan. Madcap marmot, anyone?

"Well, well, well, you stupid, mangy coyote. If, as *the king of coyotes*, ha-ha-ha, you are so wonderful and famous how is it a fat marmot could out waddle you? Uphill, too. You are not so great Borya. You pathetic excuse for a hunter. Here I sit all comfy and cozy in my warm little nest feasting on fresh clover", smacking an extremely loud bite of greens, chomping with decadent delight.

"Oh this is sooo good. Delicious. Yes, you hurt me with that slice. I did bleed, but ended hours ago. Now, while I am fat, full, and warm, you lay in the cold, hungry, and pitiful. I mock you, cur. You will never reach me. I can stay here for days, much longer than you can go without anything to eat. Phfft," he blurted in an insolent tone.

It was all the coyote could stand. He immediately dug in earnest in the tunnel. Dirt flew. He saw movement within and made for the other passageway. He anticipated the marmot's head would appear. He would grab it and twist for the thrill of the kill. Turning left, he darted toward the other opening. Suddenly, ground beneath him gave way, disappearing in a shower of loose dirt and rocks some twenty meters below.

The few roots still hanging were not a thing the coyote could bite. There was nothing he could grasp with his paws to stave off the inevitable fall. Daylight filled the gap. It was clear he would not be able to stop. He howled in all encompassing fear, plummeting to earth, landing with a solid thud on stones. He twitched a bit, then lay still. The hunter failed.

What the two did was clear. Mr. Mizithra was busy digging out the tunnel dirt above and beneath. Murray began to push it over the side of the opening to the ground below. With the soil moved, he climbed to the large roots, chomping them off so the coyote could not grip it with teeth.

With the dirt gone, roots as well, the trap was laid. All they had to do was implement their plan. Mr. Mizithra used the last strength to taunt the beast, then scurry down the burrow a few paces. He fell for it totally and died a frightening, painful death. Good, better he than we, thought our thankful marmot and woodrat.

They evaded death with quick thinking. This lesson taught the pups to never give up, use the imagination in a fix, and, at times, being madcap works. It taught the value of Murray as a friend. He was brave, innovative, and loyal. Mr. Mizithra could not imagine a better woodrat in all the Olympics. They survived by refusing to accept the inevitable outcome.

BOOK 2: HIBERNATION

Once in a place of safety, with time to think, they found a way to beat the odds. It was carried out to perfection. Sounds of the soft shower of earth and pebbles were masked by the noise of river rushing while the haughty hunter boldly taunted his prey. The young ones knew from the long, deep scar on his side and patch of missing fur, this story was true.

The coyote died. They hobbled downstream gingerly. They stopped often to let Mr. Mizithra rest and tend his wound. Food was abundant. It took time. After a trek of five kilometers and ten days, they arrived at the doctor's lodge. She immediately took the marmot to her surgical rooms. She worked on his side, assisted by two talented Douglas squirrels.

Graedy took one look at Murray limping and motioned for him to follow the chiropractor to his offices. With some tender manipulation, he relieved the aching pain, providing comfort for an exhausted rat. Mr. Mizithra's case was more severe. Once out of surgery, it was two full weeks before he was allowed to go home. The foray to Queets River was canceled. They took it easy, bypassing Bear Town, finally reaching home.

Mr. Guar Gum and Mr. Mizithra's younger four year old brother, Mr. Beef Jerky, helped him to bed. His two year old sister, Miss Sushi, tended to him for a few weeks. Food was brought in bales. He was fed all he could eat. After a long healing period, he was permitted to forage. The scar was large and deep. He had to bulk up enough for Sleepy Time or death was a possibility. On the other hand, were he to ignore recovery, the wound could get infected and kill within a day. Too much movement and the cut would bleed. He could not overdo it. Balance was required.

In time, he gained enough weight and formed a strong scab over the wound. Over hibernation, his body got back to normal. After emerging from this necessary rest period, he was thin and hale. The wound was no longer a bother. He felt great. He was ready to go for the quest for the White Mountain.

Once again finding his way to escape the wily coyote and death, he was eager to get his rat pal and seek fresh adventure. This journey had the added needs of his colony. This mission fulfilled, he now could rest at home in the Valley of the Stones. This big scar and new one on his rump from the red-tail hawk provided stories galore. His body was source to many a tale. You do not get to look like him without a story or two to tell.

Chapter 22 The Exception

Sayings which endure over time have a profound meaning, making their understanding easier to grasp. One saying is: there is an exception to every rule. This means most of the time something goes a certain way, but can, with rarity, successfully go the other. They are those who defy logic and keep the world on it's toes.

Take a bumblebee, for example. By the laws of physics it should not be able to fly due to it's wing to weight ratio. Some say the reason it flies is no one told them they can not. Such is the case of this story. Mr. Mizithra told this tale last. It had a profound effect on any who heard it. It was the story of the only known man to be accepted by the forest, the Exception.

Born south of the Olympics, he was an orphan. Raised with any number of foster families, he finally ended up with an older couple. The man was an old out-of-doors fellow. The one good thing he did was send the young man to the Boy Scouts, and every summer camp available.

The old man was getting rid of the kid. In truth, he could not have done a better deed were he to try for a million years. The youngster took to wilderness like a duck to water. He studied every subject possible. The nature section of the local library became a second home after the woods. The house in which he dwelt was existence. The couple who took him in made it plain: it was a money deal. Fine with both parties.

Human society rejected him from birth parents on down the line. He had little need for human contact. Few students in the various schools attended could recall his name. He was easily forgotten. Nothing of his appearance stood out, he was just sort of a guy in class. He was little more than filler in Thumb's scene of life. He was whom you overlooked in the background. He ditched the day photos were taken for the yearbook, took part in no clubs or sports. He never dated or went to a school dance. School was trying to get by and studying the world apart from man.

In fact, the only subjects in which he excelled were the sciences, biology in particular. There, he indulged a love of nature. Nothing else held appeal. He hiked alone weekends, exploring, finally choosing the Olympics as favorite. Once high school thankfully passed, he hitchhiked to Sequim. He moved to

Port Angeles, working seven years in the lumber mill. He made good money, spending only on needs for the future.

At a required funeral for a co-worker, his widow, kids, parents, family, and friends attended, naturally. The Exception came, sitting in the back of a packed funeral home. The deceased had a dog, who came along. Of all the people in attendance, the pup came and sat by him, resting next to the aisle. He had never met the dog, nor did he know the dead man. It was simply the dog knew who to come to when he was hurting.

This was who he was to the animal world. He lived in a modest rented room. He kept to himself, using days off to explore the hills and vales in the Olympic Peninsula. He knew the trails of the Hoh and Elwha. He hiked up Queets River, going up the east side outside of Liliwaup, and Hamma Hamma. He knew the south side for hiking out of Humptulips.

He took vacation high in the mountains, communing with Mother Nature. She was his only love. He respected and admired her with a true devotion. The woods would not betray him. This world was honest. It was not filled with greed drenched Thumbs who lied, stole, killed, and polluted in every way feasible. Animals kill to live, not merely for sport or waste time. He felt more at home up on Mt. Lena and Lena Lake than in the city. He joyfully avoided human interaction. Each foray in a green world provided more and more focus on his goal.

The arrogance of man is his clock, her calendar, as if they alone own time. An animal is more realistic, ruled by sun, moon, weather, and time of year. To the Exception, this made more sense. Life run by nature is more in flow, like a river. Man and his clock are a dam placed in the way. Their audacity is daylight savings time, hah. How ridiculous. Scratches on a piece of paper, and uneven arms of enslavement continuously winding around, bound to your wrist, rule the lemmings. These trappings animals know not. In nature, he rose with the sun, slept with the stars. Easy.

Classes were spelunking and mountain climbing. Survival in cold climes and foraging off the land were learned. He bought every conceivable book on subjects concerning the wild and how to live in it. After seven years being amongst a world he could not abide, he felt ready to leave it behind. It did not want him. He had no affinity for humans.

For some time he had been transporting supplies up to a safe spot. It was remote and would be difficult to locate. The contents would not interest animals, there was no food. He quit his job and was replaced the next day, his employment a forgotten vapor of time. He got a ride up Hurricane Ridge, camped overnight, then went to his hidden spot. It was a location discovered on an early journey in the forest.

For days he walked, enjoying every liberating step. The only sight of man was airplanes soaring thousands of meters above. It gave him pause to smile to think of how this activity loosened his bonds. He was not able to do as he wished without condemnation. He never fit in and knew the wild to be his real home. After many days he spotted landmarks to locate his stash. It would take a few trips to get everything moved in the digs.

He stayed a couple of days to ensure no one was around. For the next three days, he hauled the stash down a narrow cavern. It was deep, dark, silent. You could not see it from surrounding hills. He found it by falling off the trail. He rolled downhill, coming to rest wrapped around a sapling. Ouch. He peered over the rim in the chasm.

It was beautiful. A stream ran through. The remains of a large glacier lay on the north side. Above was a cliff up a thousand meters. Coming off it was mist. A weak stream emptied over the side. He could see a nice cave carved in the rock. On the stone laden shore he could spot places to lay out and catch the few hours of daily sunlight.

Fish were spied swimming in the pure water. He was able to locate the spot on a map. He formed a thought to live there as long as life would permit. This was his retirement plan. He went down on lines attached to stones with carabiners. Using this effectively, he explored this remote haven. Over time, he built a cabin within the cave. He set up stores.

Noting the walls lining the cavern, he set everything well above the high water mark to ensure dryness. When finished, he removed all traces of the storage bin. The final hike to his new home took several days, seeking total privacy. Hikers were all over in summer. He knew how inquisitive they tend to be. Exploring the wild is an amazing place to fill one's eyes with new sights. In order to avoid others, he took great precaution. The two times he saw others. He simply waved.

His dull clothing, baseball cap, casual manner, average appearance were utterly forgettable. He spoke little. When asked, he was out for a couple days of solo hiking. He was from Seattle. No one asked any more. He was positive nothing attracted attention. He kept calm and managed to avoid further intrusions. On the third day, he arrived. Confident after several hours of waiting to see if anyone was about, he descended the chasm walls. His only way out was the lone line, which he disguised with moss, and camouflaged the carabiner hammered in the rocks.

Over summer, he caught and smoked fish. He found wild carrots, pine nuts, and berries to can. He went to select hidden trees, fell and chopped them, stacking wood to dry. He kept a smokeless fire going. He learned how in outdoor survival class.

Weather turned, he was snug as a bug in a rug. Snow accumulated swiftly. He prepared. Any good scout would be. He had an area where snow easily swept away. Knowing no one would come for seven months, he was completely free of man, the sensation incredible. He wore a real smile for the first time in life. No other human being knew, or cared.

During the cold winter, animals do what they can do survive. The species which do not hibernate are forced to find solace, safety, and food where possible. Such is the case in the depths of winter, especially in high mountain country. Forgiveness is not available in the wild. As the saying goes, "Any port in a storm".

He awoke from happy slumber and instantly felt the presence of some thing. The cabin wall built in rock had a small window on each side of the crude door. He peered timidly out the right one. His view was nothing but wapiti. Dozens crowded both sides of the shoreline along the stream. They stood milling about, the fog of their breath thrust from of each pair of vibrant nostrils.

What they were doing there was a mystery. How they arrived was one, too. His stream came off the mountain in the form of a waterfall forty odd meters above the cavern. It would be impossible to come that way or jump in the chasm. It was a long way down, some thirty meters.

The only way they could come was downstream. He explored this region, but it was hazy and hazardous. Similarly, were he to make a foolish error and break his leg or ankle he would probably die. Caution was paramount in mind and deed, yet here stood over fifty elk.

Humans cannot speak to animals. Okay, some people talk to dogs and cats. I do. Some talk to horses, donkeys, goats. Chickens are favorites. Dogs respond for treats. Parrots can speak, but do not invent words. Dogs cannot speak. If they could, it would be the coolest thing in the world. That, and if they had thumbs. Think of the possibilities.

The man could not go out and ask the elk questions. There were so many he could not confront them. The only thing he could do was cower in the cabin and pray they would go away so he could get back to the whole happy hermit deal.

Then, the most incredible thing happened.

"You, in the cabin, human. Come out. We will not harm you", came a loud, firm male voice, "I am the leader of these wapiti. My name is K'wati, the Shapeshifter. We know you. You are: Man Who Is Not. We watched the entire time you came to our homeland".

"Other animals and birds told us about it. You see, my friend, in this region, you are under our jurisdiction. Wapiti rule this land. My family is from the far west, in rich Hoh country. The forests are thick and full of rain. Up here, it snows more and I led my gang here".

"Come out now and meet us. You have nothing to fear, human".

He emerged owning a feeling of comfort. He inherently trusted the voice of K'wati. Once outside, they introduced each other, then K'wati shared their new friend with the herd. Soon they were close. He traveled along side them. Wapiti are much larger than a man, but he was agile and healthy and could run for long periods. He ate meat, but never wapiti or deer. The wapiti explained some things which provided a much needed education. It was vital he know these things in the realm. Their rules were the law of the land.

Wapiti ruled the mountain lands. It was a loose collection of gangs. K'wati had over three hundred in his herd. Some were close in size, but his was largest. They roamed a large territory in high alpine peaks. When weather changed with winter, they moved to lower regions. This is one reason they came to him. It was cold enough to finally meet and reveal themselves. The shock wore off. He found not all wapiti could speak, so he spent most of his time with K'wati learning the ways of the mountains.

Deer are cousins, but do not travel like wapiti. They are, well, not so bright. You have heard the saying, 'A deer caught in the headlights'? Yes, well, they do not say an elk, do they? Enough said. Wapiti had a hands off approach to the wild. In other words, let predators prey. They did not interfere in other lives.

On rare, very rare occasions intercessions were made, but not as one might imagine. It is difficult enough to survive here without some elk telling you what to do. They did provide one very special service for everybody else. In quiet, secret ways, they did their level best to stave off the intrusion of man.

This is why speaking to the first man since early days, eons of years before, was such a big deal. It was the proper choice, though. He turned out to be a fine substitute wapiti even without antlers. He ran with the gang, learning how they moved in the wild. He saw things from their perspective.

More and more, civilization faded. He was happy, truly happy for the first time in his life. The method of man was not his. He could not see working in a lumber mill or an office or a store. Only in the out-of-doors did he gain a sense of comfort within. Oddly, he only fit in nature. It was home. He lives somewhere in these hills to this very day.

When he finished this tale, Mr. Mizithra would sit in still silence for a time. Then he would explain the value of this story. When man became the Thumbs the separation was vast. The space between was so far not even domestication could bridge the divide.

The pride of man will not allow surrendering of power. This rarest man did, however. He was not like anyone in history who went in the mountains, except the most ancient tribal members. He was first because his heart was pure. He was in spirit not man, but animal.

Mr. Mizithra let them absorb this story in silence.

Once lessons ended, he returned to his sleep chamber and loving wife. Each day he told one. Kids would go home to discuss the tale with parents. In turn, the folks helped their young answer follow up questions. This was planned by the elders before each winter.

He shared with the young. Time to sleep for a few more months and enjoy the fruits of marriage. Having wedded before, he missed the true joy of a wife. Love is pretty neat stuff, you know? He and Mrs. Saffron, tucked away in their love bucket, fell deeply in love. They emerged in spring, two fine young ones accompanying them in their first taste of sunshine.

The reason for stories was to benefit the colony. He truly reveled in telling them. It was his passion. Each journey into the unknown provided applicable tales to be better Olympic marmots and strengthen the body as a whole. Mrs. Lip Gloss assured him he was carrying on his parent's legacy well. Yes, it was his own unique way, but no one could reach the young as well. Mr. Mizithra had a charm about him no other marmot possessed. He used it to benefit others, not merely himself.

Chapter 23 Springtime

This was a long Sleepy Time. Snow began to melt. Older marmots knew it was time to come out of darkness into the light. Here was one necessity the young needed to learn. WHEN to emerge was as important as when to hibernate. The cycle of life was controlled by weather. In these high alpine slopes, cold and snow dictate everything.

If it did not warm up for a few extra weeks, all the more important to start eating. However, if snow began melting early in the season, they have more time above, which has it's own problems. The more time above ground, the increased chance a hunter will eat marmot for dinner.

Elders knew from knowledge passed on by their seniors. Knowing precisely when to egress from wintertime homes was sagacious wisdom from the past. Mrs. Lip Gloss stuck her well tuned nose up to the end of the tunnel and pronounced it time to exit to the sunny world and enjoy the fruits of nature. It was the first time these babes saw the world. Everything was new.

Mr. Mizithra had double duty. He had stories to tell, but the additional responsibility of his own wee ones. Mrs. Saffron was exhausted from feeding, passing child care on to her willing husband. They operated very well together. He enjoyed fatherhood to the utmost.

On the day they sprang from winter chambers, he took his son, Mr. Murray, and daughter, Miss Paprika, by the paws. He led them to the open field in their birth home, the Valley of the Stones. They were the first generation born here. These winter babies would build the new marmot colony as models of love, safety, and peace. His son was named after his best friend. This was a cause for alarm.

It was simply not done. Perhaps a daughter was named after grandma, but to be named after a woodrat was uncommon. However, since Murray was instrumental in finding their new home, was a boon to the colony, and saved Mr. Mizithra's life, it was fine. His name evoked the best memories. The son loved it and tales of the rat behind it.

Children ate their first greens. It was time to be weaned from mother, who needed to eat now. She took time, but eventually came up for a lovely meal of violets and clover. The kids were off with dad, sliding around on snow and finding new things to eat. They were fascinated with birds flitting about. Butterflies made them whistle with delight. She could see they were in good hands. Mr. Mizithra certainly did enjoy being a papa.

He was a good marmot. She was pleased with her choice. Their first winter together was steeped in love. She was happy with pregnancy and loved her two young pups. They gave her an easy smile. To see them with her husband playing and eating, enjoying the late spring sun filled her heart with joy. This new home in the Valley of the Stones was ideal.

It was the perfect place for a settlement of marmots. While the kids chomped away at a lovely patch of lilies, Mr. Mizithra received a tap from behind on his right shoulder. This caught him by alarms. He was totally wrapped up in the frolicking his children were enjoying. He turned right only to be grabbed by the loose hanging left arm and flipped to the ground, his assailant landing on top of him.

"You are getting lazy, old man. Do you want your kids to see you turn into an old softie, Mizzy?", shouted Mr. Milk Chocolate from his position of superiority.

A quick grasp of the right leg by his right arm and a tug, Mr. Milk Chocolate was on the ground with Mr. Mizithra on top, his arm firmly planted across his foe's throat. While still weak and groggy from nine months underground, he was a force. This fellow was not going to get the best of him.

They stood facing each other in mock anger, locked arms, then wrestled for the better part of five minutes before hugging and falling on the ground beside each other in laughter. These two chums missed seeing each other. They spent Sleepy Time in different colonies. Mr. Milk Chocolate did not have children as Mrs. Mascara did not conceive this first time, which was ordinary.

He introduced Mr. Milk Chocolate to his children. He got a kick out of having a son named after Murray. Mr. Milk Chocolate knew him better than any of the other marmots. When they lived in the old colony, he did not visit. Mr. Mizithra came up to his place. Murray felt intimidated by so many rodents bigger than he. Some of the young ones would pull his tail. Indignant, he did not care to be around.

Murray would much rather be around grown-up rodents than have little ones around his ankles. After a few hours getting used to the outside world, Mr. Mizithra herded the kids back down the tunnel. He helped his pooped wife to their chamber. He brought her some violet tea on a tray with an assortment of greens. She needed rest. Weaning would take a few days, so the young ones suckled mother. Mr. Mizithra stood in the entrance to the chamber admiring the scene.

His smile was one of gratitude and satisfaction. The love these two formed was profitable for both because it was honest. Neither bore some hidden agenda. In this spirit of goodness, two young ones sprung forth. He was one happy Olympic marmot. He had what he always wanted. The past and it's pain were pushed far to his background. He only had time for today, for the here and now.

Mrs. Saffron was a blessing he would never take for granted nor would she for him.

He watched as she fed them, humbly bowed his head and smiled.

BOOK 3: WAR

Chapters

1. So It Begins 276
2. An Interesting Chat 279
3. A Time For Every Purpose 285
4. What Murray Learned 288
5. Mr. Mizithra Gets Hip 293
6. An Evil Discovery 297
7. What The Skunks Found 302
8. Retaliation 307
9. Murray Learns More 314
10. Mr. Milk Chocolate's Idea 319
11. The Reinforcements Move 321
12. The Battle Begins 326
13. Surprise 332
14. Igor 335
15. The Unthinkable 341
16. Ralston Is Found 346
17. Uninvited Guests 351
18. Interruption 356
19. War Is Over (if you want it) 360
20. Party Time 363
21. Back Home Again 371

BOOK 3: WAR

Chapter 1 So It Begins

Bad things have humble beginnings. It came to pass like collapsing a line of dominoes. Everything fell perfectly in place or it would not work. Sadly, it did. What transpired was not planned or desired by the attacked. Survivors who fought and lived to tell the tale were forever changed. War is not good. It is not pretty. It should never be desired.

Warfare is ugly, evil, kills without discrimination. No one says when over it was a good battle. Many innocents are slaughtered by military action. War is good for absolutely nothing. Hostilities have been waged by animals and man forever. Cain and Abel come to mind. Watch a Cape buffalo in South Africa fight off a lion. Whoa.

Come spring, hibernators emerged from warm chambers into fresh air and sunlight. Food was of instant concern. Those who did not hole up had newbies to contend with for sustenance. They got reacquainted. Miss Bluesberry came to the bungalow to see what was what. Mr. Loganberry of the Hills went to see his skunk family to inform them of his new bride. He wanted them to meet the lady.

He had four brothers, so there were a whole lot of skunks around. A group of polecats is called a surfeit. Were they to get in the Beaver Blues Bungalow, well, it would raise a stink. Upon entering, she sat in on a jam on stage. Murray vaporized from the hazy background at her first notes. He sat next to her without a word. He pulled out a harmonica. He began to blow a painful dirge. She was blown away. It was incredible. Murray really delved into the harmonica.

They played, then took a break. He told her about Carnival Island and the happenings. His marriage shocked. Once she met Momo, it was obvious this was a perfect match. She gave him balance. He was truly happy. He found she was as well. Mr. Loganberry of the Hills was the best mate. He adored her talent. He encouraged the singing career. He had no problem with crows managing her. Miss Bluesberry's life was now a very joyful one. She owed it all to Mr. Mizithra and his open mind.

The rest of the beaver gang were pleased to see their favorite skunk. Twigs gave her a massive hug, but held his breath just in case. Bette's loving looks told it all. Time formed a strong bond between these ladies. Buford and Beulah brought her a huge plate of food and a bowl of sweet clover water. Once sated she made for the stage again when the door swung open.

Through the entrance fell a tattered crow. It was Mon. His feathers were shredded, beak bleeding. One eye was shut, torn apart. His left wing hung limp at his side. He fell on his back. Those within rushed to his side. Too late, they found Mon lay dead. Tobias darted outside. In moments he returned carrying in his arms the limp carcass of Tutti Frutti, both talent agents gone. Waves of shock reverberated. Murray flew to his room to check on his wife and family. Dragonflies were dispersed to important animals to declare what occurred. Homer and Henry went out to ask if anybody had seen anything.

For crows to be murdered was not uncommon. These two were well known and had good reputations. Crows are celebrated crooks, often harshly chastised for cheating. However, they were mere music lovers and did no harm. WHY they were slain was now a mystery in need of solving.

A group of ravens is an "unkindness", which they proved to be. It was discovered more than a dozen hunted down and pecked the crows to death. The reason was not given. News provided was Tutti Frutti and Mon interfered in raven business. The spokeswoman was Lady Satin, a harsh one. Her words deeming them criminals were vicious and demeaning.

Charges were crows infringed their territory. They were in violation of realm rules, warned multiple times. The justice was a death sentence. The questionable verdict rendered, this unkindness was perpetrated by a bevy of big black birds. The unrelenting assault proved fatal. Their deaths were sad. There was no way to prove innocence. If you intrude, you die.

Within an hour, none other than Lloyd Joseph flew from the Elwha with a contingent of crows. This was not a "murder", it was a massacre. Dozens draped trees encompassing the club, cawing incessantly. Aged Lloyd Joseph demanded to see the bodies. He examined them with their physician, Dr. Marley. He concluded both showed signs of raven attacks. The telltale clues were there. The doctor assured Lloyd Joseph it was so.

The sawbones added his disgust. The boys, Mon's family was related to his. He then spoke low with his ancient benefactor. Lloyd Joseph had well-earned respect. He was older than any crow, the true definition of honorable.

He did not hold with modern ways. Tutti Frutti and Mon held the past while embracing the now. He knew something was rotten in the state of Denmark. He would find out.

The bodies were taken by a mass of black feathers. Each corpse had four brethren carry them to burial. Crow funerals are the most somber in all nature. They do not take death lightly in the avian world. It is a sacred time when birds of their number passes. Roje and Agwe attended. Lloyd Joseph called them to his home in Whiskey Bend. This was not a request.

If you were to listen to and heed the words of any crow, it would be this elderly one. He knew something was afoot. He wanted to glean any info they had and warn about what happened to their trainers. Roje and Agwe were not ready to take over the reins of discovering new talent, but could muddle by until improved. After the service, they accompanied Lloyd Joseph and his minions down river to his mythical lair.

In the aftermath, the Beaver Blues Bungalow held a vigil for their fallen comrades. It was a shock to the system. Seeing friends in such a tattered, bloody, beaten condition was horrific. To witness his death was devastating. The immediate reaction was a heightened sense of security.

Face it, while a crow was more than welcome in their midst, a raven was not. They are brutes, much larger and more audacious than the smaller, more plentiful crow. Size is not everything. A crow has more intelligence than the bigger breed. While no one knew precisely what happened, they could not imagine anything Tutti Frutti and Mon did to prompt death. They were not that kind of crows.

Time passed, a few weeks, and life calmed. Roje and Agwe returned from Lloyd Joseph's. They trimmed a lighter tour for Murray. He would not return to Carnival Island per his wife's request. He wanted to go back to Waadah Island and clubs around the area.

This meant none of the other stops around Willapa Bay would get to enjoy his sax. They had to understand, he was a new husband. He needed a year to experience marriage before running back to the road. While Momo loved his musical talents, she needed him home. They had a new marriage to establish. It takes two to tango in a relationship that serious.

The new scouts agreed. Preparations were made for Waadah Island. Paisley and Ness specifically requested a whole week. He negotiated five days. They would return home at the end of the gig. After Club Club, he could hit the mountain circuit. He would wind up at the huge Grand Meadow Festival with Miss Bluesberry.

This suited he and Momo fine. She could attend the other shows. This would be their only separation. He wanted four days. They wanted a week. His compromise was five. Time apart was painful. They were very much in love. She had Nediva and Frema to comfort her. Hey, they could teach her how to sit in the corner and eat moldy potato skins. Alright!

The trumpeter swans arrived, a welcomed sight. Murray never really got to thank them for saving his life back on Carnival Island. They waved it off, urging him to climb aboard. He said good-bye to his family and band-mates. Donning his new kippah, armed with two saxophones and three harmonicas, "Sugar" Murray was on his way to once again shine on stage. Ah, the life of a professional musician. His new title, earned during last year's travels, suited him well.

Murray was a sweetie pie, a honey of a rat. He was beloved. He had good friends, an excellent wife, and loving family. Murray smiled in humble gratitude for his lot in life. Considering last year began at home under the log pile next to the ranger's shack, he had come a very long way. As swans soared in the azure sky, Murray sank back in Theodor's down and reveled in his good fortune. The flight would take two hours, so he closed his eyes with a happy smile of reflection and joy.

Chapter 2 An Interesting Chat

Too soon Murray felt them quickly descending. This was unusual, so he poked a fuzzy rodent noggin out of the comfortable, warm down. He saw a river. It was the Elwha. They aimed for a water pool. They landed, paddling ashore under a rangy stand of cedar. The tree was massive and very old. It bore dozens of smaller trunks jutting skyward, the tallest over thirty meters. The area was a semi-circle.

Where the Elwha calmed, lay a lilliputian, grassy knoll. It rose a half a meter from the wet stuff, over seven meters long by two meters wide. The shape resembled a crescent moon. The cedar loomed large over the pool's north end. This spot is in Geyser Valley, near where Haggerty Creek joins the Elwha. The river continued on the west side, growing swifter once past this oasis on the trek downstream to the Strait of Juan De Fuca.

Puzzled, Murray sat upright as the trumpeters trumpeted. The octet approached. A murder of crows appeared, encompassing the shoreline. Terese told Murray to get off. They would wait. The swans milled about in laggard water, feeding. The continent of crows led him along a small path. They waddled, cawing all the while. Not one spoke. They were solemn.

It was much warmer down here in late spring. The humidity was daunting, adding to the drama. As they neared the massive cedar trunk, a few flapped into the dank branches, disappearing in darkness. He was led to a small break in the lower limbs. A narrow pathway upward appeared. He ascended, winding around the ancient cedar. A place where a massive limb broke off was whittled into a platform.

Crows split, the final feathered fellow lifting his left wing, waving Murray to a branch seat on the end of the dais. He sat alone, bewildered. No one gave him any information. He was sitting on a cedar instead of wending his way north to the gig. While a rough and tumble woodrat, this mystery had him unnerved.

Swans were friends, well trusted in the animal kingdom. He mused within his own mind what was happening. The day was warming, sun addictive. Were it not for the weirdness of the moment, he would be one happy camper. The sound of silence was deafening.

After an awkward loneliness, a dark figure flapped on stage, wafting softly, slowly to the platform. It waddled with laborious pain to his side. Lloyd Joseph. He neared Murray, staring hard in his eyes. He was peering intently at this little rodent. He felt terrified, but did not budge.

Up close, he saw age on this old soul. Lloyd Joseph's legend was deep in the hills. No other crow stirred such honor, respect, and mystery. Earning a reputation with hard effort, honest dealings, and a continental air, none dare question his value. Not only did he garner love and honor amongst his own, but in the Olympics he was revered by every living beast. He was fair and honest about his own kind. Crows could be shady characters. The incident with Tutti Frutti and Mon proved it.

"Murray, my talented young friend, please relax. Welcome to my home, Crescent Cedar. I mean you no harm. The swans, especially Trygve, are trusted allies. Only this particular flock. When their earliest ancestors arrived in this country, my great-great grandfather helped them. Swans arrived from a blustery journey from Scandinavia disheveled and lost."

"He rendered assistance. Since olden days our two families have worked together, which is why they do all the flying for us. By being our transportation service, they earn certain items only crows can locate. We help each other. Our species work against a common enemy: ravens."

"Speaking of the past, it was ravens who misinformed Trygve's flock about the winds. They were happy in Norway and Sweden. One bad choice and they ended up in a freak windstorm. It blew them to America. They lost many on the way. The flock numbered in the hundreds, but the evil work of ravens caused this error. Old leaders listened to questionable sources and paid dearly. Now, they work with us. It is a feud with swans. Ravens earned the hate. It is not a wise thing to make an enemy out of a Trumpeter Swan."

"When you were at the festival in Willapa Bay, my sister tells me you witnessed the attack. Yes, I see. Trygve says they got you out in the nick of time. Raccoons were after you. By the look on your rat face you did not know this fact. They were after Winston and Beverly as well. The ruse was Paula. She was a diversion. Had she died, all the better as far as the ravens were concerned".

"They solicited the raccoon gaze, all boars, retelling Paula's joke about them. This incited a riot. They were happy to cause the havoc accomplished that fateful evening".

"You told Agwe you are not going to the festival due to your wife? I understand her parents were slaughtered, too? So sad. I do apologize for these questions and secrecy. You will soon be on your way to the Island. Delroy, Neville, and Clive were cousins to Tutti Frutti. The time we have is brief, Murray. You must tell no one of this visit, not even your wife or Mr. Mizithra".

"What I impart is chilling. It is imperative you listen. You are admired amongst crows. We are on your side in this trouble. What is to come will be difficult. We of the Olympics will rise together and defeat a common enemy".

BOOK 3: WAR

Ancient lore in the hills was: ravens were the chosen creators. They were given two boxes, one with the hills, fire, animals, the woods, etc. The other was given to a seagull. There were two ravens, one who opened the first box, the second tried to trick the seagull to open his. The seagull refused and could not be tricked.

In frustration, the second raven drove a stake in the seagull's foot until it released it. Once opened, the second box spilled forth sunlight, the moon and stars. Murray knew the story well and the arrogance ravens assumed due to the honor bestowed upon them. Rather than express humble appreciation at the choice, they chose to boast and brag. Many assumed unearned and unwarranted royal titles. Those who headed the largest "unkindness" were this sort of bold ones.

Lloyd Joseph slowly moved closer. Murray could feel his breath on his fur. He spoke quite softly and low. Muttering, he looked deeply in Murray's eyes, seeking an unspoken answer. He wanted to see how this touted fellow reacted to the closely guarded info bestowed.

Very few between Neah Bay to Seattle to Willapa Bay knew anything about what Lloyd Joseph nervously shared with Murray. He had to be able to trust this fellow he only knew from his sax and some bloody prancing about on stage. This was no endorsement of ability.

"You see, Murray, one major source of trouble for us are ravens. They are led by King Olaf. He owns satellites all over the Olympics from Gig Harbor to Humptulips to Forks, up to Neah Bay, over to Port Angles and Sequim. He made forays up here, too. This is our home, our territory yet they keep encroaching. Since we are much smaller, it is hard to fight. We have much stronger numbers, but their shear size can kill two or three crows in combat. We are not fighters, but deceivers. It is a superior brain power which made us the power we became. The ancient regal way of the raven is hard to overcome."

He explained boundaries in the Olympics were under assault. The ravens were testing all over for weakness. They exploited the advantage. Wapiti were useless. The squabble did not interfere with their stoic lives. What happened between two species of birds was irrelevant.

Truth be known, any bird who wanted something of an elk better be an eagle. They had no time to worry over a flock of flying feather flappers. King

Olaf had spies all around, aching for a fight. He wanted war. It was his evil desire to so diminish the crow population they would be forced into slavery under raven masters. The attack at Carnival Island was coordinated by the king with an evil henchman, Nevermore, whom he knighted before the cowardly assault. Sir Nevermore enlisted eager aid in the gaze of raccoons who hit the festivities with reckless abandon.

This is where Murray came in the picture. A certain rat and marmot had a previous tangle with one Ralston the Rancid Raccoon. Murray acknowledged it as true. Apparently, getting bashed in the head by the jeep tire tore off part of an ear, scraped his left muzzle and nose, and bashed his brain. He had a dent in the side of his head and carbuncle on his schnozzola from when he was whacked. The monstrous beast swore vengeance on Murray and Mr. Mizithra.

King Olaf's spies knew all about Tutti Frutti and Mon finding talent and how they enlisted Murray. The plot to kill him was hatched. It was to be coupled with the assassination of Beverly and Winston.

The position they held down south was very prestigious. King Olaf wanted his brother, Prince Bjorn to run the crow's operation. This was a major ripple in the animal world. For one species to try to displace another was serious business. Animals always battle for turf. Each critter has it's own territory. For ravens to remove crows and take over was something only a lowlife like Olaf would pull.

Lloyd Joseph saw fear in Murray's eyes. He reassured the rodent he had friends. Swans were sworn to protect him, beginning with Thyra. She was like her name said, thunder. No other swan could deliver such power. Along with Torhilde, the fighter, he had two feisty, able lady and pen trumpeter swans to watch over him.

No raven in the world could best one of these avian behemoths. He had an eight swans escort (or seven swans a swimming plus one). They would guard him from now on, subtly. It is not easy to be nonchalant when you are a huge swan. He was warned, blackness would descend on the high hill country. Bad times were afoot. He had to prepare beavers, marmots, and others for the pending doom.

After words of encouragement from the wise, wizened black crow leader, he lay a fatherly wing on Murray, bidding him adieu. Many plans had to be made. Ralston was raising an army. He had the support of Olaf. These two vile companions hatched the attack on his sister and brother-in-law. He was certain it was they who organized the murder of Tutti Frutti and Mon. It would only be a matter of time before trouble visited the Beaver Blues Bungalow and the Valley of the Stones. Paradise, sadly it seems, is never destined to last.

He had to go to the gig, so as not to arouse suspicion. He had to play as if nothing happened. Lloyd Joseph already had a division of unknown troops en route to the battle front. He had murders all over the place. It was a mass gathering of crows from all over the Olympics. No less than a dozen would gather from now on to stave off all raven attacks. Murray was shaken, stirred to his core.

The first instinct was to go home, nab his family, and find safety, but knew Lloyd Joseph what was right. He boarded the swan. He bore more respect for them. Coming down the cedar to the water's edge, all eight swans awaited. A tear in his humble eye, kippah in hand, he bowed in thanksgiving to his noble guards.

He climbed on Theodor's downy back and flew in complete contemplative silence to Waadah Island. They would blanket him, no flying off singing Ride of the Valkyries this trip. He was their sole focus. No bird or stinking raccoon was going to injure or kill this woodrat. Truth be known, they loved the little fellow. Crows conspicuously milled about. As Murray took flight, the same female crow from the Murderer's Row event silently left Lloyd Joseph's side, flying south.

He took a moment to suck it up. Club attendants grabbed his sax. He again humbly bowed, acknowledging his guardians, then entered the dark doorway to Club Club. The gig went well. He hit it off instantly.

Pockets, the short tailed weasel, was replaced with Slasher. Pockets was wanted for some unspecified trouble and disappeared. His shady past caught him. He vanished. Slasher was grand, Pockets forgotten. Graham was playing like mad. Paisley and Ness lavished praise on Murray. Once he produced the harmonica, it was all over.

They jammed for five solid days, sleep an after thought. He hit new heights of excellence. He played better than last year. The gigs after the Club Club paid off and it showed. The time was so good, he almost forgot what was going on up in the Olympics. Ever present in his mind was the pending war. He fretted about the safety of Momo, Frema, and Nediva.

He knew Olaf and Ralston would never stop. Those bent on death and destruction seem to always find a way to succeed forcing animals to fight. Those on the side of good and right are forced to stop their lives and address evil creatures like Ralston and Olaf.

Chapter 3 A Time For Every Purpose

Once he wrapped the Waadah Island gig, Murray was eager for home. Murray heeded Lloyd Joseph's warning "Trust no one you do not already trust". Chided to keep cards close to the vest, Murray spoke to one suggested by the aged crow. He must remain for now unknown (patience friends, please). His importance will become clear later.

This individual could be trusted. He was incapable of lying or deception. Any who came to him, knew nothing discussed would go beyond his ears. Were you to speak with him and word got out, it could only come from you. In this case, you would be put to death for betrayal. Such is the harshness of crossing the truly pure.

The animal world is honest and coarse. The essential thought Lloyd Joseph disseminated, the necessity of forbearance. While the initial urge was to attack, it would not be prudent. Not at this time. The results would be catastrophic for the woodrat and his family. Heeding his elder's advice, Murray had the bungalow establish a security guard outside. This task was handed to beavers.

The notion met with a positive response after the invasion by Jimmy Joe John Brown Jr. last year. The solution: place two large otters to keep out the riffraff. They set up near the entrance. Mice messengers were kept under the top of the guard's desk to run instant data inside via a handy cubby hole next to the door. They were paid in mouse food. Besides, all concerned concluded, it gave the place an air of class.

Lloyd Joseph sent crows with communications. Murray had daily walks, rain or shine, to receive correspondence via ebony evangelists. This way, he avoided suspicion. He took various routes. Each intersected with one specific area hidden to anyone. The deliverer would always be waiting. It would not leave until long after he moseyed along.

Not every day brought messages. Some days, he merely went for a stroll. It became a known habit. He told others it was for reflection. He was a bachelor so long, regular walks alone kept balance. Besides, his chiropractor, Graedy, ordered daily ambling to help his bum leg.

Momo, Frema, and Nediva were oblivious, the desired goal. Were they to know, the news could not be contained. He was the only one who knew what was happening around them. Part of the vital information imparted provided a detailed map of the pond. Why became evident once he saw the important facts.

He knew the small creek alongside the long path. He and Mr. Mizithra took this route when they came across Graedy and Molly. He knew the stream to his right. He took this when Ollie showed the way. He knew the rapids because of the journey east to White Mountain. What he did not know was the southern end of the tarn. It was shady, mysterious, and deserted.

This area was of particular concern. It was here, due south of the huge creek which led to the pond, his attention was drawn. The shore on the far southern end rose sharply from the water's edge to a ledge some seven meters high. As this ridge line ended, a small creek near six meters wide, flowed in the pond. Fifteen meters past it was another, larger stream. It was eight meters wide, deeper, faster flowing.

The island between these waterways formed a V, edged on both sides by swift moving water. Directly behind the V, a jagged rock face arose about ten meters, causing cascades to nearly become waterfalls. This space was of particular interest to Lloyd Joseph. Were it checked by Murray, that would raise suspicion, but he had a source.

It was Twigs's cousin, Ripple. Inscrutable, shy, silent, he swam at night, patrolling the shore. Twigs's family inhabited the area in numbers. Ripple was the loner of the clan. Twigs was his only real contact. He liked watching Twigs tap his drums. The talent displayed was amazing, certainly nothing Ripple could do. His claim to fame was the most extensive knowledge of the entire soggy boundary.

He knew every stone, best places to hide, find food, or swim freely. Once he filled in the map, Murray had a solid resource. It was a matter of waiting. Worry about Ralston weighed heavily. He could not allow fear to show. His biggest concern was Momo. Keeping her in the dark was best. He did not like it. It felt dishonest.

Twigs had no idea what Murray was doing. Murray simply said he was gathering data for Mr. Mizithra. This satisfied Twigs who was a drummer, not a deep thinker. As long as he could keep perfect time, the rest of the world could do as it pleased. He only wanted to tap, bang, and boom the day away. Music was his life, not a bad choice. Music brings out the best. Ripple also knew Lloyd Joseph.

This crow very carefully selected a few to help. Those inclined to a solo life were best. Married animals naturally focused on spouses and children, where bachelors could do a better job. His private female crow messenger was unwed. Ripple was an excellent candidate. He avoided does. As a young muskrat buck, he got in an extreme fight over a lady. He lost badly, causing him to shy away from muskrat mating. Thus, he became reclusive, not bitter. As such, he was invaluable to the cause.

Ripple met Murray. It was simple. He came one night to listen to his cousin, a common event, hanging thick as the crowd thinned. He helped with the equipment. Soon the club stood empty. Ripple and Murray met our mystery critter. The trio moved to the back booth out of sight of any who might venture in late. They went over the map in great detail. Momo was sawing logs in back. They were quite alone.

Murray knew the other three watercourses bleeding into the pond. He was of great use. They could use his knowledge to their advantage. He was going to have a leadership role. His ready wisdom on this battle front was excellent. It was during this session he learned some terrible news. It was vital to know in spite of the pain.

Being given information you cannot share is rough. The urge is to let others know important things. He knew a lot, too much for his little rat mind to comprehend. Learning more and more about the goings on of Olaf and Ralston quaked him with fear. However, not being able to tell anyone drove him crazy.

Ripple became his close confidant. He was the only other creature who knew what was going on. Focusing on music, Murray channeled the emotions to his sound. Using the harmonica more and more, he delved deeply in the blues. He and Henry would duel, the band jamming behind them. Each time, of course, Bette had to finish the battle by ripping out psychotic riffs. Momo began to add flute to their sound. Oh, wow, that about tore the place to pieces. The delirious sound made the Beaver Blues Bungalow a favorite.

Still, the news he received that evening put him on a course which set everything off and forced Ralston's hand. Murray found he was a doer. As such, he wanted to help, especially his friends. Never taking himself too seriously, he bore an easy demeanor. Playing music suited him like a fine, handmade tuxedo. On stage, he shined, glowed, flashing brilliance.

For a humble old woodrat, his hidden talent paid off. He was in one rather unique position. Lloyd Joseph's choice of this little fellow was excellent. He did not make it until peering in Murray's heart. There the aged one saw what he sought: a spirit of gold, loyalty in stacks, boundless love. Hey, not too bad for a little mountain rat.

Chapter 4 What Murray Learned

The revelation that fateful evening dealt with Mr. Mizithra and the Valley of the Stones. It shook Murray to his core. The big galoot was his best friend. To know he and his colony were in severe danger filled his spirit with dread. Realizing he could do nothing caused anxiety. He could tell no one. He could not go to help. By the time he reached the Valley of the Stones, it would be over. Time was the enemy. Murray was beside himself.

A war does not usually begin on one front. Multiple spots are chosen for weakness. Probing various places ensures finding areas for penetration. The first blow was struck at the Carnival Island Festival. It was a precursor to the northern conflict. The second found it's way to a couple of errant crows. It was the third which aroused the natives.

What happened in southern Willapa Bay at some rowdy festival did not make a dent in life in the Olympics. Sad as it was, the deaths of two talent scouts did not evoke any response, save those who dealt in booking acts and finding new performers to exploit.

Most thought their deaths to be over some transgression. There was no reason to doubt the explanation Lady Satin delivered. Frankly, no one dared question her due to a consistent, surly demeanor. The third front occurred a couple of weeks after the marmots came out of hibernation. The colonies got into organizing their new homeland. Each had specifics for their own needs. Mr. Milk Chocolate's peeps were new to each other. It was necessary to assign tasks, find elders, and assure all concerned of their value to the whole.

Time was needed. Since they did not need to go anywhere to forage, a committee formed. Mr. Milk Chocolate and the others worked hard. In short time, the issues happily ironed out, they implemented changes. A great deal of digging went on as couples expanded homes and made way for young. For Mrs. Lip Gloss and her brood, it was a need to welcome new members, like Mrs. Saffron.

They were well established. Everybody knew their jobs. She ran a well-oiled machine. Due to this asset, they advanced quicker than the other colony. The result was Mrs. Lip Gloss authorized a few to explore the high ridge. The route they descended into the Valley of the Stones was a wobbly, narrow path.

Their initial urgency was getting to the new homeland. Sightseeing tours were not on the agenda. Mr. Mizithra held a tight leash on the moving mass. This was necessary. Transporting an entire colony was a huge undertaking. They coordinated six mobile groups a long distance in unknown country. They had old and young, too. No lives were lost.

By necessity, they settled quickly for hibernation not long after arriving. It was a rough winter. A few elderly, young, and a handful of newborns did not survive the long period underground. One sad thing they weighed was precisely this result. They knew it would filter more than the usual amount. While more difficult, in the long run the residual would be better off.

Survival of the fittest is harsh reality in the wild. No man, fewer predators, and a perfect location made the decision easier. They did not wish to move. It was an arduous trek. Now they were safe from the Thumbs. No traces to follow. Befuddlement was the achieved goal. They got their wish. The cold trail was never found by a human being. The same cannot be said for the forces of evil. Raven spies above, Eastern gray squirrel scouts below, the new location for Mr. Mizithra's family was soon discovered, info passed to Ralston. He wanted this stinking marmot to kill him, then kill him some more.

Vengeance was his only goal. Coupling with Olaf, he found a bosom buddy in evil. After he was whacked in the head, his sole purpose was to slaughter Murray and Mr. Mizithra for stealing his Pacific Northwest banana slug meal. Time passed. He raised forces to defeat those he despised. He heard of the failed attempt at the Beaver Blues Bungalow by Jimmy Joe John Brown Jr., so Ralston sought him out to become allies. A promise of turning the club over to Jimmy Joe sealed his treachery.

Others joined. Eastern gray squirrels were belittled as an invasive species. They are, and wished to defeat and eliminate the much smaller native Douglas squirrel. Voles were loathed for their vile ways. They had no shame, no decency, and were deservedly shunned. Many clubs refused them service.

Ravens, as noted earlier, got the same treatment. They earned their scorn due to interaction with other animals. Ravens were the only species who thought themselves royalty. Certainly a jerk like "King" Olaf was not considered regal by anyone in the Olympics. Starlings, another invader, were added due to their numbers and blind allegiance to ravens. Theirs was a deep, long relationship going back to the days in Europe when they joined forces many times all over that continent.

Pilgrims surviving the first winter in the Valley of the Stones were given permission to ascend the pathway to the top ridge. A specifically selected group went. To qualify you had to be hale, at least three years old, trained in traveling. The last qualification demanded taking classes previously taught by Mr. Gingersnap and Mrs. Cupcake.

They were taken over by Mr. Baklava and his lovely bride, Mrs. Marsha Marsha Marsha. These two were renowned adventurers years ago, well before Mr. Mizithra. They did not travel as much or as far, but knew the ropes. He filled in with fables and lessons on the four whistles. They did the rest. Marmots wishing to venture up the hill had to pass muster with these three requirements.

One of the lucky few was Mr. Burrito. His family made it safely through winter. His only concern was his mother. Mrs. Jazz was ill on arrival, but recovered nicely. Her husband, Mr. Perry Mason, doted on his wonderful bride until she was healthy. Mr. Burrito was three, going on four. He wanted to marry and would this year. Many of his friends wed right before Sleepy Time, but he was too worried about his mother to focus in on females.

He took the required classes, Mrs. Marsha Marsha Marsha noting he was by far the best student in class. He was a master at traveling safety. He knew how to spot hunters from air, land, or water. Predators were known to hide in reeds and cattails. He was excellent at locating the best spots for digging emergency tunnels. The fellow could dig with the best. He was old enough and his health was never a worry.

Mr. Burrito and friends began early morning. The sun had not hit the valley floor when they began marching uphill. The chance to explore the immediate area exhilarated him to no end. In many ways, he was like his Uncle Mr. Mizithra. Mrs. Jazz was his younger sister. They took time to find the proper pathway. Snow was still piled up in places. They knew to function as one rather than venture off alone to their own destruction. If you make a mistake on a mission such as this, you endanger not only yourself, but your brothers and sisters in arms.

The trek upward was pleasant, flowers and clover plentiful. The best thing was the panorama. It improved each time they paused. The ten, all from the same colony, huddled at these stops. The troop was five females and five males. All were single, so yes, Mrs. Lip Gloss was matchmaking. What better way for shy ones to find that certain special marmot?

Mr. Burrito took her bait, eying Miss Coupe`. She returned his ocular advance with a wink and smile. Up top, they took in the incredible vista. They decided to venture to left or south. They were to travel a kilometer at most.

Whistling was important. To ensure safety, they formed five couples. Mr. Burrito paired with Miss Coupe`. He took the lead. She liked his decisive nature. He was handsome to boot, to a lady marmot. He was smitten with her, making this hike enjoyable.

The sun-glossed day glowed. They had a marvelous time, returning late to their tunnel. Once inside, they were met with Mrs. Jazz, Mr. Perry Mason, Mr. Mizithra, and Mrs. Lip Gloss. They were surrounded, bombarded with questions. Young Miss Ballerina was missing. She must have followed her older brother up the path. She was absent.

The entire valley was hunted stem to stern. Unless she fell over the side by the Misty Stream, the only answer would be following her sibling. She was a wanderer. She did on the way to the Valley of the Stones. None of them had seen her. It was too late in the day to search.

Worried, come sunrise, an exploratory force of Olympic marmots ascended the hill. They fanned out in groups of five. The search spanned until sun set. Darkness halted their efforts. They camped on top, digging a snug tunnel and chamber for thirty weary rodents. The next day proved fruitless as well. She was gone. Even Mr. Burrito had to admit it. Life in the wild does not support dreamers. Reality is simply and sadly what it is.

She did not adhere to the rules. She did not consider the risks. The goal of seeing what the world had to offer her eager eyes proved fatal. Sadly, the son had to return to his parents and deliver the worst news. Their lovely young daughter was lost. She, the sole survivor of her litter, was gone. Her parents and brother mourned. Both colonies united in grief. It was a most sad event, scarring their first outing. Once word of this tragedy got out, it shook the Olympic animal kingdom.

Sadly, a foolish young marmot played perfectly into the awaiting paws of evil Ralston. His minions spotted her. Raven spies hid in trees about the valley. Once seen, they sped to him with this vital data. Ralston expedited his closest ally, Alston, an equally nasty character.

He was distantly related to Ralston through his father's line. Alston was a no account bum who squandered his life. Rejected by his family for evil, he vanished into these woods. His sin: he actually tried to become a human's pet. There was no other choice. He was a loser according to tradition. His family had to disinherit him.

He reached out to Ralston after he was smacked by the jeep. His thought was two evils are better than one. Dispatched, he quickly found and slayed the young lady. He took the body back to the agreed upon meeting place.

Ralston thanked him, paying in glittering loot. Alston took his prize. He was one happy degenerate. This story was related to Murray. He suddenly realized he knew more than his best friend, but could not tell him. Were he to spill the beans, the enemy would be alerted and begin war. Murray's side was not yet prepared. He was working on it, but time was necessary.

Luckily, no one questioned any of his doings. He was a bachelor so long, no one knew his regular schedule. He could maneuver safely. Murray learned to be cautious, hide activities, without looking like he was concealing a thing. Then it became a game, one he mastered.

Though under constant surveillance, he nonetheless managed to carry out missions for Lloyd Joseph effectively. Contacts were made, the circle in on the action was small. Some in the loop did not know other parts of the new army. This was by design. Were someone caught, they could only furnish limited usable information.

He prayed for Mr. Mizithra. The music tour had him going to play as planned, but not for a long while. Were he to simply appear one day, it would be disastrous. Ravens would spot him in an instant. All sorts of pain would result. He was forced to possess knowledge he could not impart. This tore at him. One burning thought: how could he live with himself were his keeping quiet result in his best friend's death?

Murray did not enjoy this position. Lloyd Joseph knew this and sent messages of comfort to the lonely warrior. Daily, he aided the cause, at night Murray wailed away on sax and mouth organ with a fury.

Even his band-mates noticed the change, but attributed it to a new marriage and not being able to take off with his old buddy. He knew it was the best way to expel internal stress. Whatever, he played more than ever. Sadly, war was in it's genesis and he knew it. What worried him most was what was coming next.

Chapter 5 Mr. Mizithra Gets Hip

One thing about our hero, it was very difficult to pull the wool over his eyes. He was in far too many near death scrapes and escaped bad guys to know when it did not feel right. Spies are notorious for not realizing they are being spied on. In this instance, he saw the mistake. Mr. Mizithra learned a very important tell years ago about crows and ravens.

Both species really love shiny stuff. To distract ebony feather bombs, wave something lustrous and watch the fun. They go bonkers for a silvery flashy bit of anything. This is why they love metal instruments. They can worship the lustrous metal object to music. It's a twofer.

Due to his past, Mr. Mizithra never felt wholly safe above ground. He spotted a glint in a distant tree south. He viewed it without looking directly for an hour. Slowly turning his head, two other ravens spread out in other trees, each in eyesight of the other. The one spotted foolishly brought a shiny toy along. He surmised ravens had diabolical purposes.

When Miss Ballerina vanished, though it seemed a simple case of a bad choice, he knew by instinct evil transpired. Knowing how bad this was, he went to Mrs. Lip Gloss and the elders. His report ruffled their fur, both layers. They knew and trusted his opinion. One thing demanded was the info stay in this tight group. Agreed, a new vigilante force began.

Mr. Mizithra's Marauders were formed. A division came from each colony. What was called for was madcap but the only solution. They had to work at night. Unbeknownst to them, Ralston was one step ahead. He enlisted the help of long-legged bats to keep nightly vigil. Mr. Mizithra trained his forces. They went at night, digging small caves connected by a massive tunnel system to guard.

This series of interconnected routes ensured safety. They had many escape paths. Included were a few to throw off bad guys. It was perfect for their needs. The bats were semi-effective. Early on, they were heard flapping. Tunnel digging ceased. They set about improving the path to the valley from Grand Meadow. This threw the bats off. Soon they departed. Nothing was going on of interest. Bats are easily bored, so begged off the job to go hunt bugs instead.

Employing the chamber dug when seeking Mr. Burrito's sister, they soon had a solid place to repel any invasion. The new defensive position, Fort Ballerina. She was never found. They knew her fate and honored her memory as one of their own. Miss Coupe` saw Mr. Burrito change after his sister's murder. He did not blame himself for her having followed, but the incident set his firmness of purpose. His only goal now was to find and kill Ralston and those responsible.

The elders decided it was important to have constant guard above. A schedule was assigned each colony. No one objected. It was done secretly. Ravens kept daily vigil, but night belonged to marmots. The garrison was constantly occupied. Usually eight or ten Marauders occupied to watch. An excellent addition was a whistle, an improvised version of the VCR.

It was a relay, impossible for ravens to comprehend. Knowing it unwise to communicate with others, Mr. Mizithra set a thought in place. Risky, yes, but success meant a great deal. Keen's mice are an alpine friend to marmots. They look sort of like a hamster, white underbelly and soft brown up top, fuzzy, cute. A family of them moved in the Valley of the Stones with the marmot's permission.

Both species were on friendly terms. It was agreed with no protest. The requirement from Mr. Mizithra was two of their very best for a secret mission. It would take a long time for a mouse. Were they caught, torture was sure to follow. They had to be willing to die before being caught. To find such a pair did not take long. The mice were very happy in the valley and wanted to prove value to the rodents chums.

Zum and Zax were chosen. Zum was the fastest lady mouse in their nest. Zax was the strongest male. They were sent to Mr. Mizithra one night, after dark. He gave them charge, sending them on the way. It took the rest of the night to get up top. They were met by guards at Fort Ballerina. They slept, then darted off at night silently, well hidden.

No path for these two. They made good time through a lichen laden forest. Within a week, they were at the bungalow. Murray was located. He took the subject matter to his contact and sent a crow off to Lloyd Joseph. The worst had happened, a young one was slain.

Zum and Zax rested two days before heading back. The additional day would throw off any trailing the duo. They were given an alternate route home. Ripple carried them on his back across the pond, depositing them on the south side of the rapids. From this remote site, they started the journey home. They arrived safely, hailed for the effort. They bore a message for patience from Murray. Exhausted, they rested a long time.

He and Mr. Mizithra invented a code during their first trip. From the note provided he knew war with Ralston was in it's early stages. He was told the sad fate of Miss Ballerina. The last note was most difficult: keep up your vigil. Above all, patience. Raccoons were organizing with ravens. Others were joining. They needed to do more, but what?

The fort was reinforced as best they could. Mr. Mizithra had a good thought and put it to action. Soon, with a bit of effort, the entire fort was lined with pointed sticks. Booby traps were placed at false entry points. They fashioned spears by tooth whittled pine branches. The only nasty thing was pitch got in the teeth. Amassed weapons were stockpiled for defense of the Valley of the Stones.

The ravens continued to be fooled. They kept watch on a dull pair of marmot colonies. Their only joy was wrestling matches. They grappled away, but added being hams to the mix. It is like the difference between real wrestling and the sideshow: professional wrasslin'. They did not don gaudy spandex outfits, but did begin to make it more of a spectator sport.

When tossed overhead, the one landing would let out loud spurts of noise. Holds became very flamboyant. Mock wrath was intensified. The ebony spectators got a kick out of the show, clueless they were being taken. The marmots enjoyed the entertainment value. Long after the war, they continued to wrestle in this ridiculous fashion. It was just so much danged fun, hence the appeal for wrasslin'.

Meanwhile, rodents made active preparations. Were any big animals to get to the valley floor, they wanted to ensure it would not leave alive. Defenses were improved. Even during the day they dug and moved rocks underground. The ravens were clueless. Working day and night, they soon assembled a strong, multilayer defense.

They were proud of the effort. Improvements would be realized, but these were superior in every way. Some of them were actually considered offensive weapons. Mr. Mizithra desired his colony to be safe, prepared, and angry. He got everything desired. They believed in this cause.

Once again word went forth. This time Zum and Zax were to remain at the Beaver Blues Bungalow until further notice. A later pair, Zil and Zyp were sent on a completely different route. They never made it. No bodies were found. Nothing changed, but the fear felt was wise. Ralston may have gotten one to talk. Over time they accepted the duo did not spill what they knew. More than likely an owl got them.

Another pair, Zor and Zau went another way, arriving safely at the bungalow. They hid in Murray's home. No one knew they were there. Since they were such little rodents, everything moving scared them. Were a beaver or otter to sit on one of them, curtains. Timid by nature, their duties added fear.

Stashed in deep chambers gave them as much safety as possible. They awaited the coming storm with dread.

Chapter 6 An Evil Discovery

A cordial summer's evening made perfect by excellent music, Miss Bluesberry practically floated home from a night of singing and playing. She was so happy. The lone downer: her husband was not home from his trip. He had business to conduct and was adamant about getting his four brothers to come back and meet her. He wanted them to hear her sing and pat him on the back for finding the perfect, exquisite lady.

Warm air wafted her along. She hummed all the way to her humble cave. It was a stone's throw from the blues club. She had the best of both worlds: close to work and an excellent cave in which to hibernate with Mr. Loganberry of the Hills. Arriving home, she grabbed a snack, a sip of honey clover water for her voice box, then climbed in a bed of soft leaves and was soon out for the count.

She was the only animal to get any rest that particular evening.

Come morning, she awoke, yawned, stretched, then headed out for an early stroll. On these jaunts she would find food, then wend her way to the Beaver Blues Bungalow. She reflected on the joy her life possessed. Mr. Mizithra stood up for her. Since that day, life was magical. Meeting Mr. Loganberry of the Hills was best. He loved her completely. Caught in joy, she thought it best to head to the club. However, when she turned the corner to the entryway, a sight of horror annihilated her.

The front area otters guarded outside was demolished. Their bloody bodies were thrown in a heap, corpses shredded to pieces. Black feathers were strewn about the area. Dead crows, more than twenty, painted the entry red and black. The bungalow door was torn off the hinges. She saw it floating in pond water, splintered.

The carnage increased. The tarn was filled with the dead: beavers, muskrats, otters, woodrats, squirrels, birds, and mice lay motionless, buoying up and down. She peered over the edge of the dam to witness the carnage. Blood pooled near the deceased. Bugs were busy feeding on bodies, buzzing them with zeal. This was a meal, food to insects.

She could see more destruction near the doorway. Yet, it was only when she turned the corner and stepped inside the whole scene became clear. The entire interior was razed. Water from the pond was leaking in a few places through the solid wall of wood. Years of building this crucial structure saw the distinct possibility of the dam breaching.

It's impact would decimate the whole region. The flow of water would tear apart as the torrent fell down. This was done on purpose. Fighting and death amongst animals is common, but this was much more. This was a statement, unmistakable to those who found the massacre.

As an observer, Miss Bluesberry witnessed the statement embolden itself. The stage was torn to pieces. Instruments were thrashed. The piano was gone. It was found at the bottom of the pond, no longer whole. A huge rock sent it to a watery grave. Guitars were busted up, the standup bass had a dead beaver shoved through.

This, sadly, proved to be Homer. Later, she found Henry, so badly beaten he was only identified by the woven clover wristband he always wore. Dead western heather voles lay everywhere. They were the only one of the four varieties who lived up high and evil enough to join Ralston and Olaf. The bar and lovely display of bark were gone. Every place she looked death and destruction met her eyes. Her lovely orbs filled with tears. She ran for a long time. Terror filled her mind.

Why?

"Miss Bluesberry", came a voice from above her.

She looked up to see Lloyd Joseph peering from a low pine branch.

"You see this horror? It was Ralston and his army. They attacked late last night. For hours, evil poured forth until all lay dead. You lost friends. You will find more are gone. We are at war, lovely lady. Go to your cave now, please, to ensure your safety. In a short while, a visitor will appear."

"He will fill in the goings on, but your life is too precious to be lost. Please, dear lady, heed my words. Your cave is a haven of salvation. Do not leave, stay and await the visitor. I have much to do. Do not fret, sweet sow, the forces of good will defeat these armies of evil. They are legion, but we have right on our side."

Lloyd Joseph flapped away. Ebony wings made a tremendous sound as a hundred crows guarded their leader and general back to the void. In a moment the area was silent. The information invaluable, she instantly felt very alone. For the first time since she was a young kit, fear and isolation visited her.

She missed her husband, missed friends. The conversation cleared up a few things, forcing her to accept reality. This wise old crow's words had good results. She concluded a need for clear thought. She had to react. Dead friends needed avenging. The power this polecat possessed was greater than she knew, but would discover later to her own surprise.

She responded immediately. Due to the situation, she was a bit lost. Running without thought, she needed to gain bearings. This was done in a state of shock. Miss Bluesberry lay in her cave weeping for the ache in her heart. Holding in pain is foolish. Stoicism in light of the scene of death and destruction is not wise. Her mourning was deep, real, painful. It is usually best if you experience tragedy to talk about it with family and friends. By not doing so, the mind twists thought. This can lead to horrific results.

Some societies do not mourn well. Others get all the nasty out of heart and mind. While expression does nothing to bring the dead back to life, it allows the living to go on. The notion of only crying for a passed loved one out of the public eye is antiquated and ineffective. In regions where outward signs of an inward grief are permitted, you find a much more relieved populous. Mourning is real, difficult, personal, sad.

On advice from Lloyd Joseph, the club was avoided. It was not a good place. She was fortunate to avoid injury. The air was different from morning. Thoughts swirled about her mind. A portion was mad, a portion sad, and a part of her thoughts confused. The incident made no sense. The evidence was not a fight, like last year.

She missed that hullabaloo, but saw evidence the next day. This, this was far more. The bungalow she knew was gone. It would take a great deal of work to rebuild. With Henry and Homer dead, others succumbed to savagery, reconstructing the club would be a huge undertaking. For it to become what it was took decades. In one night, it was a pile of trash.

BOOK 3: WAR

Miss Bluesberry lay supine in bed, quivering, weeping, thinking, thinking, weeping, quivering. Outside, deadly noises rose. Leaves rustled. It was not wind. The sound increased. It was not one animal. It was many.

As a skunk, her first instinct was to stick her rump up to the entry, raise her tail, and spray the first fool who dared stick their nose inside. Tears wiped away, anger engorged her heart.

In a moment, whomever wished stink would enter. Scent glands, a skunk's formidable defense, were on the ready, quaking for action. Noise increased. Louder and louder it grew, nearer and nearer came paw prints. A harsh anxiety gripped her, like stage fright. Come what may, she dug down inside, clutching the sides of her bed. Okay, come, she thought.

"Hey, honey, I'm home. Please, put your sweet tail down, darling. My brothers came to meet you. We will give you a minute to get ready.", Mr. Loganberry of the Hills' timing was perfection.

She turned, ran, and hugged him for all she was worth. He was knocked over. Male skunks are 10% larger than ladies, but the speed and power she came with was so strong he went down. Words exploded from her lips. He led her out. Not pausing to introduce her four brother-in-laws, Miss Bluesberry poured forth her horror. This was too troubling to stay trapped in her mind. It could not remain pent up.

The fear of his approach sent her over the top, filling her mind with images of terror. She did not, as instructed by Lloyd Joseph, talk about meeting him. This was a sin of omission, but the reason for secrecy was necessary during war. Enemy eyes and ears lurked in every corner. You could not say where a raven, vole, or Eastern gray squirrel lay in hiding.

Mr. Loganberry of the Hills listened intently, his jaw slackened in complete stupefaction and revulsion. Lovingly, he did not interrupt his wife until the incident was completely vomited out. The tale told, he took his brothers, leaving one behind to guard his distraught bride. It was clear after experiencing this repulsive ordeal, Miss Bluesberry was in no shape to defend herself.

She was all aquiver. Normalcy was blown out of the water. Since they did not know the whole story, only what she witnessed, he would take his brothers over to investigate the scene, then report back. She was to rest, stay in bed. Her brother would tend to her needs with tenderness. If he did not, Mr. Loganberry of the Hills would clobber him.

"My darling doe, we go to the devastation, but before that, may I please introduce you to my brothers. As you know, I am the eldest. Next, the twins: Mr. Elderberry and Mr. Elderberry Blue. We simply call him, Blue. After them is: Mr. Salalberry. He is the one who will stay here with you. He is by far strongest. No one beats him in a tussle. Last, but certainly not least: little brother Mr. Salmonberry. Since he is the kit in the surfeit, mom and dad made us swear he would not get hurt."

The brothers introduced themselves to their new sister-in-law. A mass of polecats was what came to mind. They were big boys. He gave his love a gentle hug and kiss, lingering on her lips an extra moment. This is the reason she loved him so, he truly cared for her.

No one else was ever so loving and kind. She reflected on her good blessings. All about swirled death, yet she was afforded a wonderful boar to fill her with comfort and love. Mr. Salalberry bid his brothers adieu, then set about helping Miss Bluesberry relax. He brewed a cup of her favorite honey clover tea and put food on a tray. She ate. He went outside to keep guard.

"It is always wise to take a gander at what is what. I do so at odd times to keep spies off guard. They get sloppy. Mistakes turn the tide in war. Back home, I'm sure big bro shared, we fought battles. We had two other brothers. They were slain in a vicious conflict with another skunk family. We won, the price very high".

"Had they been victorious, we would be slaughtered. Afterward, we told Mr. Loganberry of the Hills to go find a wife. He needed to make his own way. I am so glad he met you, sister, so very pleased. He gushed about you so much we had to beat him to shut him up", Mr. Salalberry uttered with a smirk.

The love the five brothers shared was skunk deep. They form strong family bonds. It was the case for these fellows. Their parents did a good job. They came from a strong, solid surfeit. Mr. Salalberry got another cup of tea, then bade her to get rest. He was going to do some reconnoitering. With enemies afoot, proper defenses required erecting. No beast would hurt his family while he drew breath.

Miss Bluesberry acquiesced. Her head hit the clover pillow. Fatigue overtook her and was quickly asleep. Knowing she was well cared for by a guardian permitted slumber, easing a woeful mind. The shambles of the club, scenes of death, coalesced a sad psyche. Her husband's love, and his brother gave time to recover.

Trauma is real. Witnessing devastation, seeing blood, dead floating bodies, slain friends, and band-mates destroyed her. For Lloyd Joseph to find her was the first positive after fleeing the club. His words of comfort and encouragement did much to calm. He provided a task upon which to focus. His was good advice.

She was delivered from confusion. He put her on the road to recuperation, sending her home. Her husband's timely appearance was icing on the cake. His care, his brothers' strength in numbers, and laying in her own bed gave this lovely sow time to sleep. It was the right thing to do. It helped provide rest for her weary brain.

When you face dramatic, painful incidents, love is the first thing needed. You require comfort, one to ardently listen to your ails. This comes in many forms: spouse, lover, family, friend, clergy, are good resources. First, you need to accept you are in turmoil. Many deny it to their detriment. None of us is perfect.

If you can accept that, you begin to address and overcome a soul's trauma. Second, hopefully you have someone to open up to and share your inner feelings. Counsel of some sort helps. Seeing a dead relative is rough. Witnessing a car wreck and it's aftermath can ruin your life. If you behold war, well, enough said. Third, if good advice is given, it is up to you to employ the information.

Knowing and applying are two very different things. You cannot learn to swim without getting in water. Dwelling in your own misery is like a pig laying in it's own filth and slop. Addressing and dealing directly with pain is agonizing. If you do not work on it early, with time it can be impossible to recover.

Chapter 7 What The Skunks Found

Leaving an overwrought wife in his brother's care, Mr. Loganberry of the Hills led his other brothers swiftly to the site of the trouble. It would take but a short while. They spoke no words knowing the task at hand. Serious looks arrested their faces in anger. Mr. Loganberry of the Hills did not appreciate somebody causing such pain to the love of his life. He and his siblings were a force to be reckoned with. The pace set was quick. They marched fifty meters from the cave site. Mr. Loganberry of the Hills then stopped, raising his arm to stop his entourage.

He lifted a paw to his mouth, shouting a skunk holler. White Striped Skunks make a chirping noise. He aimed it towards the huge wall of fir trees separating their love nest from the rest of the wild kingdom. They wanted privacy. Instantly, they were surrounded by a handful of crows.

He spoke in low tones. After a short bit, they flapped away. He continued towards the club. He paused before turning the corner to the bungalow. He chided the squad to prepare what they were about to see. It was going to be extremely nasty. Brothers in tow, they made quick time to the hideous scene.

They found what she described outside, inside, and bodies in the water. While the brothers cleaned everything, Mr. Loganberry of the Hills ventured inside to ascertain what happened. This was a crime scene. A battle was fought here as any fool could see, but no war was declared. The episode was frozen in time. He went about finding clues and deciphering responsibility.

He was certain, the culprits could be sought and dealt with. He hollered again. A few crows flapped inside. They were uncomfortable to see so many of their number dead. He ignored their worries, issued new messages, then continued his work. The crows gladly left to deliver their epistles. One thing to know: Mr. Loganberry of the Hills appearing when he did was no fluke. Known solely to Lloyd Joseph, this was the extra military unit coming to the battle front. This crafty crow, apprised of the battles the skunk won, pointed Mr. Loganberry of the Hills in Miss Blueberry's direction purposefully.

The polecat was recruited years before his skirmish at home. He was an excellent soldier and leader. Once defeated, skunk remnants became of part of a growing surfeit. He was a bachelor leader. This needed to change for many good reasons. Chief was it would give him someone to love and protect. Sending the unwitting clod towards the bungalow was a stroke of genius.

Lloyd Joseph's spies knew where she was. The crow aimed him in the right direction. Results were practically written before they met. He was correct. Both were delirious. The general of Crescent Cedar had the ideal leader in the perfect spot to guard vital positions and launch a withering raid. The enemy struck first, but would not long enjoy victory.

The plan Lloyd Joseph drew began once King Olaf initiated his reign of terror. He slowly poked and prodded soft spots in the crow kingdom. Quickly, he saw the intention and called Murderers Row together. Boston Charlie did his job. He sent Mr. Mizithra and Murray to the Beaver Blues Bungalow. While no other critter knew Murray's talent, Lloyd Joseph did. When the rat passed the hat back home, a crow or two listened to him.

Unbeknownst to Murray, an alerted Lloyd Joseph came to hear him play. Nestled in a thick spruce, he focused on the rat on sax. The lilting notes rested sweetly in his bosom. The old crow was moved to tears. It did not take much to make the dominoes fall in place.

A nice spot to rest, knowing the woodrat would have to play for his supper, and the souls in the club made simple a plan. With Murray in place, the first phase was complete. The second was Mr. Mizithra, whose bravery and mighty deeds echoed throughout the Olympics, to "find" Miss Bluesberry. Knowing she was distraught at the rejection, a noble fellow like this marmot was perfect to get her in the club and on stage.

Her talent was well known. Lloyd Joseph heard her singing on an escarpment one sunset. The voice emanating from this sweet sow was overwhelming. Each note broke him further and further until he could offer no resistance. He knew what to do. Sending Mr. Loganberry of the Hills, unaware he was fulfilling a plan, completed the picture. In the long run, he was a happy sap. The players were ready.

There was sadly one thing he did not count on in the enemy's battle plan. Lloyd Joseph knew the onslaught was emanate. He knew not when. Ralston was a crafty so and so. He had one weapon employed the general did not know. The tragic outcome: the Beaver Blues Bungalow Massacre. Cowards attacked without warning, slinking while their victims rested unawares. Those who perpetrate such evil, history never forgives.

The vile blight of this pusillanimousness cannot be erased. As a witness to evil, the skunk's conclusion: this was not done to skirmish for territory. It was a mission of insane hatred. This was genocide. No animal on either side was found alive. It was evident any who survived the fight were slain without mercy in the anguished backwash.

No provision was given for the injured to recoil and recover. His search led throughout the entire facility. The kitchen was in shambles, food looted. The area was desecrated in scat from various species. It was foul. The entire room was useless for it's previous function. No one could use it to make a meal. It was a complete loss.

This is where the dam leaked. Water pooled at his feet. The rear of the kitchen led to Buford and Beulah's quarters. Here, oh so sadly, he found them. Backs to the wall, they fought to the end. Husband and wife lay as one. Other animals lay strewn in death masks on the floor.

One was the notorious Jimmy Joe John Brown Jr. He was dead not from his attack on his aunt, but a sharpened pine branch buried in his back, shoved through his heart. Blood coagulating underneath him proved this. The telltale scent of raccoon was on the handle of the spear. Ralston assured Jimmy Joe John got his reward: the club was his for the rest of his pathetic life, the useful idiot.

Next, with dread, he went in the living quarters. Here again were the dead. Rooms were emptied. Those with animals found them in death struggles. At the end were rooms for Murray and his family. He took a deep breath and opened the door. Unoccupied. To his relief, then fear, the little woodrat, his wife, mother, and grandmother were missing in action. Signs of four mice were there as well. No bodies of the little rodents were found. He was an observant polecat, a good detective, and leader. He thought before he acted.

Here was clear evidence of a raccoon attack. Large, fierce, sly they took out the entire club. It was a total bust. A curious thing were clues he had to analyze further. They did not make sense. Were they a part of the attack or aftermath? He pondered while assessing the damage. Repairing the breach in the dam was about all one could expect. The Beaver Blues Bungalow was history.

Bette, Twigs, and Tobias were missing. No sign was found. The one he expected to find was not there; Ripple left no sign. Mr. Loganberry of the Hills concluded his investigation and stepped outside, disgusted at the entire incident. His brothers fished bodies out of the water. The otters, crows, etc. were lined up.

BOOK 3: WAR

The enemy, Eastern gray squirrels, Western Heather Voles, Long-legged Bats, and two raccoons were tossed in a heap off to the other side. Scavengers could pick at them, but the valiant dead were buried with full honors. Soon, a familiar figure appeared, perching on a branch next to Mr. Loganberry of the Hills. It was Boston Charlie. His head hung in sadness at the desolation. He was a part of Lloyd Joseph's inner circle, though few had a clue.

Given an urgent message, the Belted Kingfisher flapped swiftly in the direction of Whiskey Bend. It was time to coordinate the response. Fortunately, Lloyd Joseph was prepared. Most of what took place was expected. It was sad lives were lost. Had he warned them, more would have died. The rub in battle is the need for proper timing. Were the club informed, surprise would become a full attack. Spies blanketed the area. Anything out of the ordinary and Ralston would alter plans.

The outcome would be the same, the bungalow was doomed, not sacrificed. The difference would be even worse. A handful were alive. Where, Lloyd Joseph, Mr. Loganberry of the Hills, and Boston Charlie knew and were prepared to retaliate. The mystery critter knew, too. He fled the club in time to effect evasion. He shared data with Murray alone. The woodrat knew what was coming and made proper precautions. He was to be captured, held for ransom to lure Mr. Mizithra to the meeting place where both would be slain by Ralston.

The one thing found was a hidden note. Murray managed to leave it. Knowing what was coming helped. The mystery creature left shortly before the attack, but prepared Murray for the terror. Armed with this precious news, he knew to leave word behind. The purpose was keeping Lloyd Joseph informed. Were details omitted, desolation could follow. It was vital he get the message. In an instant, the handy, mysterious female crow nabbed the note and flapped off, blanketed by dozens of others.

With the declaration of war, received by Lloyd Joseph from King Olaf after the attack, no chances were taken getting communications through. Ravens were known to kill lone crows. They killed Tutti Frutti and Mon in an outrageous way. Crows knew it was no holds barred. As one, they bonded. War mode engaged. Win or lose, it was not because crows were unprepared to repel Olaf and his jerks. If it was a war they wanted, war they would receive.

Mr. Loganberry of the Hills returned to the bungalow's residue. He stirred about for a long time. He turned hearing flapping feathers. Lloyd Joseph flew swiftly to his side to examine the evidence.

The polecat led him about the battleground, ending at the back of the stage where the worst sections of the dam were marred. Pond water pooled deep, debris bobbing up and down. It was a sad, loathe-filled scene. They went outside. Lloyd Joseph was visibility shaken by the carnage. Though the bodies were underground, feathers, fur, and blood painted the floor and walls in a mind altering mural tainting his ancient mind until life's terminal rattle.

"It is clear enough to me, what do you think, General?", came the polecat's query.

"It is difficult to agree, but it is so. I cannot believe he is so warped to go extreme. This is an inexcusable weapon. How could he? To be so filled with hate to begin this commotion. He did not need to sink so low. Our crow kingdom is threatened, so be it, but to use a bear as a weapon is too far. I personally sent Boston Charlie with a message to the ones who need to know what is occurring in the hills and valleys."

"A bear, why would he bring such a beast to our battle? This, this is disturbing. He cannot be forgiven. Face it, this fight is to the death. It will either be Ralston and Olaf or us. We can no longer exist on the same mountain range. War is declared, Mr. Loganberry of the Hills, or should I now refer to you as Captain Loganberry?"

"The forces built last summer before falling in love with the Darling Doe now come forward. I am pleased you already thought to call them. They are to assemble within the next few days here, at the edge of the pond. In time, we move south for full impact".

"I contacted Graedy and Molly. They will set up a hospital for any who survive. They are clearly murdering the injured, but we can extract our own when possible. Remember, in war, sympathy for the enemy only spells death for your side. War is absolute. Okay, Captain, to your duties".

Chapter 8 Retaliation

The investigation concluded, Captain Loganberry made a beeline home. While his brothers guarded outside the cave, he went and saw to his love. She was refreshed, saddened still by her witness to such evil. Her life would never, could never be the same. He told of the death of Buford, Beulah, Homer, Henry, of Murray and the others missing.

He assured her they were not dead. Ralston and Olaf declared war against Lloyd Joseph and crow kingdom. Any allies were included. The goal was the death of Mr. Mizithra and Murray. The other aim was to take over the entire mountain range, subjecting humbled crows to raven supremacy. King Olaf wanted serfs to attend his royal raven masters.

Captain Loganberry explained a plan would soon hatch He hugged her, urging patience. All would come to pass in due time. He had more skunks on the way. His sources formed a skunk brigade. Once they were here, things would change. The plans Lloyd Joseph created were brilliant.

They lost the Beaver Blues Bungalow Massacre, but would exact revenge on the knaves. They were low cowards and would suffer for reckless demolition. From all over the Olympics polecats heeded the call. Raccoons were enemies, Ralston a major pain. Crows made great partners versus ravens, who felt superior to every animal, even wapiti.

Skunks were rejected, but here was a place they were accepted. Miss Bluesberry had fame in these hills, a fan club and adoration. Not once did a raven attend her shows. So, the choice to rally around Mr. Loganberry of the Hills met with great aplomb. He received overwhelming support. Now, now they came.

In every war, some seek only to get out of the path of destruction. Such was the case here. Several species begged out of the melee. Those who selected this option made clear to both sides they were not involved. Fine, get out of the way, let the slaughter begin was Ralston and Olaf's message. Lloyd Joseph understood, everybody needs a Switzerland.

Seagull, other vole varieties, mole, chipmunk, and most bird species begged out. If a majority said they did not wish to choose sides, animal law meant no species member could cross lines of neutrality. Were this to occur, the truce would be off, the species subject to combat. This was an old, honored tradition. It kept combatants honest. Here, though, Ralston tossed the rule book aside, making his end all the more essential.

Lloyd Joseph, from Crescent Cedar castle, issued an army of crows a thousand strong, making for the Valley of the Stones, to contact forces, and coordinate combined efforts. This was a two front war with a third to come. The news for Mr. Mizithra and his two colonies was met with relief. They felt isolated. It is never good to be cut off, like those at Masada, the Alamo, Wake Island or Khe Sahn.

Mr. Mizithra's Marauders prepared. Utilizing crow communications versus dragonflies sped the news. Besides, it was much more secure to send a flock of crows to stave off any raven raiders. Hundreds now encircled the Valley of the Stones, ridding the trees of any ravens. Eastern gray squirrels were put to death whenever spotted. This was for all the marbles, winner take all.

Mr. Mizithra knew his best friend and family were taken hostage to lure him to a deathtrap. A call went to Olympic marmots. Any who could fight were urged to enlist and come. Were Mr. Mizithra to lose, all the marmots would be eradicated. Opting to defend made sense. Resolute, a mob of marmots massed, marching towards the Valley of the Stones from all points of the compass to join the Marauders. Mr. Mizithra had an excellent reputation. The plea reaped positive results. Elders of many colonies authorized fighters to go to war against raccoon and raven.

They moved by night, hiding from prying eyes: vole, bat, squirrel, raven. The gathering would take time, but Ralston was in no hurry. He underestimated his adversary. Here, cock-sureness proved a failing Lloyd Joseph exploited.

With Major Mizithra, Captain Loganberry, and Lieutenant Boston Charlie, the old crow waited for everyone to go to assorted destinations. An essential communique arrived via crow to marmots. Major Mizithra read it, nodded, conferring with staff. The meeting lasted a good while, then he gave an answer to the awaiting murder. Off they flew, over one hundred crows. This was a special message and needed to safely arrive at the Crescent Cedar. It did.

The Major was in 100% percent agreement with the General. He would implement Operation Surprise immediately. His new forces were gathering. Soon all would arrive. He counted marmot soldiers at nearly five hundred. This was a huge number.

This constituted a majority of marmots in the remote species in the Olympics. Ralston had no idea what he was about to face. Back at the ranch,

Captain Loganberry assembled the surfeit. Lieutenant Ripple materialized, his news invaluable. It confirmed the conclusion Lloyd Joseph reached. The plan was ready to spring, parties needed to assemble. Captain Loganberry would lead.

A fantastic result was the transmutation of Miss Bluesberry. She armed herself to the teeth, standing shoulder to shoulder beside her husband. Now a soldier in his squad, she never imposed her marital status. She begged to be treated as any soldier of the army. He gladly complied, a happy boar was he, oh yes he was. She was the best danged sow in town. He had to be certain to thank sly old Lloyd Joseph after this was over. She was, however, the only enlisted soldier he ever kissed.

Some of Ralston's assault forces, the lily-livered leader not with them for the attack on the Beaver Blues Bungalow, rested on the V island between the streams Ripple mapped for Murray and Captain Loganberry. Here with prisoners were Alston and the rest. The beavers of Jimmy Joe John Brown Jr., unaware of his assassination, were with them.

In the mix were ten black banded villains. This gaze were the worst of the worst. These were the total bottom of the barrel of the raccoon world. Each was shunned by family and had no friends save the likes of Ralston. In their midst were dozens of voles, watching the woodrats, beavers, and lone muskrat gleefully poking and prodding the captives. The voles were vile. Night was pitch-black, four days past the attack.

Raccoons bragged to an unkindness of ravens there to gather news and revel in victory over stinking beavers. They taunted Bette, Tobias, and Twigs about their music. Insulting their dead mother, father, and brothers added pain. The words were harsh to hear, but caused an awakening like a sleeping giant. They were ready to strike with power. Hate could wait, a reckoning was in progress.

They were ready to fight, even peaceable old Twigs. He was not a beaver, but had many as friends and would not desert them in a time of great need. They could not communicate with Murray. Raccoons insisted Murray and his family be kept separate from the others. There was no sign anywhere of Zum, Zax, Zor, and Zau. The Keen's mice must have hit the road and gone off to hide. Who could blame them?

The southern end of the pond was unlike the other three. It was ridge-laden, idled water full of stands of cattails, shrouded in darkness. Trees clogged the shoreline, long lichen strands dipping into the water. This was the aphotic portion most animals avoided. Amongst muskrats, for example, only Ripple was known to swim it's mysterious waters.

The sun passed so the farthest southeast portion never saw sunlight. The ridge at back was high enough to keep the arena perpetually dark. It made it ideal to hide out and await further instruction from Ralston. The raccoons fished and lollygagged, while voles ate and mocked the rats. They would not even give poor, old Frema a bite to eat. An ominous, smoky haze came to rest over this end of the pool, chilling the air.

A glint of light flashed from the opposite shore. The ravens' reaction was instantaneous. They winged as one toward the shiny metal object. It was an addiction. They flew to the bright light in a flurry of plumage, flapping across the pond, each seeking to be first.

The noise was deafening. Dozens of pitch colored wings rising at once in a fury made great sound. They darted over water towards the light. Raccoon, beaver, and vole watched, commenting how foolish they were to chase glittering items. They saw no joy in this beguilement.

Distraction it was. At this punctilious moment, backs turned to the island, the ravens missed the real action. From black shadows, five rafts materialized, each pushed by two beavers, sides blanketed in muskrats. On each float was a weapon faced directly at the enemy. The attackers were now under direct retaliatory assault.

As rafts neared shore, from the back of the island, a ridge rose up the treeline encompassing the top, a handful of phantoms descended to the ground. Via crow messengers, Mr. Mizithra arranged, coordinated with Lloyd Joseph, a little surprise for the bad guys. Lieutenant Nicolo and the Night Ninjas hit, they now had not five but twenty. Murray and his family were immediately taken out of harms way on vine ropes. Ninjas in training were up top, pulling comrades to safety.

The rest of the Gliding Ghosts flew from all angles, scattering beaver and vole. Only the raccoons did not fall for it. Alston was too savvy for this old trick. Beavers dove in the water, voles in tunnels. In the water, Jimmy Joe

John's army was eliminated by Ripple's navy. Beaver rebels numbered about twenty after the fight at the club. Jimmy Joe John Jr. died with a dozen of his forces. Muskrats, otter, and beaver submarined the traitors. They were killed, every last one. War is hell.

The secret weapon was the rear end of a skunk pointed at raccoons on shore. Alston's eyes blew. They instantly scrambled for the ridge. The best they could hope for was to shinny up the cliffs to safety. Turning around they saw Bette, Tobias, and Twigs slip in the water disappearing in reeds. Lieutenant Ripple hushed them ushering all to safety.

Their plot busted, the only thought was escape. Any struck by the stream of stink a skunk cannon shot would be in bad shape. They found the bottom of the ridge. Oddly, four little Keen's mice stood over the top. One, Zum, thumbed her nose. The quartet faded in shadows. The raccoons shook their heads in confusion, then began to ascend the cliff.

This was their last choice. The mice made it, running to Mr. Mizithra who sent one hundred Marauders to the location Lloyd Joseph assigned. Only Zum's and her squad knew where to go. Lieutenant Ripple had a crow message of what was afoot. He was smart enough to figure out the cryptic wording. By the time Twigs and the beavers hit water, aiming for the reeds, he was waiting to escort them across the pond. The days of going back to the bungalow were over.

As the beaver neared shore, pushing skunk artillery, the muskrats climbed ashore. They were witness to the payback. One hundred angry Olympic marmot soldiers attacked Alston and his wimps. A few marmots were injured, one died. No raccoon, beaver, or vole escaped. Voles were dug up by marmots and tossed in the pond to drown for their vile crimes.

What, pray tell, happened to the fifty odd ravens who dashed across the waters for a glint of steel? As they neared, a lone figure sat perched on a limb. In his beak Lieutenant Boston Charlie held a flimsy strip of tin foil. It flickered in the air, sending forth shards of glittering light. The mere sight of this sent them rushing with a zeal to steal foil from the stupid little bird. Suddenly, a blistering, oncoming shadow materialized. With no time to turn, the shade closed nearer and nearer.

It was Lieutenant Neville, Lieutenant Clive, and Lieutenant Delroy, enlisted from Neah Bay to join the war. Eager to prove their worth, they volunteered to lead the attack on ravens crossing the pond. They led five hundred black battlers on the surging raven squad. In minutes, every raven lay dead either in water or on shore. Lieutenant Boston Charlie smirked, tin foil fluttering in his beak. He spat it forth, diving for a fish.

The response to Ralston and Olaf was decisive. None of those who participated in the Beaver Blues Bungalow Massacre lived save the bear. Jimmy Joe Johns' rebellion resulted in death for a number of beaver. Voles were missing hundreds. A number of Eastern gray squirrels were gone. Ralston the Rancid Raccoon's Regiment was down. Alston was lost. To lose so many early in the war did not bode well. Olaf lamented the loss of his air forces, but had more of that currency to expend.

The next phase was to attack the Valley of the Stones. This was going to be a huge battle. The bad guys had an army marching up the road next to the rapids. They would be coming to the battle site soon. Once they were in place, Ralston planned to hit the marmot colonies in force. He had hundreds of raccoons, voles, squirrels, bats, and the biggest bad boy of them all, a black bear to even up the odds. His other forces had to come up the road, then he would be ready to attack. He waited for them to arrive, finalizing plans for the demolition of Mr. Mizithra's world.

Blind, undaunted by complete failure, Ralston saw victory in the assault on the club. He attributed the subsequent loss at V Island to poor leadership on Alston's part. He cursed his very name, forbade it to be spoken forthwith, assigning a new second in command, his other cousin, Charleston. Not as nuts as Alston, he was a formidable fighter. Charleston was vicious in combat, unforgiving, but not the planner Ralston was. He was happy for promotion, vowing to kill his share of marmots.

Since the ravens were no longer able to give updates, they relied on the fact marmots had no where to go. If they could not leave, all Ralston and Olaf had to do was wait for the proper moment to invade and rid themselves of a mutual enemy. Blinded by hate, ruled by ego, both were the worst type of leader. These weaknesses Lloyd Joseph counted on to defeat this overwhelming foe.

Ralston sent Charleston from his rocky lair in cliffs near Grand Meadow. This fortress was well hidden, surrounded by dark, dense trees, guarded by Eastern gray squirrels, voles, raccoons, bats, and ravens. No other animals knew it's location. None dare venture to locate it. Those who tried, died, bodies never found. He wanted Charleston to lead the support from the road by the rapids. They were to follow the river past Gilligan's Island. He was to lead them to Grand Meadow.

From here, they were to join the fray at the Valley of the Stones. It would be a tasty disruption for his victims to see one hundred angry masked warriors rain down from atop the crest of the ridge. They had no defense against it.

Were they to stop this army, which was inconceivable, their secret weapon, Igor, the black bear, would end the battle with a distinct finality. Charleston dispatched, Ralston waited for all to fall into place. Spies abounded. He was apprised of happenings on a constant basis. He began to move forces towards the focus. They had to take the ridge. No problem. Next, lay siege to marmots. They could test weak spots, then descend in a wave of death and destruction.

Murray would die, having escaped due to Alston's bad leadership. The lousy woodrat would not be so lucky next time. Once the marmots were defeated, woodrats dead, crows laid low, he and Olaf would rule the Olympics with an iron paw and claw. He wanted to dance for joy. The dent in his head made him cockamamie.

War hit the next stage. His reinforcements would tip the scales. Evil joy swelled his black heart.

Chapter 9 Murray Learns More

Humboldt Flying Squirrels pulled them out safely. Murray, thankful to the bottom of his fuzzy toes, wanted to know what in the wild world of animal sports was going on and how everyone else was. He and his family witnessed the Marauders' demolition of the raccoon gaze who reeked such havoc on the blues bungalow. In a flash, the island was secured.

Lieutenants Baklava and Bubblegum, officers appointed by Major Mizithra, reported to Major Murray. Sent by his best friend to rescue Murray and family, they were fresh troops enlisted from all over the mountains to fight the forces of evil. They were a hearty bunch, proud Marauders, glad to rid the homeland of scum like Olaf and Ralston.

They gave Murray news. Captain Loganberry, one of those on a raft, provided the status of forces in the Valley of the Stones. Miss Bluesberry was a weapon as well. She hugged Murray for a long time, happy he was alive. The new marmot soldiers took great pains not to be revealed massing. The one hundred taking the island were a shock.

With no survivors, Ralston was blind to the truth of the battle. He knew nothing of Marauders. The Keen's mice effectively diverted them to the scene of the kidnapping. In sad retort, Murray imparted the invasion by Alston and wicked army. They saw most of it, heard more, then when found, extracted swiftly, sequestered on V-Island.

"We were on stage, jamming away like usual, when suddenly the front door was torn off it's hinges and tossed in the pond. The violent sound of death echoed from outside. The otters were slain. The next thing was a mass of black fur blasting in the club. The bear made for the bar, tearing it to shreds as raccoons poured in through the ravaged entry. Next, ravens flew in by the score. Eastern gray squirrels and voles came next in droves. Homer and Henry were killed instantly."

"Instinct pushed me to the hallway to Momo. The noise had mama and grandma already in the room, clutching young Momo. She had been through trauma of the attack by raccoons on Carnival Island, witnessing her parents die. Now she was under assault by the same species. We did not know what was going on, why it was happening, but huddled in fear. I had no weapons to use in defense."

"The sounds of horror filled our sad ears. I knew Buford and Beulah were in their chamber, resting from an enjoyable evening. The assailants were particularly destructive. Any who surrendered died with no regard for injuries. 'No prisoners', they told us. We were found and taken to the island for ransom. Mr. Mizithra was to come. He and I to be killed, our heads on pikes to show every creature in the hills who was in command."

"Raccoons said the bungalow went first so reinforcements could get past here undetected, where they were passing through to the Valley of the Stones. I overheard one raccoon scoff at killing Jimmy Joe John. They considered him a willing idiot easily duped to believing he would own the club. The beavers' party house gone, critters would avoid the area. Now his forces could pass through to battle sans hindrance from General Lloyd Joseph's forces. Secrecy is important, that much I do know."

Captain Loganberry and Major Mizithra's Marauders filled in the rest. Lieutenant Boston Charlie joined the mix. The message was not expected. Instead of racing to the Valley of the Stones to join the fray, General Lloyd Joseph had other plans. Major Mizithra was on board, fully aware, and at work on his plan. Major Murray was first to take a flight via Trumpeter Swan Air Force.

No other trumpeter swans were involved, skirting the Olympic Peninsula, west over the Pacific, past the mighty Columbia, coming to rest in Young's Bay, exhausted. This they willingly did for the cause. Crows were good friends with swans. A solid shared history of cooperation made both species excellent partners. Theodor's eight were there, ready again to defend their best woodrat buddy.

In the darkest hours of night, Major Murray was flown by the octet from high places. They did not take him to Crescent Cedar as expected. Lake Crescent, home to the Ladies of the Lake, has a surface area of twenty-one kilometers, and 190 or 300 meters deep, it is still unclear.

Local tribes revered it in folklore. It was here the flock took him. The three ladies and Lloyd Joseph were waiting in a deep stand to the far north end of the lake, west of Piedmont. The enemy was hip to the fort on Whiskey Bend. This was the best site to meet.

"Major Murray, while we admire your saxophone playing, this is a war. We need you to join the fight. You must build a rodent army to help. Ravens are trying to rush us like a tsunami. The insanity of Ralston adds the nastiness necessary to wage such a battle. He will never surrender".

"Know this for certain, young woodrat. You are going to see more of what you saw at the bungalow. What we must impress upon you most of all if you are to succeed, is you must not ever allow emotion to make your decisions. This spells instant death. General Lloyd Joseph has immense confidence in you and the marmot. Do not, I pray, let us down".

After words of encouragement, swans flew him to the pond an hour before dawn. Long-legged bats who saw them return assumed, a lazy and costly conclusion, he visited the Crescent Cedar. Dutifully reported news, thankfully incomplete. A lack of firsthand information about the enemy's plan was a blessing to the good guys. No one coordinated with spies watching the tree. The trail went cold. Major Murray took the news well. He had to work quickly. The enemy was massing daily. The lady woodrats, orphaned beavers, and Twigs were sent to a place of safety. Bette had plans. Her anger could not quell. She and Tobias got immediate approval for their idea. She was promoted Lieutenant by General Lloyd Joseph, Tobias sergeant. Twigs, unable to join beavers, went to aid his cousin.

Muskrats were given an interesting task. Once Lieutenant Ripple and Sergeant Twigs had sufficient forces, they reported to headquarters, then headed into position. This dividing of friends was sad, but quite necessary. Each had a function and was glad for the opportunity to show value to the cause. Were they to survive, there would be time for reunion.

Until that day, a good-bye hug would suffice. Each departed for assigned posts, eager to fight for home, family, friends, and a way of life. Once Murray got hip to the goings on, his astute wife, without betraying herself, saw him start organizing increased safety methods. The two otter guards out front, contract with odd characters, and walks alone added up to some internal struggle.

He assaulted his instruments way too aggressively to be normal. The stress of what he knew and could not share was torture. Murray was honest with his wife. It was anguish not to be able to let her know. He obeyed his wise leader, but it was difficult. Momo was observant by nature. Silently, she worked behind the scenes to help her husband and the battle for the Olympics.

When they returned from the Carnival Island Festival, as a young bride, Momo took great pride in her husband's talent. She noted all who heard his playing enjoyed it. The mountain's woodrat populous were contacted via dragonflies. A fan club formed. Sugar Murray's Maniacs numbered in the hundreds.

She kept them updated with his schedule at the bungalow and away gigs. In this newsletter, she shared bits of their marriage and personal info. He was a hit. She intended to help make him a success. She was so talented on flute. Momo jammed with the band and thrilled at the music. With a solid army of rodent enthusiasts, it was a simple matter to turn them into partizans. Shedding dragonflies, she utilized crows, sending hundreds to every colony, horde, and mischief. A response blossomed. Over 3,000 woodrat warriors heeded the call.

An equal number of Keen's mice joined ranks due to the efforts of Zor. These Momo presented to her husband. Major Murray was stunned and amazed. General Lloyd Joseph knew. Crow messengers reported back to him. He now saw forces swell. Rodents were sent to the Valley of the Stones to help. Labor was needed for defenses. They were proud to serve.

For his humble part, Murray wanted desperately to see Mr. Mizithra and look in his eyes. They were supposed to reunite this year, but under much more conducive circumstances. Musing about coming to play his sax left a wry smile on his serious face. It seemed trite compared to the current situation. Music would wait until this slice of evil was overcome.

With Momo, leaving his mother and grandmother again, though this time to battle, he made for the Valley of the Stones with a steely resolve. Ralston and Olaf had to be defeated. It is not easy to hide such a force with forests full of spies, but they had alternate methods. Major Murray led his army up a secret, arduous pathway to Grand Meadow.

They came out at the far northwestern shore of the lake. It was very remote. Here they waited. Major Murray now took time to organize his forces. He promoted Zor and Zax lieutenants, Zau and Zum sergeants. The leaders selected more to lead soldiers. Two woodrat standouts were promoted to lieutenant: old Moshe, musician from the Grand Meadow festival, and Asher (the happy one).

A well respected woodrat, Asher's positive adoration of his troops and willingness to leap in the fray made him a natural choice. These two selected ten sergeants apiece to form squads, platoons, and companies. They had to fight as a solid unit or the enemy would easily pick them off one by one. They trained like mad. Time plodded until some days later when Major Murray heard the hushed call of his contact.

It was Sergeant Twigs. He and a select handful of determined muskrats led woodrats and mice through dense, moist undergrowth to a clearing on the edge of the pond. Thick legions of cattails hid their transportation. Convoys of muskrats each took a woodrat or several mice on board and swam in complete silence to the far eastern shore.

They awaited orders where to proceed. It took time, but the entire army got ashore. Major Murray now split his forces. The Keen's mice had their orders. He conferred with the two wee lieutenants, then sent them on their mission. He gathered forces, explained the plan, then made for the Valley of the Stones.

The initial task complete, muskrats moved to another position. Here they were to hole up, out of sight, and await the next phase. Again, they had to move at night, sleep during day. It took a long while relocating to the place assigned, but once there they were able to rest and recoup.

Traveling such distances was not common, so they had to adjust. This they did with hard determination. Ralston had to be defeated. The entire animal world in the mountains hung in the balance. They would do what they could for the betterment of the whole. Nobody could ever find fault with these noble allies.

Chapter 10 Mr. Milk Chocolate's Idea

Forces gelled towards the Valley of the Stones. Major Mizithra built a mighty defensive wall around Fort Ballerina to stave off invasion. The raccoons were strong and had to be thwarted. The rudimentary tunnels and chambers were inadequate for something this intense. Ways were suggested. Ralston had bat, squirrels, and voles as well. Captain Milk Chocolate had some ideas about the vile voles. His colony's decimation was greatly aided by the voles burrowing into tunnels and chambers. This surprise gave him ideas how to stop them. Major Mizithra had others give him help. In the end, they built a strong fortification system, complete with a myriad of escape passageways.

One of the first things was to increase the amount of decoy tunnels. These would stop Eastern gray squirrels. Upon an excellent suggestion, some of these traps were designed to kill any who entered like a finger torture toy works. Once in, if they tried to pull back out of a dead end, they would find themselves trapped. Soldiers inside stab through the hole on top of the tunnel with a sharpened pine branch, slaying the bum.

Booby traps led to tunnels with small stakes driven in the ground, sharpened, and covered. A squirrel stepping on this was rendered useless in battle from bleeding paws. For bats, large tapestries of tree lichen, dense in this moderate rainforest, were strewn from tree to tree. They would have a difficult time getting through.

Bats are not too brave, so should they stop a number, bats would flee in terror. Captain Milk Chocolate had an excellent idea. Invaders got in his colony because of soft walls. Voles are excellent little diggers. Taking time, striking victims completely unaware of impending doom, they poured through tunnels by the hundreds killing without discretion.

When woodrat forces arrived, he put them to work with his marmot soldiers. They laid large flat stones on the bottom of the trenches. Walls in front had pine stakes driven 45 centimeters deep in a line. Stones filled walls on the sides and bottom. Escape tunnels were lined with long limbs joined with pitch. Voles hate pitch and would avoid contact. They had stakes pointing out towards the oncoming enemy. It was decided to add a few surprises for the jerks.

Major Mizithra, viewing newly erected defenses, offered hearty approval to his Captain and troops. While they would not stop all the raccoons, the smaller fighters would succumb to death in great numbers. He implored Captain Milk Chocolate to set about making spears for defense. Wrestling alone would not cut it in war.

They installed a few covered holes dug deep to trap a raccoon, then slay it with rocks. This was to the death, no one was hugging the foe after this like marmot wrestling. It was kill or be killed. Rodent forces would not let raccoons and ravens eliminate them and subjugate crows to slavery. Weapons would surprise the scum. He would prepare as best feasible. Major Mizithra was a fighter.

Murray's fuzzy kippah topped the horizon. Major Mizithra was in the valley playing with Mr. Murray and Miss Paprika. He told the kids to sit right there, then darted towards his chum. They met and hugged. It was a long time since they had seen each other. Both were now married. Murray met his namesake and sweet sister.

War was real. They needed to discuss plans.

Woodrats were on the ridge assisting marmots. They were excellent at cooperation. Work went quickly. Time to finish activity truncated. The raccoons were on the way with a black brood from the dark forest south of Grand Meadow. It was in this dense fauna Ralston lurked until King Olaf gave him the sign to advance.

Nothing was going to stop his revenge. He knew Murray and Mr. Mizithra were together up at the Valley of the Stones. The two who denied him a Pacific Northwest banana slug dinner would pay with their lives for such insolence. His spies knew not the number of woodrats, but counted many. They did not know about marmots either, who stewed nervously underground awaiting the coming storm. Those in chambers and tunnels prepared castle keep, were the worst to happen. All Ralston knew was he was on his way, King Olaf issuing the order.

He had reinforcements sneaking through the newly claimed region of the demolished bungalow. Why they razed the club became apparent. It had to die to provide cover for operations to move into attack position. The key was to arrive undetected. To accomplish this they moved at night, quietly along the path between road and rapids.

Raven spies above and Eastern gray squirrels below were very limited in their ability to pierce defenses and see the goings on in the valley. Their chances to view the defensive work undertaken was better, but not much. Crows blanketed the skies. Marmots and woodrats sent constant patrols against invasive squirrels and burrowing voles. They wanted Ralston and Olaf blind.

Any slight advantage was a blessing given the larger size of raven to crow and raccoon to marmot. Were they to breach the bulwarks above, the enemy would pour downhill, along a narrow, windy pathway, then kill the entire population and win the war.

Not one marmot, woodrat, crow, otter, muskrat, beaver, or Keen's mouse let this fact escape them. This was a war to end all wars for these animals. They had to defend home, hearth, and a way of life. Ralston and Olaf be damned, they were not going to lay down without a fight. If they wanted war, they would have it!

Chapter 11 The Reinforcements Move

Defenses were modified. Ralston sent word via an unkindness of raven messengers to move the massive, undetected army up the path by the human road. They used every method known to avoid crows see them arrive. For the reinforcement's path, rapids soon slowed, turning to their right or south. They were to move along this course, crossing water at a calm spot, then attack the enemies' right flank.

The shock of such an astronomical amount of raccoons, Eastern gray squirrels, and voles would overwhelm this side of the fort, creating a gap in the wall. Thus exploited, others fighting could join the downward flow of troops. This would ensure swift victory.

Mr. Mizithra and Murray were to be spared, toys for Ralston. His hatred for them knew no bounds. They were to be beheaded, heads up on pikes displayed for all to see. Word arrived. The reinforcement's leader, Cheston, Camp of Soldiers, a large, strong, natural leader, gathered company captains together to advise the plans. It would be dark in an hour. To their great good favor, it was going to rain all night with gusty winds. This would help shield their approach.

The road laid them open to crow's prowling eyes, eliminating the surprise attack. Moving in total darkness was wise. To have rain and whipping Olympic Mountain Range winds would send any crow to shelter from the storm. No land animal would venture forth either.

Perhaps an otter or muskrat could keep an eye on the road from the pond, but rain would cut down on this possibility vastly. They would hug the wall, running into Gilligan's Island. Once past this deserted site, a turn right and they would disappear in complete black cover in the forest. The only danger was this road. It was of utmost importance they get past the pond, rapids, and road to the thick timberland before dawn.

Once the sun arose, rain usually tapered off. They would lose the blessed cover. Cheston informed the officers to keep moving. If anyone complained, they were to be beaten. This alone would stave off whiners. Down to the weakest vole they had to complete their mission. Cheston was a powerful warrior. He was the natural choice for Ralston.

No other raccoon was closer in strength and reputation than he. With Cheston leading the reinforcements, victory was assured. Logistics coordinated, they set off to join the war, no communications permitted. Were a raven messenger caught and spilled the beans, all would be lost. It was imperative they arrive by surprise. Shock troops would overwhelm an already weakened defense.

Ralston reckoned marmots and woodrats could last for a few days at most. He could throw wave after wave of troops from above, under, and on the ground. Raven air forces were ready, a secret weapon added to ensure air superiority. Eastern gray squirrels amassed a huge regiment at Ralston's lair. Cheston's mob would arrive precisely at the proper time.

They marched with the head of the war, General Thackelberry. An old, rodent, he hated native Douglas squirrels, Townsend and Olympics chipmunks, and marmots. He alone rallied these rotten rodents to unite with Ralston and fight for the annihilation of those despised. He gladly threw his hand in alliance when approached by Olaf. He and General Thackelberry had an agreement. He took the ground, the royal raven could have the skies.

The old soldier pulled every string, called in every debt owed, and assembled a mass of squirrels large enough to impress Ralston. He sent half to join Cheston led by General Thackelberry. The other half would join Ralston and move forward to engage the enemy at Fort Ballerina, then on to final victory in the Valley of the Stones.

As his choice for this half of his forces, the aged General selected his old friend, Colonel Duisburg. He was a well-scarred battler, known and feared from one end of the Olympics to the other. General Thackelberry knew no better Eastern gray to lead forces into battle. Ralston knew Colonel Duisburg from past joint ventures. Approval was instantaneous.

They joined the voles, led by Princess Hera, first and eldest daughter of the vole king. This aristocrat spent no time as a dutiful wife, instead fought along side her father after her mother's murder by a goshawk.

King Xenophon quickly remarried. Hera remained by her father's side, a warrior feared by all. Voles were led by their leader, the young and brash son of their king, Xenophon (soldier and mercenary). The king literally rose from a lowly place as a mere private to general then royalty.

He was brave, a military leader with a genius for strategy and tactics. His days of running off to war were long gone. He ruled, his children carrying on the proud legacy. Other vole species refused to join his fight. King Xenophon warned any who chose neutrality, he would return after the war to settle with them. He sought to rule his kingdom of diminutive burrowers and enslave those who defied him. This was one mean little guy.

His son, Prince Vasilis (magnificent knight) led the mass of voles at the rear of the column up the road. They did not like this weather, yet he permitted no complaint. Death instantly befell any who dare whine about conditions. He told his vole vanguard any objections would be viewed as a direct insult to him and the king. Vasilis had great respect. When his father went on to rule it is this son, not the seven older ones, who became the greatest warrior in their ranks.

Vasilis was part of his father's seventh litter by queen Anfisa (flower girl). Mad, she would only deliver litters in a bed of petals. As offspring of the king, he was a distant prince, but rose as did his father from the bottom. He fought along side his older sister, Hera, so impressing her above all his older brothers, she added him to her armed forces. As the seventh son of the seventh generation, he was considered special.

He, Thackelberry, and Cheston began to trudge towards the fight out of the dank, dark hovel in which they congealed. Raccoons led the way, silently marching a fast moving column through dense woods. Once out of the foliage they spied the desecrated Beaver Blues Bungalow. The deep of night and nasty weather aided camouflage. They shot past the pond, staying far from it to avoid detection from spies in the water.

A patrol of young raccoon scouts went ahead, checking the point before these troops arrived. They wanted to be certain no living being found the forces. The journey's worst portion was the climb up the path by the rapids. They were exposed. What if a truck came by? Were Thumbs to see such a site, they would surely stop and destroy the whole surprise.

Aware of the timetable, Ralston began moving forces out of his rock lair south of Grand Meadow. Happily, he was shadowed by a black cloud of ravens escorting him to warfare. Rain and wind played havoc with the avian air force, but King Olaf was leading them personally, so they were not about to stray from formation over the raccoon army. A determined ground force moved south and east towards the Valley of the Stones. The ultimate battle was to destroy the marmot and woodrat enemy.

Completely insane Ralston the Rancid Raccoon, almost foaming at the mouth, led his ground forces with a greasy, grimy grin. His cold heart and evil soul focused completely on the task at hand. He would not be detained any longer. Everything planned was in place. Alliances with the ravens, bats, squirrels, and voles were paying off.

The few beavers with his forces, were led by a beaver who did not like Murray or Mr. Mizithra. He was not from Jimmy Joe John Brown Jr.'s colony either. He was Zanzibar, Bette's spurned suitor. She rejected him shortly after Murray arrived and began to play sax with the band.

He and Miss Bluesberry were adequate to turn the tide. Bette did not really want to marry him. With new band members, she ended the relationship. She told him the band was her husband. He came in one night, watching her rub up against that stinking woodrat. It was enough to cause hate and green jealousy to arise in his soul.

He vowed death to the rat. His dam was far from theirs. He was not around for the attack by Jimmy Joe John Brown Jr. He was, however, in the force joining the raccoons' raid. He was also not part of kidnapping Bette, Tobias, and Twigs. He went to Ralston's lair instead, to gather his colony's contribution to the campaign. He brought a mighty force of fifty beavers, male and female, who desired to conquer Bette and her army.

As Ralston's forces moved towards the battle sight, others made the way to the pathway up the road. The journey uphill would take hours. Converging with the leaders before this ascent, Cheston chided the small ones to keep up. They could not afford voles to drag at the back. Prince Vasilis angrily charged the raccoon to set the pace. He was not about to be insulted merely because he was smaller than the masked bandit. Cheston nodded with a sly smirk.

He had a difficult time offering anything in the way of respect to these little buggers. They had a purpose, to burrow under the works and in hollows, but were virtually useless in battle. It took a dozen to take down a woodrat, let alone a marmot. A raccoon could swat woodrats away like flies to get to two marmots at once. He was an egotistical jerk.

The trek uphill had scouts way ahead, making for the turn in the road, past Gilligan's Island. Rain and wind made crows remain grounded. No way they could fly in this stuff. The same was true for any Douglas squirrel or woodrat. It was too nasty to worry about spies seeing them, provided no one from the water spotted them. Traveling via the road made the time and distance go much more swiftly than when Murray and Mr. Mizithra came up along the rapids.

The dark of night kept them faded in shadows. The only thing which could stop them now was the voles slowing things. Cheston was kept up to date on their movements. Prince Vasilis, true to his word, was directly behind Eastern gray squirrels in perfect formation. The surprise force wore the same disgusting look painted across Cheston's ugly mug. He loved battle. He loved the taste of blood, to see his enemy vanquished, to eat his kill. Cheston was a massive beast and fought to prove domination. Ralston promised a kingdom of his own after the war.

It was Cheston who drove the stake through Jimmy Joe John's back. He thrust through the dupe, then finished off Beulah and Buford. He lay scat on the kitchen floor before exiting the demolition derby he fostered. Others took the prisoners. He sent Igor to Ralston, while he went off to organize the disruption for the marmots. He took ultimate joy knowing they had no idea what was coming to their right flank in two days.

Chapter 12 The Battle Begins

Spies arrived, bearing horrible news. Ralston and his minions were closing swiftly southeast. Braced for the advent of day, Major Mizithra directed Lieutenant Burrito and Captain Milk Chocolate to defend Fort Ballerina. Male and female Olympic marmots and woodrats filled the trenches. Green trees, sagging with an air corp of black crows, awaited battle. Lieutenants Delroy, Clive, and Neville eagerly scanned the skies for the huge unkindness of ravens to blacken the sky. This would be an aerial display heretofore unknown.

No air force so large gathered to do battle until King Olaf called for this daring plan. Those in troughs got encouragement from leaders. Each marmot held a spear. The rows were solid. They felt no voles or squirrels would get through. Raccoons were quite another thing. In an effort to thwart the raccoon menace, Captain Milk Chocolate supervised the erection of wood walls, as tall as possible. By marmots lowering beams on moss vines, woodrats could line them in place. Now, the only thing visible was the occasional opening filled with spear points and angry rodents.

Some walls reached four meters high, the shortest just shy of three. This would force raccoons to climb to get to them. The plan was to have small holes in the wall to harpoon intruders. They could pick off some, weakening the invasion. They could stave off a number of raccoons, stop Eastern gray squirrels and voles. They felt groovy with these defenses.

They counted on crows doing their job. Thousands filled the trees, squawking and cawing loud encouragement. If they were to defeat these ravens, they had to stay strong together. No crow could shirk duty today. They had a trick or two up their sleeves. One thing, they had practiced formations to combat the larger foe.

The flock mocks fleeing, suddenly dodges, allowing another murder to swoop in, then massacre the unkindness. The lieutenants trained others daily until it became second nature. Guards flying overhead would caw a warning to stop and fool Olaf's spies. Knowing the enemy was on it's way had the young, untested combatants stirring at the unknown.

Just then, the sky darkened from the west. Air yet dripped moisture from the previous evening's stormy weather. Ravens swarmed at great speed towards the marmot's left flank. Below, a ravaging force of voles and Eastern gray squirrels emerged from the dense underbrush. Captain Milk Chocolate wisely cut shrubbery back ten meters to expose any enemy approaching. Once in the open, any advancing enemy would be spotted. This would make firing at them effortless.

Raccoons came directly behind, with long-legged bats above. It was still dark enough they felt comfortable flying to fight. As smaller enemy soldiers approached the ramparts, a shower of spears and rocks assailed them. Rodents fell in death. Still they came, undaunted by a few dead. As they neared the bulwarks, voles began to dig. Eastern gray squirrels made for tunnel entrances.

The raccoons halted, assessing the wall. This they did not anticipate, but upon conferring with Ralston, came up with a plan. A few tried to climb up, but were stabbed with spears in the belly and fell curled up in dire pain, some with spears sticking out of their guts. Soon they had enough holes spotted in one section, they could implement plans to scale the rampart. The largest raccoon hoisted a fellow up top, followed by two more in succession until the last one could see over the side. She signaled back to Ralston the Okay, then turned back to give a report over the wall.

With her data, they could exploit any weakness in the divider. Just as she got turned around, a massive shillelagh hit her smack dab in the schnozzola. Orlaith, Irish for Golden Princess, head of the otter force, on guard with marmot and woodrat allies, took her old granddad's club and belted the sow right in the gob. She flew to the ground, out cold, blood flowing freely from her mouth. Teeth on the left side of her face vanished.

Voles dug ferociously, soon halting. Digging was normally 20-25 centimeters down. They hit solid pine. This meant a much longer time until they could breach the walls and overwhelm the enemy. They dug down 45 centimeters. They finally hit bottom. Word went back up to the raccoons. They were in position to dig through the floor.

When Ralston heard how deep they dug, he was concerned and affected. Defenses were better than when he wiped out Mr. Milk Chocolate's colony last year. He was impressed voles so quickly reached the deeper destination. He mused how fun it would be to witness the vole vanguard rising out of the soft soil to shock marmot, otter, and woodrat. He would love to see them and the squirrels come to take out many.

Eastern gray squirrels dove quickly to contribute to the fight. This largest squirrel in the Olympics hit passageways with zeal and strength. Many were along this left flank. Some fed around skirting down. This was the far southern end of the rampart, the very end defensive spot. The wall wrapped around the edge, allowing it to slope downhill. Visible tunnels at this end numbered eight or ten.

A large scurry of gray invaders made for these, whilst their fellow soldiers filed in along the rest of the bulwarks' empty holes. The thought was if enough could get inside the trenches, defenders would be forced to flee. They could kill more in fear and retreat. Either way, these rodents' efforts would disrupt marmots and woodrats so raccoons could pierce the wood embankment and slaughter the miserable foe.

Squirrels who made for the lower tunnels on the sloping side of the wall had an unexpected fate. These holes fed the same way. They ran for a meter, then turned directly down on a blind corner. Squirrels raced for this end.

Quickly, they found their mistake. Too late, the tunnel spilled, falling straight to the valley floor.

Large stones greeted those who fell, dashing nasty, little gray heads against solid rock, killing all who fell. None made it back to warn others. One after another Eastern gray squirrel met the same fate. It was another Captain Milk Chocolate innovation.

Not waiting for word from voles, Ralston assumed they would emerge through the floor at the coordinated time. He called his forces to storm the ramparts. He selected a focal point then shouted to move forward. Seeing this from above, Olaf knew to initiate his attack from above. He swung his unkindness into action, swooping down toward the near treeline filled with smaller crows. They gained speed, shocking the crows who burst out in a million different directions.

This was expected. In a flash, they were one. A triad of lieutenants brought three seemingly disoriented groups together. The now organized mass murder, if you will, of over twenty-five hundred ebony warriors quickly turned left, rising high. They shot to the precise place the ravens had to be. They could not turn.

However, as trained crows were about to turn tables on these foul invaders, from the left above came a blur of black and green. It was a tight chattering of European starlings employed by Olaf as a distraction. This intruder from another continent made it's grubby way across North America, eventually ending at the very end. Here they continued their gypsy ways. This chattering was about five hundred strong.

Their rigid formation hit crows as they made the final turn at the ravens, throwing them off course. This allowed time for ravens to recover, who now made for crows with angry caws of vengeance. The European starlings, for their part, headed swiftly, cowardly for fir trees on the far side of the valley. Here they watched the fight, could join if needed, and stayed safely out of the way. They were too small to fight a crow and win.

The mission Major Mizithra sent Lieutenants Zor and Zax and the rest of the Keen's mouse army on paid dividends. The only hold outs who wished to serve, but not fight, a sort of conscientious objector, were the Douglas squirrels. It was not that they did not hate raccoons killing some of their own.

They were not fighters, preferring to be of service in the medical field. Most trained by Graedy and Molly were this species of rodent. They did not like to fight. A scurry was usually found playing darts, arguing surgical techniques, not how to best an Eastern gray squirrel in battle.

Zor, a persuasive lady Keen's mouse, issued them an ultimatum. They accepted with pride. Gathering forces under General Douglas MacArthur Squirrel, they snuck in shadows and stealth for the Valley of the Stones. Squirrel forces numbered five hundred. The mice had three thousand. This was their moment to shine.

As European starlings landed on branches, they were swarmed by defenders in the form of mice and squirrels. Shock troops were shocked, falling in death, feathers wafting in the aura. Within moments, every European ally lay dead on the slopes of the Valley of the Stones. Once finished with the small stuff, they turned to the swooping raven air force as it broke into three formations to match and overtake the triad of crow ones below.

To do this, one of the unkindnesses turned close to the very stand of trees European starlings found as a last resting place. As ravens did, rocks reigned down from within shady branches. They were pelted by bullets with deadly accuracy. All those nights swilling cattail tea and shooting the breeze, Douglas squirrels learned to zero in on a dartboard with focus. Ravens paid the price for skill. As this unsuspecting force fell dead, it's number, once one thousand, mustered a mere two hundred and fifty.

This was all they had left to attack the twenty-two hundred strong murder. With such shocking devastation, reduction in numbers, and advantage against a shaken foe, Lieutenant Delroy's squadron did a fine job mopping up a full one third of raven air forces. In a flash, they joined the crow air corps. This united force now focused in on one target: King Olaf, above the fray, encircled by royal protectors, the Blackguards.

Word was sent to the front via messengers. He could see everything. To his horror, a third of his forces lay razed by trickery at the treeline. Whomever it was, they countered his twist with the European starlings. This meant the marmots knew about them already. Moving this secret weapon in position was done with utmost care, yet somehow they knew.

European starlings headquarters was in a very large field in which they could feed freely. They moved about in chatters of a hundred or so all over their little realm. Unknown to them, trees which surrounded this field were home to a hidden force. It was the home of the Gliding Ghosts.

They filled Mr. Mizithra in on the foolish starlings. He told General Lloyd Joseph. Keen's mice convinced the Douglas squirrels to fight. A coordinated counter attack from trees was a success. European starlings, five hundred strong and a seven hundred and fifty unwanted raven unkindness rotted on the slopes of the valley.

Dozens of Eastern gray squirrels lay dead, stabbed in the back of the head with a sharp jab. Tunnels bulged with corpses. They could not turn around, sitting until slain. Rocks collected their brethren, who swiftly fell from delusory, hidden passageways.

Voles finally made it directly underneath the trench floor. The signal was given in a dozen places. On a count of ten they would dig through, sending hearty warriors through the hole to surprise defenders. They counted, then dug the last dirt to expose the bottom of the trench and effect a shock. Each lead vole struck solid rock. The noise, expected, was met with a shock of their own. Orlaith and her shillelagh had friends. A holt of fellow angry otters, equally armed, whacked on flat stones under their feet as hard as possible.

Impact waves burst voles' brains, who died in droves. From above, marmots cast rocks upon the vole tunnels, trapping them until woodrats could scurry in and finish them. The air force weakened, his smaller forces failing to penetrate defenses, Ralston sent for reinforcements to come break through the rampart on the enemy's right. Raccoons and remaining ravens could maintain a sturdy attack until Cheston moved in place. The tide would turn. He would be victorious. Those in the trenches felt safe, but did not stop trying to kill the enemy.

Raccoons attacked the walls, seeking weakness. Bats flapped in the belfry, trying to drive the enemy batty, but were pelted with stones. They departed, telling Ralston he was on his own. They did little to help. Their cowardice was expected. He dismissed them out of hand.

Thus far, the battle had not gone as expected. Their spy network was ineffective. Walls were a surprise. Animals in trees counter attacking the shock troops was not anticipated by either leader. Zanzibar had a force of beaver busy chewing through the wooden wall. If they could clear a large enough hole, raccoons could pour through and stave off the rodents until plenty could enter to slaughter them. He ran into pitch everywhere.

Thus finalized, a nice jaunt downhill and the rest of the marmots, woodrats, and foolish allies would perish at his hands. This gave him monstrous hope. It was only a matter time before the mass Cheston was springing on unsuspecting marmots arrived to win the day. Ralston, totally lost in his own egotistical mania, was so happy he could burst.

Chapter 13 Surprise

One day and the night before Ralston hit Fort Ballerina, troops of Cheston, General Thackelberry, and Prince Vasilis, trudged in pelting rain, driving wind, and a totally ebony night. They marched in shock and awe against an unsuspecting foe. This was a miserable, challenging mission.

Each leader was proud of this opportunity to terminate the enemy. The army would be sufficient to breach the ramparts, changing the tide in their favor. They did not know about defenses up front due to the communications blackout. Ralston's and Olaf's' logistics were sound. The idea of having a fat force come on the blind side was a wise war plan.

Point guards conveyed word from up top. The pathway south was free and clear. There was no sign of any enemies. The sun would soon rise, spreading light upon their hidden attack. No time to lose, Cheston sent word for the scouts to make immediately for Ralston and give the good word. They could be at his side in hours. The raccoons could take swifter paces after crossing the stream. They would rest a day, then pounce.

Once across, no impediments lay in their path. Squirrels and voles could enter the battle after the attack Cheston was going to lay on the marmots. All they had to do was get to the top, past this old pipe, and off the road uphill. The forest lay before them, across a wide stream. They would ford a short distance from the curve ahead. They would assist little guys across, then speed the next day to the war zone carrying voles on their backs. Cheston was foaming at the mouth in heated anticipation of what gore he was to find at the front.

The point of point guards is the concept of ensuring the large force behind them does not face unseen attacks. Yeah, well it is a nice theory. Often a large force is hit when an awaiting army allows the small fry to slip on by. They then take the big fish. The guard came to clear the path for Cheston and his troops, the only oddity: an old pipe.

The rim was canopied in rocks. Lichen hung thick in front. A trickle of water ambled along. Nothing across the road alarmed. Not wishing to chance crossing and being hit by a vehicle, they lazily lingered. The place looked long abandoned. They stood deciding whether to cross the road. A noise around the corner came as the rapids turned to slow. They saw a stir in the water and zipped to investigate. One of them signaled all clear to Cheston, then ran with his buddies to solve the mystery of this splash.

Cheston moved quickly, realizing the advantage of being away from the road and rapids. Once in the forest and across the stream, they were close to the quarry. He burst with joy at the prospect of the savagery he was to visit on these upstarts.

Nearing the pipe at Gilligan's Island, an even louder sound shattered the silent night. The moss lining the front of the tunnel flew away as rocks burst like cannon shot, blasting Cheston and front row of raccoons with deadly fodder. Some turned to flee, but what emitted next entirely solved that problem.

A stream of water came out of the pipe at a rate of speed so forceful, it cleared raccoons off the road into awaiting arms of swiftly descending water. The H2O cannon did not discriminate. Eastern gray squirrels and Western heather voles tasted the spray. Those who did not were met with an army of marmot, woodrat, otter, and beaver who fell upon them, killing any survivor. Cheston and his entire reinforcement army lay still on the road. Dead bodies bobbed up and down in pond water below, an odd payback for the Beaver Blue Bungalow Massacre.

The scout force found nothing of the dissonance. Confused, they thought it sounded like a stone falling into water. They turned to head for Ralston and their good news. A louder sound of the gushing pipe hit from behind. Instantly they turned to go back and see what was what.

Marmot, otter, beaver, and woodrat befell them dispatching the lot. Ralston decreed a total blackout on communications. He never knew the troops reckoned on to shock the marmots and woodrats were caught in Operation Surprise. The disruption planned for rodents' end hit a real snag. No backup was coming. Ralston would soon discover this sad truth.

Day two and his soldiery was found faltering. They stalled. Voles and Eastern gray squirrels were utter failures, long-legged bats a joke. He watched in the soggy, gray sky as a huge, determined murder of crows flew at King Olaf. His Blackguards flying firm, other ravens flapped vigorously to his side. They were unable to quell the crows in this mass murder, who defeated the unkindness. King Olaf finally admitted defeat, endeavoring to flap off.

He neared trees on the close side. Lieutenant Nicolo and his Gliding Ghosts flew in from all sides. Olaf could not fly out. Gliding squares of fur mauled him to death. Plummeting to earth, his regal corpse bounced off rocks. It rested like all other dead. His body looked like everybody, so much for royalty. The few remaining ravens sounded retreat and were pecked out of the battlefield until gone from sight.

Major Mizithra sent forces to Gilligan's Island to effect a surprise. General Lloyd Joseph knew of the build up from spies blanketing the area around the bungalow after it's demise. He knew there had to be a reason to begin war at such a specific place. The only logical conclusion was so Ralston could run troops up quickly to reinforce his flank. With the club razed, his forces could pass through undetected.

He thus ordered Major Mizithra to send Lieutenants Baklava and Bubblegum to this grid point. Forty went from the Valley of the Stones with this plan. Muskrats under Lieutenant Ripple came to the pipe, then burst through clearing the opening. It collapsed over the past winter. They improved it, lined it, placed a mossy facade in front, then blocked the entrance on the Gilligan's Island side to keep water dammed.

Marmots, otter, and muskrats dug out the abandoned boat exposing the old homestead. They came up with a little surprise for the jerks coming to catch them unawares. Earth and rocks filling the old boat were hoisted out and piled up towards the road and rapids. This built up the walls of cover from approaching combatants.

Once empty, it filled with water from the spring at the back of the chamber. By digging, otters exploited the flow, increasing it tenfold. It gushed, reaching it's high point. Restrained liquid awaited true freedom. Cheston and his motley crew caused water to surge. The gusher's force knocked enemy soldiers off paws to death in rushing rapids.

Sensing something wrong, Ralston sent a gaze of raccoons back to find what the holdup was. It had been far too long. He was at a stalemate. Defenses were weak. Dozens of marmots and woodrats lay dead inside the walls. They were breaching the dividers in places. The siege was under way. Raccoons proudly taunted those on the other side.

Under strict orders not to speak back, raccoons ran into a wall of silence. No one responded. They ceased. Psychology only works if you let it. Major Mizithra knew too much to let him get fooled in a word game with raccoons. It would win over a rodent. It was how it had always been.

His messengers returned, winded and excited. They saw scenes of death. The victors were in the process of heading to help those inside the ramparts. Ralston knew it was time to do what he hoped he would not have to; it was extreme. Every other option was either gone or no longer possible. Ravens, bats, voles, squirrels, were gone. His backup was washed away and slaughtered. He was outsmarted at every turn by an old crow sitting in a lousy cedar grove down in Whiskey Bend.

It was now or never. Ralston's hatred caused him to get smacked in the head by the jeep that day. Marmot and rat still lived. He could not stand for them to exist while he drew breath. Nothing would deter him from causing their demise. He knew the next choice was unforgivable. He no longer cared. His sole purpose in life was to get a hold of those two rodents and kill. He made the choice. He sent a small gaze off to the cave.

Chapter 14 Igor

Raccoons made haste to Ralston's evil lair, retrieving the ultimate weapon, a black bear with attitude. He was a friend to Ralston from way back. It is true raccoons and black bears do not usually become bosom buddies. A bear will eat a raccoon. It will eat anything. They met when Igor was first on his own. At two, his mother prepared to have another litter. It was time for he and his sister to strike out on their own.

Kira died her first winter alone. She did not gain enough weight to live off her fat reserve. Anorexia does not cut it in the Olympics. Ralston met Igor in his third year on a high spot. Ralston had a marmot dinner, when a starving Igor approached. Rather than fight, Ralston cut a hunk for himself, offering the majority of dead rodent to the hungry bear. This formed profound trust and friendship.

Igor stomped on all fours through the flora at an extraordinary speed. The raccoon messengers could not keep up with him and were soon out of sight. He wanted to be of service to his friend. He would only call Igor in an emergency. Igor was not bright, vocabulary meager. He knew Ralston needed him. He was on the way.

Skies clouded. Wind tossed leaves pellmell in arbitrary whimsy. Raindrops drenched soil, clamoring to strike Mother Earth. Lightning blasted a blackened sky with bold, yellow thunderbolts. Fear gripped defendants. Approaching the wall, Igor saw his chum, hastening to his side. Ralston spotted his grievous comrade and called his forces together.

Before the bear arrived, Ralston gave new orders to his diminutive warriors. The few beaver, Eastern gray squirrels and voles remaining were told to stand down and get well-earned rest. They were to go eat and drink in the rear. For now, they were not needed. Truth be known, they did not want the bear witnessed.

Exhausted beavers, squirrels and voles did not need to be told twice, immediately heading to rest, eat, and drink. Sleep and tending wounds were on the menu. The battle was costly for them.

Most forces died in horrible ways. Vole brains burst with shillelagh shock. The Eastern gray squirrels had a case of back pain. So many were speared right through, they abandoned the tunnels completely. It took a while, but they finally quit losing soldiers through side tunnels. It was an ingenious weapon, costing only a bit of thought and digging.

Igor was quickly led to the weakest spot in the wall. To the revulsion of those in direct sight, a mass of black fur fell on the ramparts, tearing timbers out one by one. Troops were ordered to evacuate, Major Mizithra and Major Murray last to depart. They set a few booby traps, then made for the pathway down.

Here they had to hamper the enemy. The problem was the immense weapon Ralston now employed. It was lowest of the low to use such an animal in any turmoil. Species were free to enlist others, but hunters were strictly forbidden. Did Ralston wish Mr. Mizithra to employ a cougar? It was never done, never before, but a bear was shredding their strong defensive bulwarks.

Through the first layer, he tore relentlessly at pine stakes, ripping them out of the ground. He tore his paw on a stake driven in the bunker's roof, pausing to lick the wound. He leaped back to work, tearing out the back wall, exposing the footpath down to the valley and victory.

Igor led the invasion, raccoons aplenty happily surging behind. Due to the narrowness of the path, it was single file. Trees above swayed in the wind, rain dripping off summer leaves. Flashes of nature's spotlights filled the sky with color. Moving forward, those same trees pelted them with a steady flurry of rocks. This did nothing to the bear, but took out a few raccoons. They were stuck in the way. Many fighters were slain or injured.

At the steepest portion down, Igor stepped in a nicely covered hole, filled with sharpened stakes. He tripped and fell the rest of the way down to a pile of awaiting stones. Raccoons hastened to aid him, avoiding the same opening. This caused them to leap over the gap. On the other side were more stakes driven in ground, sharpened and covered.

This took out more, stalling the army's descent. Igor was at the bottom alone, injured. His paw was stabbed, next his ankle twisted in the hole, foot stake stabbed. He fell to the rocks, facing an angry mob of little guys. He finally struggled to his hind feet lifting his front paws high in the air and growling his nastiest black bear growl. Blood flowed down his face to his muzzle, dripping in mud, mixing with rain water.

The descent and subsequent pain made him goofy. He fought back. Dozens of marmots and woodrats assailed him, but his bulk was too much. He waded through them, tossing defenders off to the side. He moved to the last vestige of marmot safety: castle keep.

Mrs. Saffron, other mothers, the young, the old gathered to wait out the battle. A strong battalion of male marmots stood vigil as a last gasp were Ralston to breach the wall. Some of the strongest males were held here in reserve for this precise purpose.

BOOK 3: WAR

No one ever imagined he would sink so low in the gutter to have a bear do his dirty work. His army lay in total defeat. Olaf's air force was obliterated. To use a bear as a last resort was really too much. Regardless of the outcome of this war, Ralston's days were numbered. Even K'wati and the wapiti could not ignore his intolerable actions. He would be excommunicated from the mountains. Shunned, he would be forced to the lowlands, there to remain, hunted by man.

While marmot and woodrat tried their best, it was to no avail. The bear was simply too big, too strong to stop. He pawed at tunnels to find a nice snack of young marmot meat. Raccoons engaged marmots in battle. Crows swarmed, aiding as best they could. Raccoons slowed. Their only hope for victory lay in this nasty ball of soggy black fur. He was tearing it up with zeal.

Defense was on it's last legs. They were only able to hurt his injured paw, feet, or sprained wrist. Blood surged from his wounds. Adrenalin of battle boosted his power. In a few more scoops of dirt he would reach the deep chamber and extract supper. No marmot had been able to hurt him. To get to the last bit, he came around to his left, nearest the ledge of the valley. The battle flowed around him, but he was ready for marmot jelly.

Making a vicious point, Igor reared on hind legs as a lightning bolt struck above his head. This was his final swipe at the cover for terrified inhabitants in the chamber. Bright lights lit the beast's drenched face to Major Mizithra. He was a few meters away, trying to get to the bear to kill it. The lightning exploded, taking his breath away.

There, on it's hideous black nose, was a huge marmot tooth sized hole. In that timely flash, he knew. He saw. It was the same bear who ate his daughter, his precious Miss Cake that sad and fateful day. Reason left his mind. The only thought occupying his thoughts: kill this piece of waste. The opponent mattered not at all. He was possessed with hate. This is not wise, but nothing could turn off this spigot!

His eyes filled with dense rage. He flew at the bear's head. Igor did not expect this, looking at his opponent in the face. Though a moron, he instantly knew this was the one who bit him, father of the little marmot girl he ate with his mother and sister. Mr. Mizithra tore again at his eyes and face with the fury of a marmot gone mad. He did not think of the battle, only revenge for his family's horrible end.

Igor had enough. He tore Mr. Mizithra from his face, biting at him, receiving a marmot's swipe across the muzzle. He threw the stinking varmint down to get a better grip and kill the thing. He no longer cared what Ralston wanted. This was the one marmot Igor hated. All bets were off. This meat was his!

The safety chamber lay completely exposed, it's marmot contents spilling out. Some joined the fight. The rest huddled young and elderly to safety on the southwestern rocks. Here, a few tunnels were left for the taking. The bear could not move these massive rocks. Only Ralston's small rodents could get through. They were on rest and relaxation.

Out of the chamber, a guard came, spying bear and marmot.

It was Sergeant Guar Gum, his old friend. He held no responsibility, refusing to become an officer. He volunteered to guard the weak and wounded. He was not afraid to fight. Mr. Guar Gum was as strong as his friend. Wrestling gave evidence. He and several others not inclined to take lead in the fight would serve best guarding the most precious of the colony. It was not a job looked down upon, rather an honored necessity.

The bear chased Major Mizithra through the pond. He crossed some rocks. Igor climbed in hopes of diving on him. Major Mizithra was a dead marmot. The split second before Igor launched on his intended victim, another marmot flew out of nowhere, hitting him off balance. Sergeant Guar Gum stayed with his foe, the weight added to the bear's head caused both to plunge over the side, to the abyss one thousand meters below.

Sergeant Guar Gum gave his life for his best friend. He could not measure up. He was never jealous of Mr. Mizithra. They had been friends since childhood. It was impossible to have bad thoughts about him. He had one life to give and did so without a second thought.

Falling to death, he knew the sacrifice worth it. It was so for him, his friend, and the colony. Both hit ground at the bottom of the long fall. The shock of having two such bodies hit was immeasurable. Igor fell in front of a diminutive waterfall in a small chasm and stream trickling past the Exception's cabin. Yes, the Exception was still alive, in his free world. He knew nothing of the world above.

He knew a misty flow made rainbows in the sky from time to time. Now, it was raining bears and marmots. Utilizing his way of contacting wapiti, he sent an emergency signal to K'wati, leaving home to see why it was showering such large raindrops. He looked up to the ledge so high above, wondering what was the cause for falling animals.

Back on the Valley of the Stones, Igor's death spelled doom for the raccoons, who instantly made for the pathway away from the fight. They ran in fear, again pelted with stones as they tried to escape. The poor ones at back were slaughtered by angry mobs of woodrat and marmot. Ralston directed forces from a nice spot to view the battle in battered ramparts. He saw Igor go over the side.

He witnessed Sergeant Guar Gum's personal sacrifice, giving his life to save Major Mizithra. Why this stinking marmot incited such loyalty was beyond him. This was because Ralston shed his soul a long time back. Anyone meeting Mr. Mizithra was struck by an innate good nature, desire to help, and marvelous stories.

Anyone who met Ralston ran to get away as quickly as possible. He was pure evil. Ralston fled, path to his lair cut off by an army of muskrats, led by Lieutenant Ripple and Sergeant Twigs. He and his few remaining soldiers swam around the massive pond at Grand Meadow, down the rocky cliffs, disappearing in underbrush. The battle for the Valley of the Stones was over. Losses were deep on both sides. A number of marmots and woodrats died. Keen's mice and Douglas squirrels paid a toll.

The murder of crow murdered was immense. The only solace one could take was they won. General Lloyd Joseph selected his leaders well. Major Mizithra and Major Murray proved the proper choice. Igor died. The sacrifice Mr. Guar Gum made was the most humbling event in Mr. Mizithra's life. He was never able to reconcile what his friend did. It saved his life. In the heat of battle a most loving episode sprung, leaving a humble marmot bewildered, thankful, blessed.

Chapter 15 The Unthinkable

Ralston fled to avoid capture. He had no conscience. Escape filled his dented head. Sending his beleaguered forces on ahead, Ralston the Rancid Raccoon earned his moniker. Faithful troops headed to a rallying point behind the old club, above Crescent Cedar. This piece of garbage used warped thinking to devise a way to stall those in pursuit.

Lightning usually hits trees. Some ignite, causing small fires. Humans send fire teams to put blazes out lest Smokey Bear come chew them out. He found dying embers in the hollow of a split hemlock. With hands of dexterity, he wrangled a long branch midst the flames until lit.

He used two paws, carrying the limb to dry undergrowth on a hill covered with dry summer grass and stubby arid firs. Placing the burning branch on grass, it ignited instantly. Fire rose, reaching low limbs of pitch laden boughs. In a flash, the dry timberland danced in flames.

Ralston committed the inconceivable: setting fire to home. The bear was beneath contempt, but this was too much. Once wapiti heard of his transgressions, they had to cast him from the Olympics forever. Never had anyone exploited a bear in combat. And now this, this travesty. He took the most uncontrollable of all, fire, turning it on his enemies to escape. So audacious, no provisions, no thought had ever addressed the eventuality someone would do such a thing.

Flame accumulated, whipping to the top of short firs, wispy sparks floating off in soft, summer breezes. In seconds, a fleet of dragonflies appeared, dashing in every direction to warn the forest's inhabitants of fiery death. In the woods, fire gone awry will spread due to two things: air and fuel. If you combine these, fire will not stop. Fuel feeds it, air allows it to live and breathe. The combined elements spell doom.

It is not like animals have a fire brigade. An otter or beaver can go underwater, avoiding flames, birds fly. For anyone else, it is impossible to outrun combustion for any but the most agile. Deer and elk may zip off to safety, but mouse, chipmunk, no way. This was not from man, it was from one of their own.

Unthinkable.

The need was plain. K'wati ignoring this was not good. Dragonflies surrounded their elk king. They went to any and every leader warning what they witnessed. Ralston was killing dragonflies who bore red hot news. He crushed all roadblocks, a hunted fugitive. He turned his back on what constituted his world. He set fire to kill what he could not own.

The lowest amongst us do the same. When an egotistical maniac does not win, the resulting devastation is predictable. Unable to grapple with truth, others suffer including followers. Those siding with he and Olaf cast loyalty in a fatal direction. Many, far too many, paid with life for literally no reward. They fought and died for Ralston's ego. This ought to make you consider who you follow, leader or maniac.

If you have ever been in a forest fire, one fact remains, you can never tell how fast or where it will travel. A wind whipped wheat field will go up in flames so fast, you can not outrun it. Many die in this type of fire by flame, not smoke. It is a forest fire where smoke plays the most havoc. A pine tree, for example, will burn within, pitch an excellent fuel source. When heat inside the wood structure reaches a high temperature, it explodes into a billion fiery toothpicks.

These act as fervid arrows, igniting more fires. If a dart strikes man, the pain is indescribable. The other fact: flames ascend. Scaling trees over thirty meters high, it sent sparks as messengers to spread the wealth. Fire likes more fire. This is how they move so quickly.

Without any barrier, sufficient air and fuel, Ralston did the worst thing possible. The forest was in real danger! No longer content with it's genesis, flames encroached high firs on the bluff. Were these to ignite, the fire would spread unabated. Unchecked, with plenty of dry tinder, this would eradicate animal life for a huge distance and time. Ralston's plan was insane desperation.

He wanted to terrorize those he could not defeat. In war, survivors are often the worst, carrying personal vendettas to civilian life. In a time of peace, these hatreds manifest in forms of bigotry and revenge. He was already a marmot and woodrat bigot, now Ralston added vengeance to his list of heinous talents. The worst thing about punishing those who defeat you is harsh methods hurt innocents. In this case, in his warped mind, the raccoon said all were guilty of not letting him kill them.

Just to let you in on something you may not realize, there is not one fire department in the animal kingdom. In the Thumbs world are found men and women who fight fires. Some specialize in the forests. Others fight fires in cities. Here, in the deep woods, humans jump in hot spots to stop flames. This is a noble thing people do for the whole world.

Others fly airplanes, laden with water or fire retardant to quell flames. This is hazardous work. Lamentably, fires are set by humans who disrespect wilderness. This sort of person is dangerous. To destroy beauty is a sin against creation. If you do not believe this, ask an animal. Your dog, cat, or gerbil agrees. Fire stinks, kills indiscriminately. None are safe. Heat whips, snagging birds from flight to death. It is no respecter of life.

Go in a forest after fire is quenched and see what is left. It is so sad. You see soil, burnt black, every plant dead. Animals and birds unable to leave, lay strewn in death poses throughout the forest floor. If you ever witness this horrible scene, it will never leave your memory. One stupid move, lighting fireworks in the Cascades in summer, flicking a lit butt, destroys in hours what it took nature hundreds of years to create for you.

The sin of one ruins whole forests for fellow man. They end an entire civilization for animals, who have no place else to go. Enter a home, set it on fire, you go to prison and pay for any lives you end in smoke and flames. What, pray tell, is the difference in destroying the world, the homes of animals?

Ralston saw his crime light up before in front of his twisted, evil face. The size and speed were a shock. He did not realize what he did except it stopped a detail pursuing him. The reality of sin exploded in living color. He ran as did all animals. The difference: he set it. Too late, he knew what was happening. With the speed it moved ever closer, he would burn in his own device.

When you decide to sin, ripples flow, echoing the future. A fool, he ran in fear towards his awaiting troops. They saw the flames shoot up. No one imagined he would commit such a travesty on their world. Following a bad leader is unwise, now an evil result: a destroyed forest, dead animals and bird. Nice job, Ralston.

Birds flew with reckless abandon to escape smoke. Ground animals had smoke and flames to contend with to evade death. Ralston's troops were antsy to get moving. He emerged from oncoming flames, covered in ash falling like snowflakes. In fire, it's every critter for themselves.

For example, once a dragonfly completes a mission due to fire, they are done. No more messages, they get the heck out of the way. Fire respects nothing, no living being, no man built structure. Flame does what it wants. Only a lack of fuel and air stops it. Water does this nicely.

The Olympic Range is the only moderate rainforest in the United States of America. Driving through on good, old Highway 101, hiking up in it's green paradise, you see one thing for sure: it rains. Oh yes, it has an average of 149 days of soggy. Then there are the clouds.

Traversing paths to the heights, you meet long strands of lichen. It is primeval. You can imagine dinosaurs roaming around it's murky ponds and damp, green stuff. Moss, ferns, grasses, make it wet much of the time. Summer is the only time you can count on dry days.

Here, trails swell with day hikers and the occasional camper. Most pitch tents on the Pacific coast. The view is awesome and campgrounds abound. License plates from all 50 states, yup even Hawaii, and Canadian provinces attach to campers, cars, RVs, and trucks along Washington State Highway 101 in an annual invasion.

This coolest of all cool roads wends it's way past the Columbia River at Megler Bridge, north to the end of the United States. It whips around east to Port Angeles, eventually making it's scenic way to the Canadian border. Hello, eh. This is one seriously awesome drive, folks. Take a couple of months and drive it at least once from start to finish, life changing.

Spreading with intense speed, hot air, fuel plentiful, combustion chugged it's merry away. Spotted by fire rangers, flames would be hard at work until they were able to fly a plane or two over. Or get people to jump, Pulaski in hand, and start the arduous chore of putting out a fire. This is not an easy task.

Those who fight fires in our forests are to be admired. A fire takes no break or worries about a sick loved one, they just keep burning. Those who willingly enter smoke and flame are heroes, crazy, but brave. They save not only the forest, but the lives and homes of animals and birds who can do nothing to stop heartless destruction.

In 1950, the Capitan Mountains of New Mexico caught fire. It was a nasty one. In the remnants of smoky ash, forest fire fighters found a dead mother black bear, carcass smoldering. Further scrutiny disclosed an amazing shock: dying in smoke, she wrapped her entire body around her son: Smokey Bear.

A campaign arose from this incident, sparking a nation wide awareness of the fragility of forests. The spot is marked with a plaque, if you should ever choose to climb up to the place. It is high in elevation, but a point of interest for those who love and respect the forest.

If you live in a big city, forests are foreign. Too bad. Going in one, you immediately gain a perspective and respect for what nature provides. Appreciation grows from encounters in the out-of-doors. Hiking along a mountain stream opens a mind to magnificence. The smell of nature alone is a tonic for the soul.

Care for this vulnerable environment demands we take the lead. No bear, raccoon, or chipmunk will, they cannot. It is important you honor our outside by caring for it. Fire and trash are man's pathetic legacy. Caring for and protecting green, growing things is our way to pay back and respect our leadership role.

The pyre soared, spread, killed. Nothing animals could do would stop it. They ran to safety. Exhausted, many succumbed to smoke, dying in a fiery furnace. It would be a long time until Thumbs could quench it. Suddenly, winds whipped again, hard. Clouds blackened the sky, filling the air with moisture. Rain pelted thankful animals. Blackness left behind painted a wicked pattern of death and destruction over the forest floor.

Birds and animals lay dead from smoke inhalation, many burnt. Young birds died because they could not fly. Baby animals were burned alive at home, mother and father dying to protect children. Imagine fire in your home, your parents and guardians. They would do all they could to help you. This is so for animals, too.

Rain quashed budding flames. The fire died. Ralston ran. Coward.

"Rainbow One, this is Home Base. Do you copy? Over."

"Ah Roger, Home Base, this is Rainbow One. Terry, this is Chris. Lee called in, too. Rain extinguished the fire we spotted up above Whiskey Bend. I looked long and hard through the glass and saw the flames go out. Animals and birds scurrying all over, but it's out now. It looks like a lightning bolt started this one, like usual. Over."

"Roger, Chris. Lee confirmed the same. One good thing about living here instead of where I used to work in Idaho, rain is our best friend. It never rained in Idaho when I saw fire. Here, we have nature's help. Okay, stay on it until you are certain. Over."

"Ah, Roger, Home Base, out."

Chapter 16 Ralston Is Found

Scum like Ralston cannot hide in light. The storm ended, drying out trees and ground. Paw prints a gaze of raccoons leaves are fairly easy to track. Lieutenants Zor and Zax were on them soon after the flames died a mucky death. They reported where he and his followers holed up. They were on the back side of the Great Wall of China.

He ran downhill from Grand Meadow with the tittle still loyal. This would lead below the V-Island pond. The area was riddled with caves. Hiding, he could recoup and find a way to rebuild his forces.

Ralston came near the caves. He met the welcoming committee. The southern and eastern region of the pond was lined with skunks. The surfeit, under command of Captain Loganberry and his Stinky Hill Mob, spread along the shore. Each polecat raised it's tail, scent glands ready for battle. These weapons were unbeatable. It was gas warfare.

Were Ralston to attack, a squad of skunks would converge and stink him to death with deadly spray. Thwarted, he turned back to the rapids. This led to a scene of ultimate horror. Cadavers of Cheston, Thackelberry, and Vasilis lay amongst the dead on land, riverbank, and the water. A few bodies lay strewn on river rocks, dotting the rapids. Here, power mad, Ralston saw the results of his hate. Many lost their lives in his foolish revenge over a missed Pacific Northwest banana slug supper. He paid little heed. The remaining militia were shaken.

Skunks herded them up the path, past the defunct water gun, off road, switchback bordered with skunks. Captain Loganberry's call for polecats met with huge response. They came to aid the cause. They forced raccoons to a place of refuge. The enemy saw a break.

Swimming a swift, narrow stream, the disgraceful cowards spied a rocky outcropping. The other side of the stony spot was full of rushing water, although a much wider river. A man made wall lined the entire rocky hill's length. Here Ralston's jellyfish perched, quivering.

Trapped on this spit, the standstill began. Raccoons were safe on this large hunk of stone. A few scrubby trees gave modest cover. Forces under General Lloyd Joseph could not get them from the air. It was too easy to swat any crow foolish enough to try. Animals on Lloyd Joseph's side could not cross the stream. Rushing water would take out a beaver or muskrat by virtue of speed.

Raccoons would kill any swimmer before damage ensued. Escape, impossible during day, was planned for slipping away at night. Once off the island, a spelunking network of caves lay lower down the mountain. They had to keep vigil unless marmots tried something funny. Ralston urged those not on guard to get shuteye. Once they split, it was every raccoon for themselves. The battle came to this. He was not about to suffer anybody foolish enough to be caught.

Hours passed, the summer day cooled from storms was, nonetheless, long. A few tufts of nasty, black smoke from the quenched fire lingered in midair, admonishers of his crime. Ralston and his army slept. In time a guard woke him with a great alarm. He brought his fearless leader over to the side of the island where water met it. He saw the watermark, it was very low. Were this confusing twist to continue, they would be exposed to marmot and woodrat forces. Foes would attack and overwhelm.

Raccoons astir, fussed back and forth, anxiously awaiting their leader's direction. His mighty forces were down to a handful of nervous Nellies. A guaranteed quick, decisive victory whittled to a few exhausted troops, stuck on an island in midstream. The water rushing by was drying up. How was this possible?

Enter Lieutenant Bette and her beautiful beaver battalion. Teamed with the providential arrival of Lieutenant Ripple and familiar Sergeant Twigs, muskrats did heavy lifting. Beaver engineers put the dam together. After dispatching Zanzibar and his resting minions, they made for this site, knowing the task assigned was vital to victory.

BOOK 3: WAR

Combined, a basic structure was soon in place. Exploiting the handy labor, the dike took quick form. She knew her stuff, brother Sergeant Tobias, coordinated a diversionary dam. In a short breath, the lock was done. This rerouting structure led the flow past the Great Wall of China, leaving them high and dry. The man made wall, proved a detriment. The top of the rampart was too high to leap. The only escape was the back side. Other routes were covered with marmots and the army.

Now Lieutenant Delroy and Lieutenant Clive attacked from the air. Lieutenant Neville was slain in the battle over the Valley of the Stones. The living black arrows shot darts of death on raccoons, bombarding with sharpened, ebony beaks. Pecking enemies drove some to run off the Great Wall of China, dropping to death in a dry riverbed.

The only avenue of escape was the sloping end of the island. Night was distant. Ralston tried making like Houdini. Battling an onslaught of crow pests, he had several masked bandits chuck big rocks at the closing forces. This would make marmots retreat. They could slip off the back, downhill, safely away from the fray. The sole problem: water dried so fast the island was now surrounded by Lloyd Joseph's animal army.

The only way was to plunge off the side of the Great Wall of China. The rat in front was Major Murray. His woodrat forces crept up the right side of the rock. Major Mizithra, Marmot Marauders left, tailing rodent companions, narrowed his eyes. The scar above his eye shone in the light piercing out of summer storm clouds. Wrath embraced him.

Ralston, seeing the surge, lost all reason. He rushed, screaming like a banshee. In so doing, he was instantly covered in an ocean of claws, teeth, and fur. His fury did not match his foe's righteous anger. Woodrats, marmots, and the rest fell on his remaining troops, leaving none alive. Ralston died, bloody body pitched off the Great Wall of China.

Bette and the other beavers tore the dam apart, muskrats carrying off debris. Water surged, lifting the lifeless corpse of the evil perpetrator of this unnecessary warfare. In a flash, water rising swiftly, Ralston was downstream, never seen again. It was a hollow moment. In an instant, all wept. War ended. Battles ceased. The enemy vanquished, they released pent up fear and anguish.

Once removed from a violent, disturbing episode, body and mind react. Realizing you are no longer in danger produces odd responses. Crying and hugging veterans, the animals did their best. They relieved the fear of death by consoling and being comforted by others.

It is easier to recover from trauma if you can relate your pain to someone who understands what you went through and can sympathize. The memory of such events lingers in the mind. Thinking you are over it can be attributed to a desire to stop dealing with pain. Understandable, but this only shoves the problem back down, to gain strength inside your mind. It is best to get the whole nasty episode out in the open, give it a good look, then work a realistic way to handle the aftermath.

With the rancid raccoon dead, they immediately sought survivors. The Olympics were in a state of bereavement. Families were back at the Valley of the Stones or hiding about hills and valleys. It would take time to get back to normal. Damage from fire, while minimal, was significant.

Others were displaced due to blackened destruction. New homes had to be erected. Of course, nothing again could be as it was, but they could rebuild. The work they did at Fort Ballerina was incredible. They could fix the breach in the wall and fortify other areas. The chances of another war was not realistic. The army rested for a while to recuperate.

Ravens lost their king and were in complete disarray. Lady Satin led now. Raccoons were depleted and would cause no more violence. Eastern gray squirrels, Western heather voles, long-legged bats, and European starlings took significant hits.

All, save starlings, left to recover. European starlings, paying their contribution to war no mind, stayed on their home field. They assumed it would blow over. These dimwits were not known for intelligence. Crows, in retaliation for foolhardy insolence, chased the entire species off the Olympics. They were forced to dwell only in the lowlands.

Their spontaneous celebration was intense but brief. Word came, the General was on his way, so await further orders. Skunks huddled. It began in their midst. Miss Bluesberry let loose a capella. Her voice rose above the din. All sadness, pain, and remorse melted away. Twigs banged on a flat piece of wood with a couple of sticks. Tobias began to thump his tail in time, the rest humming with the singing sow.

The Darling Doe lifted the spirits of those who fought the battle. Without her ax, Bette, a tear in her eye, went to Miss Bluesberry's side to sing. While known for six string prowess, she had a set of vicious pipes inherited from her loving mother. These ladies sang, arm in arm, tears running down their cheeks for the lost. As with songs of victory, it was a combination of joy at overcoming evil, harkening to heroes passed.

The General arrived. Leaders assembled. He gave news. The good news: Ralston and Olaf were toast. All who sided with them were dead, wounded, or on the run. He told the fate for the European starlings. This got a shout of joy from his troops. Truth be known, none of them cared for the stupid birds. They were glad to see them gone from the Olympics. Raccoon worries ceased. The population was decimated. The same was true for voles and Eastern gray squirrels. Ravens were going to be licking wounds for a long time.

Crows, an affable bunch, already reached the olive branch to Lady Satin. No one wanted war. King Olaf was an egotist. His defeat did lead to one big positive: future generations of ravens no longer referred to themselves as royalty. His was the last line of ravens to make this claim. Most of his relatives died in the war. Those who did not cleared out, making for safer confines in Seattle.

The bad news was atrocious. With widespread devastation, fire, and mass murder, the clean up was monumental. Add to this one thing no one considered: a lot of dead bodies lay about. Ralston was gone, but the Valley of the Stones was riddled with cadavers.

Lifeless corpses bobbed up and down in the pond at the beaver dam and large lake at Grand Meadow. Dead animals lay all over the battlefield. The problem General Lloyd Joseph explained: scavengers would appear. While they need not worry about vultures, bald eagles, and other raptors, it was ground forces they needed to address.

They smell the scent of death and head towards odors. Meals would be torn asunder, the best parts picked over. The next wave of dead eaters would come, followed by the next and the next, until bones were picked over by worms.

Once coyotes, bobcats, vixens, cougars, lynx and bears caught a sniff, all would converge on survivors. Hunters cared little if a meal was alive or dead. Veterans caught in weakened conditions due to combat fatigue, were quick meals. This would be a scene of carnage for their depleted army. Rather than receiving necessary convalescence, they were subject to further torment.

The young, elderly, and wounded would face certain death. Nothing could stop them. Marmots had no way to conceive such things. No provision was made, how could it? They did not wish war and did what they could to survive Ralston's insanity. He and Olaf played havoc with lives. Now, both dead, the surviving remnants had to face another evil.

The General dismissed them. The mass of warriors headed for the Valley of the Stones with all haste. Whatever else, they had to get to loved ones before it was too late. Even now, carnivores salivated. The smell of fresh dead tantalized nostrils. Dead, yes, fresh kill lay close by. Yes, they sniffed, yes, odors came from…the Valley of the Stones. Knowing there had been a battle with Ralston and Olaf in command, the predator league calmly awaited the outcome.

The carnivore class would be happy to clean the butchery. Judging the power of scent, many bellies would fill in the Olympics tonight. There would be enough with so much blood spilled, no need fight over meat.

Looking skyward, a vulture kettle (vultures in flight) brewed over the valley. It would become a vulture wake (vultures eating a carcass). They wanted to arrive soon enough to get the best flesh. Every meat eater within sniffing range heeded the call. There was no time to waste. With all speed, predatory animals gathered for their scavenger party. Their feasting numbers blanketed the valley floor in bloody, black feathers.

Chapter 17 Uninvited Guests

Mr. Mizithra and all concerned traveled as quickly as possible for home. They sent word via crows for survivors huddled at the Valley of the Stones to prepare for the worst. They made it past the deadly forces led by Ralston and Olaf. Recovery was postponed for another onslaught. Now animals feared most were coming to feast on the dead.

Already, skies filled with ominous shapes of birds of prey circling. The ramparts lay in complete disarray. There was no way to defend the remnants. Their passageway system was destroyed. What few tunnels and chambers remained intact were for the wounded, pups, and the aged. A handful of guards was it.

Those under Mrs. Lip Gloss and remaining elders, began to dig by rocks on the far southwest end of the valley. Here they could stave off some larger beasts. Vixens could get in with little trouble, but by then perhaps the army would return and save them.

The missing heroic elder, Mr. Savoy Brown, passed defending his colony against dark forces. A raccoon fell on him. The madcap marmot knowing it was curtains, thrust his spear in the foul beast's belly, sending him on a hell bound train. Both died, but the invader killed no more of brother Brown's brethren.

Crows confirmed what the kettle above spelled: death was on it's murderous way. Marmot was on the menu with woodrat appetizers. Soon, the edges of the ramparts echoed with muffled sounds. Things were creeping around inside abandoned trenches. The bulwarks were deserted shortly after the battle's conclusion. Lieutenant Burrito and others left behind by Major Mizithra took immediate care of the wounded, then started in on new tunnels as fast as possible.

Per the elder's commands, they dug straight back, a smattering of diversionary hollows to throw bad guys off the scent. This was to be as deep a burrow as possible. Where rocks lay, they were to go both sides around it. This would provide ways to make excellent escape tunnels. Once back at least ten meters, they were to create chambers.

With a short time to prepare, woodrats were employed to haul out dirt. They were not diggers, but they could carry soil. This they did under direction of the sweetest lady in the land, Momo. She organized the leftover Sweet Murray's Maniacs into a bucket brigade. They took freshly reaped earth out so quickly, marmots increased speed. Woodrats suffered huge losses in the battle at Fort Ballerina.

This first defensive post meant a great deal. They dedicated a mass of their number in resistance. No task was shunned by the brave soldiers. Knowing what was at stake, they came from all over the Olympics to serve under Sweet Murray. The only thing asked in return was for a free concert once it was over. He promised, massive smile on his face.

Enough dead critters from both sides lay in a mural of blood about trench and tunnel to occupy the pariah for some time. Larger, more evasive mountain hunters would come later. The early worm, so to speak, got the delicious marmot or woodrat meat. Raccoon, squirrel, and vole, too. Many snapped all they could force in their jaws and made off into black woods. They feared any who would come kill them for taking meat.

These earliest scurried off quickly. Vultures turned from a kettle to a wake. Wafting odors hailed a convocation of bald eagles. These noble looking scavengers circled, observing the large wake held below. Few in number, they knew it imperative to land in the precise spot to get the most meat. Setting down, the vultures would fly away to feast elsewhere, complaining the whole way.

The convocation agreed to a nice place full of dead racoons and marmots along the lower portion of the pathway. The symbols of the United States of America dove for the valley floor. As a mass of white and brown feathers, yellow beaks, eyes, and sharp claws, they landed in a huge circle, blanketing the target. They pulled at dinner with claw and beak. Once opened, innards were eaten. Sated, each bald eagle would flap away, filled to the brim with fresh kill.

Other, smaller raptors made the way to this place. Even little kestrels made the feast. They found a bonanza of Keen's mice, voles, Douglas and Eastern gray squirrels, and birds. Everybody feasted. Any living creature on the mountain was safe. No hunting necessary. The shear magnitude of butchery was staggering. One could select the Valley of the Stones, Grand Meadow, or the pond below it all where the Bungalow used to be.

Next, clearing small timers, coyotes showed. They wasted no time in the trenches, but made for the pathway downhill. It was like a buffet line with any animal you chose. Big or small, male or female, young or old, take your pick. Raccoons down to little mice were available in the After Valley of the Stones Battle Smorgasbord.

Don't forget the crow, raven, and starling portion, delicious. Were a coyote to tire of mammal flesh, a nice bird would do. You have several species from which to choose. Take time, no hurry. It's not as if your meal is going to run away. Ha-ha-ha. Wily coyotes dallied, feasting on carcasses of dead warriors. Word of battle reached them as it began. Their issue was locate the Valley of the Stones, grab kin, and dash as quickly as possible.

Focusing in completely on eats, coyotes did not see the enemy enter. A clowder of hungry bobcats, brown killer eyes glowering at the fools below. Normally, as coyote well knew, bobcats hunt at night. The scent of fresh meat lured them. Professionals in silence, they crushed the coyotes, taking the pack completely by surprise. More than a dozen were slain, dead bodies tossed down the hill in a bloody heap.

The remaining cowards skedaddled out of the region. They did not stop running until the ramparts were out of sight. They counted over twenty either dead or wounded. It was unwise to seek the injured. They vaporized, most with full bellies for the trouble. Due to superior forces they left a perfect kill zone. They were upset more could not be eaten. A meal like this was rare.

The bulk of good meat was chewed by coyotes. Bald eagles feasted on meals at the bottom of the pathway. The bobcat clowder slipped down to the valley and began picking through corpses to find the right one. The twenty odd splendidly coated felines found the vacant marmots digs. The scene revealed exposed passageways and living concavities. Toms and queens sniffed the refugees' trail.

Empty dormitories reeked of fresh marmot and woodrat. The maze of tunnels kept leading to dead ends. Being thorough, they soon realized no one remained. No scent of living marmot or woodrat was found. The journey uphill was clogged with dozens of dead raccoons, marmots, and woodrats. Nothing could escape. They came down the slope and would have detected the scent of the living, but did not.

Meanwhile, Major Mizithra and the others were speeding towards the Valley of the Stones in a desperate attempt to save loved ones. They knew, crows sent constant updates, by the time they arrived bobcats would be eating the living. Add to this sad news, the cougar population, the most solitary group in the Olympics, were gathering. This was so rare there is no name for cougars in a group. They would arrive come sunset.

Bobcats were forewarned to be done with dinner before sundown or they would be torn to shreds. A bobcat weighs 9 kilos or so, a puma male can grow to 100 kilos. A cougar kitten can handle a bobcat. The returning army sped along, informed bobcats were on top of their families. The victorious returning Marauders bore faces of fear. There was no way to make it on time. It would take them two days. They had no way to stop the slaughter. They could only hope to save some.

Grimly, they trudged on, Mr. Mizithra not bothering to eat. The entire colony's marmot fighting force was down on the necessary weight to survive hibernation. Ever present in each marmot mind was the need to gain grams to survive harsh winters in the Olympics. They had no time to eat. The lives of his family, of every other rodent on the trail hung in the balance. A habit of late, weather turned sour come afternoon. Cooling quickly, as is habit in alpine peaks, wind warned of rain to come. It came.

Skies darkened. They slogged along silently, each contemplating to occupy their mind. Fear of the unknown, death for spouses, children, and elders at the claws of feline phantoms, could drive a rodent mad. They focused on what they could do, not the unseen. On and on they plodded, paw prints splashing puddles away. They marched home determined. Back in the valley, a queen, Esmeralda, with an excellent nose, sniffed out the tunnel entrance filled with living marmot and woodrat. She called her clowder. They schemed what to do to extract fresh, living meat. They preferred this to those already cold.

Her protuberance of distinction said the meal lay deep within. The macho toms thought of pulling down the large stones, digging to the holding tank. It was not as if they had anywhere to go. Ha-ha-ha. As with all feline species, bobcats play with prey. This seemed the best idea. The other thing was locating escape passageways. The other queens went to ferret out these tunnels.

Once one was located, the queen bobcat would make a howling sound and squat in front of the doorway. It was now merely a matter of time. The contents of the underground meat locker would spill out. They would be more than happy to oblige, consuming each yummy rodent.

Those who yanked the largest stone out found digging easy. The soil underneath it was soft. The glacier left stones behind. Where it rested was less firm. The toms made quick progress. A marmot and woodrat picnic was on the way. By calculation, it would be a matter of minutes. Cougars were not expected for hours. They could eat, burp, take a nap, and carry off meat for later long before mountain lions arrived. Bald eagles were about done. The vulture wake had a steady supply of hungry friends circling in as word of a mighty feast got out.

A squeal was heard in the grave the bobcat's were digging. The first tom reached a tunnel. The fresh smell of living, terrified rodents greeted hungry nostrils. Mrs. Saffron, her young, the colony, Momo, the woodrat refugees, Keen's mice quivered in the dark. All eyes focused on a beam of light filling the hollow in the chamber.

A bobcat's muzzle came in view. Sniffing had some looking for the exits. These were filled with salivating bobcat mouths. Rescue by the returning army was too late. Death would take them to departed loved ones. They would leave life behind. As one, they hugged. Love and prayer was all they had, so they shared it.

Chapter 18 Interruption

The bobcat clowder closed to kill trapped rats, mice, and marmots. The ground began to shake. They imagined a mighty earthquake were passing. Rodents within said good-bye, hugging their young. The terminal temblors would extract them. Bobcats could feast at leisure. However, the quaking was not from fanged felines. It was not earth moving.

K'wati and his massive wapiti gang arrived. Not bothering to look, warriors leaped over dented fortifications and down the steep grade. The clowder scattered in moments. With no where to go, they froze in place, cowering beneath fir trees hemming the corners of the valley.

Exposed tunnels were abandoned by predators. K'wati came to rest on the floor, immense rack of antlers proving why he led. His rack was largest by far. His chest was massive. He stood out not only because of his distinguishing girth and presence, but something clinging to his back.

It was the legend, the Exception riding with the king. He leaped off the wapiti's back and made for the last of the bobcats. He held a large stick in his hand, two and a half meters long. This he swung at the cats who hissed and pawed. They burst forth, wapiti parting to give them a way out of the valley. Tails between legs, dispersed kitty cats scampered off to evade the huge elk.

As they neared the top, the sky darkened suddenly. The twenty odd bobcats stood spellbound. The speed of airborne assailants was too fast to move. To the rescue, called to emergency service, came the legendary 300, the Royal Canadian Mounted Geese. This armed flying fortress rarely left their northern reaches. Flapping to the Olympics meant crossing the Strait of Juan De Fuca and it's treacherous winds.

The plea came via General Lloyd Joseph. They respected the ancient one, heeding the call. Their brave leader, Inspector MacDonald, led his Mounties across the divide and into the fray. They bore the insignia each prized who served. It separated them from other Canada geese. Instead of the standard slanted strip of white across the neck, the select few, this 300, painted their necks Mountie Red.

Flying geese are a skein. When they close up in a tight formation, it becomes a plump. This was the fastest moving plump in the history of plumps. Diving at great speed, with no intention of pulling out, Inspector MacDonald led his pilots into the clowder with death on goose minds. On the ground geese are a gaggle.

This gaggle was hard at work, tormenting bobcats who fought back with ready claws and hissing orifices. Twenty bobcats are a force, but not when facing 300 determined Canadians. These fliers were well trained. They did what wapiti could not. It is not in their complex nature to kill. As vegetarians, it was difficult to put to death a living being, proving why they were given rule. Canada geese have no such issues.

The cost to the gaggle was fourteen killed, a small handful injured, but able to fly. The clowder lay still. After a tally of able bodies, Inspector MacDonald, raised a wing to his forehead in a dutiful salute towards King K'wati and General Lloyd Joseph, perched in his antlers. He wanted to fly home immediately. They were not a social bunch.

The obligation to General Lloyd Joseph complete, they made it west of Port Angles by dusk, home the next morning. They abhorred the sight of the valley. It was a terrible place, marred in every corner with a litter of dead animals and birds. They had duties back home and losses to replace. This was never a challenge, every good Canadian goose wanted to be a part of this elite gaggle.

BOOK 3: WAR

They departed on a wing and a prayer of thanks from King K'wati and General Lloyd Joseph. As the Canadians vanished, he flapped to earth and turned to converse with his old, well-respected friend. K'wati knew of the war, naturally. As by his nature, he did nothing about it. Ravens and crows, raccoons and marmots, who cares? It had nothing to do with them or the balance of the Olympics for that matter. Ravens and crows, fought each other since time began.

Raccoons were low criminals. What did they expect from dishonest ruffians? A marmot, oh please. A big squirrel and a couple of woodrats, really? To his way of thinking, this was business as usual. The reports he received did little to arouse his fur. He was well aware of the impending war and probable results. General Lloyd Joseph was far too clever a crow to be beaten by a boorish, clod like Ralston.

No, it was none of those things. It was the Exception calling K'wati for help. The man had never done this. Wapiti stopped in for visits all the time, especially those who could talk. He was fun to converse with and loved elk with a passion they did not know man could muster.

He sent urgent word via dragonflies, when it began to rain bear and marmot from the ledge. He knew of a colony settling there last autumn. That simpleton, Ralston, and his egotistical partner in stupidity, Olaf knew they moved here. They set the wheels in motion for attack. When animals rain down it is time for things to cease.

The impact of Igor falling a thousand meters to the valley displaced rocks on the riverbank. The dead corpse lay, a pile of black fur, oozing bright red blood from every pore. By the time elk arrived, the Exception had moved up the cliff from his home to avoid the scene any longer. He climbed on K'wati's back. They made for the Valley of the Stones.

Arriving not a moment too soon, he chose the dramatic entrance. K'wati had a sense of flair, if nothing else. Had he not, chances are some marmots would have been taken. He was a ham, but all loved him. Lloyd Joseph urged survivors to come out to the safe light. Mr. Beef Stew, hero above, defender below, lay injured, his surgery performed by Douglas squirrels. Mrs. Saffron, arms about her children, arrived, followed by Momo, then Mrs. Lip Gloss.

Wapiti milled about, smiles of satisfaction painted across elk mugs. Slowly survivors emerged. Of the original seventy-seven members of both colonies, forty-two remained, though a few were hurt. In the ones who egressed, it was eighteen, the rest were on the road. Other marmots who arrived to help began as a company of 300. The rest went to help Major Mizithra at the Siege of the Great Wall of China. Others fought in the Battle of V-Island. Now they counted one hundred ninety-two alive and wounded. Losses were high.

Momo surfaced, Sugar Murray's Maniacs in tow. Those not with Murray, staying to defend the home front numbered 1,500. Their forces counted 978, a devastating reduction. Of the 1,500 who fought with Major Murray, 1,244 were returning.

His forces were strong fighters, coming home victorious against a much larger foe. It swelled little rodent hearts with pride at their success. Crow losses were well over 1,000 souls. The Keen's mice lost 1,008 of their volunteers. Douglas squirrels, once 500 strong, now had 398 answer roll call. They were good combatants and hiders.

Skunks lost none. The stinky ones were fortunate. Beaver losses were 189 on both sides. Muskrats lost 92 in heavy fighting. Humboldt Flying Squirrels under Lieutenant Nicolo lost three, four injured. Ravens were big losers. Of 5,000 soldiers Olaf raised, 145 made it out alive under the leadership of humble Lady Satin. They fled high climes fearing retaliation. Otter losses were but eighteen. Molly and Graedy headed to Grand Meadow, creating a makeshift hospital. Dozens of Douglas squirrel pupils, no longer soldiers, eagerly volunteered. They were in the Valley of the Stones. It would take a long time to heal those marred by violence.

Hours after conflict in the Valley of the Stones, Mr. Mizithra's fuzzy noggin peered over the top of the ramparts. Elk had gone to retrieve his forces, carrying them on their backs to home. The scene to greet him was unforgettable. Dead animals lay everyplace from both sides of the battle. Partially eaten bodies filled the trenches and pathway to the floor below. Much of the living area was filled with a huge herd of elk. The rest of the living were crow, marmot, woodrat, etc.

The elk dropped them off on the valley floor. He gave a huge whistle and aimed at his family with haste. At his heels came the returning warriors. He sorted through a maze of wapiti legs to find his wife and children standing by Momo chatting with an immense elk. Standing next to his own wife, standing right next to her, a man! One of the Thumbs stood right there all nonchalant and everything. He stood there, no one worried. Stunned, Mr. Mizithra froze, his yellow toothed mouth agape.

"Honey, you're home. I love you, Mizzy. Come and meet everyone. You have to meet this human. Honey, it's a man. It's okay, really. Come on, and shut your mouth. I love you, dearest. Come now, hurry up".

Chapter 19 War Is Over (if you want it)

Once the stupefaction subsided, Mr. Mizithra, a civilian once more, was shaking paw to hand with a human being. The Exception, tall, lanky, and hairy, sported a beard, long locks, and human duds: pants, boots, and a T-shirt which read: Keep on Truckin'. He spoke to Mr. Mizithra who understood the words. How it occurred was a mystery of K'wati's doing. Disbanded, the army relaxed.

Stories began to fly, exaggeration close behind. Fear of bobcats was contrasted with victory over Ralston. Disbelief came when they learned of Ralston's setting the fire. Igor was savage. The story of Bette and her beavers and muskrats building the dam was super. The saga of tunneling cats was terrifying. Of course, the ultimate was wapiti to the rescue.

Whilst all engaged in frivolity, K'wati spoke solemnly with Lloyd Joseph. The two bent heads together for some time. When they unbent pates, all about them was hushed silence. Without realizing it, their body language did not appear happy as with the rest of the survivors. Instead of rejoicing, they were extremely serious.

This drew attention to the two leaders. Once their conversation was spotted, word went around the crowd. Soon, inhabitants of the Valley of the Stones were watching intently unheard words being muttered. Lloyd Joseph listened, nodding his ebony noggin. Seeing everybody staring, he spoke to the audience.

"Brethren, we won. Hold applause, please. With the salvation K'wati and his gang provided, the bobcats were stopped from a scrape too close for comfort. I am joyful Inspector MacDonald and his Royal Canadian Mounted Geese came to the rescue, too. Ralston and Olaf have been vanquished. Please, no more shouts of joy, though I understand. K'wati informs me of the law of the Olympics. He asked us to realize what they did is the rarest of the rare. Bears and marmots raining on their friend instigated this action. If Sergeant Guar Gum had not knocked Igor off the ledge, all in the Valley of the Stones would perish. His memory will live forever in our hearts and legends. He was a heroic, selfless marmot."

"We must immediately terminate this festive occasion. According to the convention, pumas have every right to feast as other carnivores. They take turns, cougars come last and stay until they have their fill. Only vultures are permitted when feeding because of some inter species contractual engagement. What it is, we have no idea".

"K'wati says cougars are near. We must evacuate this place until such time as feasting on dead carcasses ends. This will take days. Bears, due to Igor, are forbidden this feast as a penalty. We will ride the backs of wapiti to our place of refuge, Grand Meadow. Provisions have been made to accommodate us for temporary shelter".

"Food is plentiful. You must begin to eat like the dickens to gain the weight for hibernation. Elk will carry you to Grand Meadow. It will not take long. There, a huge celebration. Everybody who participated in war is welcome to join the fun".

"One weird thing, something King K'wati did not understand any more than I, is the reason cougars begged off. They have every right to do as they please. The laws of the mountains have held these tenets since the earliest days".

"Survival is survival. It is not under our purveyor to restrict the eating habits of any creature, be it carnivore, omnivore, or vegetarian. We must respect each species as they do ours. This is only right and fair. It has worked out well over the ages, so no reason to change it for this incident, however nasty".

"Their reason was related to the King and I from their leader, Baldur (prince, purity, peace, beauty), the young mountain lion king. He was the favorite son of his father, late, great King of the Cougars, Einar. Baldur, learned of the presence of an obscure woodrat named, Murray," every eye in the valley turned towards the little fellow wearing the tattered kippah hanging his blushing furry head, "and said they would give time to take us out of the valley before they came to feast. It was because of Murray they would do this, only him, explained the royal cat. It seems he had a debt to pay, but would not elaborate".

"In any case, we have no time to waste. We have a lovely spot for you. It will take all summer to recuperate. We have some nice caves, tunnels with existing chambers, and trees for those who live up high. At Grand Meadow, provisions have been made for a fabulous party to celebrate the incredible victory over that stinking raccoon. Now, gather the young, elderly, and wounded. We will get them to safety first. An adult needs to hop on wapiti's backs to hold kids. Ditto for the elderly. Those injured need a Douglas squirrel medic accompanying them with a strong adult".

"Hurry now, everybody. Let us get out of here and respect the kind deeds of cougars. King Baldur is young, but wise. His father, a noble cat if ever there was, raised him well. Other sons came and went, but Baldur shone above the rest. His bride, Queen Eir (mercy and peace) is a prize. She is well revered by the rest of the forest denizens. Sure, she eats animal flesh, but she is not cruel in her killing. She does not toy with her food as some are prone to do".

With haste they loaded the elk. Young, old, and ill left. The healthy and hale got supplies loaded to Grand Meadow closing this unbelievable chapter. Chamber after chamber was searched and emptied to ensure all was found. After cougars, other scavengers would utilize the valley as a restaurant with a spectacular view.

Who could blame them? A summer at Grand Meadow sounded like a vacation, a place to rest and rebuild lives. It meant eating non-stop for the next month and a half to gain tons. The delay due to war had a lasting effect. This was a diabolical after effect Ralston relished. He was a real nasty raccoon whose memory was blotted from Olympics history. It is the cruelest of the cruel who inflict such harm to enemies. A bear and a fire his low legacy of evil.

"Say, old buddy, old pal, explain how you know a cougar, the King to boot?", curiosity got the best of him. He had to ask.

They were in the final stages of loading. Mrs. Saffron took the kids and Momo. Few remained. The place was empty. A handful of wapiti were available to load and head to Grand Meadow. As leaders, these old friends felt it necessary to be last to leave. After all, they found the place.

Mr. Mizithra felt almost guilty about his discovery until Mrs. Lip Gloss took him aside and gave him a fact. Had they remained in their former colony, Ralston would take them out in a flash. No, he saved them. He obeyed orders. She beamed at his success and selecting him.

"Oh, I did ze cougars a good turn, old chum. A trifle, it vas nichts, I assure you. Now let's get ze last of zis loaded und git. No matter vat a cougar says, a marmot or rat makes a nice schnack. It is a sad moment. Ve von, my friend, ve beat an evil bandit. You have a tear in your eye, sir, but ve vill return".

"It is wise to leave. Let ze meat eaters feast. Ven ve return zis place vill be left alone. Zey vill have no reason to come back. Ve can build better shelters, deeper tunnel works, more vell reinforced chamber valls. Now, let us go. Get on your elk und ride like ze Valkyries. Ha-ha-ha."

With moves as graceful as Fred Astaire with Ginger Rogers, Murray side-stepped answering the question. Told he could never blab, Murray remained a rat of honor. No one brought it up again. The secret remained intact. His honorable self could not permit his present self to shame the word given and memory of so noble a cat.

For his son to honor Murray was humbling. By being a good guy, he received an unanticipated reward. Forever, he would recall Einar, the fall, rescue, and his son's repayment of the good deed. A dilemma presented itself. He did right. Riding on a wapiti's back, Murray dug in the fur, held on, and shed a tear. He was one fortunate little woodrat; yes, he was.

Chapter 20 Party Time

Mr. Mizithra and Murray were last to Grand Meadow. Coming over the tree-lined knoll mounted on wapitis on a fabulously sunny day in the Olympics, eyes met a large field, massive pond, and rocks at the back. This was a magnificent spot. Every time, from the very first, the place took one's breath away. It truly was a grand meadow. The dead were gone, a tribute to the efficiency of those inhabiting the region.

The thick greenery from whence Ralston emerged, darkened the south end. That was past. Milling about were beasts from wapiti to Keen's mice. Areas for shelter, guest quarters for various annual music festivals, packed with woodrat and marmot refugees. Crows flocked trees. The lake was so full of ducks, one might walk across without getting wet.

A tad bit of smoldering off in the distance was the final clue a war was waged. Ralston's fire did little to hurt anyone save those nearest. His reign of terror reaped no lasting success for raccoon or raven. In the morning's light, only a moist, blackened patch of burnt grass and a few small trees would remain to recall his evil.

Within a year, greenery would return, trees repaired. All evidence of a vile life would disappear downstream as did he. Caves to the right of the pond, at back, held others. No one seemed crammed or upset. In fact, it was downright syrupy. Yuck. Regardless of species, they got along. The stage was in place. By a coincidence full of deja vu, a Belted Kingfisher flapped his way over to them, lady lookalike by his side.

It was Boston Charlie and Matilda. Greeting the heroes, he was so pleased, he chattered away. Both had the finest accommodations, side by side, of course. Wives already had homes readied. The kids were playing by water's edge. The band would start soon. Horace and Boris were ready to help. Miss Bluesberry was going to sing. She was backstage.

They threw a band together. Ishmael, Deidre, and Ronaldo were on hand. She was getting them ready. It was hoped Murray could play. He had no sax. He had not seen one in forever. Boston Charlie had both alto and tenor backstage. He had a harmonica. Murray went, kissed Momo, grabbed her, flute in hand, then made for his old singing partner.

They had not seen each other in a long time. He had to admit he missed her over other singers. Sadly, Bette, Tobias, and Twigs were unable to attend due to work undertaken at the old dam site. Their parents and brothers died. Beaver siblings honored family. Celebrations would wait.

Mr. Mizithra went to his wife's side. They hugged a long, long time. It was over, all safe. He was scraped, bloodied, and bruised, but intact. Though shaken by the near bobcat meal escapade, Mrs. Saffron was fine. The hug over, she led to where the kids were playing.

Ralston's revenge failed. Jerk.

They sat on the lawn and ate it. They ate lilies, they ate clover, Flett's violets, water bulbs, anything in sight. The lovebirds ceased cooing words of love; it was time to feed faces. They were happy to oblige hibernation's urges. Each ate until rumbling stomachs were quelled. They lay on the beach to watch the show. It was going to be a huge extravaganza. They called the kids to come shove their faces full of greens, and watch Uncle Murray and Auntie Bluesberry entertain.

Over to one side, a crowd full of activity beckoned a look. It was the medical area, swarming with Douglas squirrels and otters. On further inspection, it was their doctor/chiropractor pair hard at work healing wounded. M.O.S.H. (Molly O'Grady's Surgical Hospital) was the place to get better, It's nickname: the Mosh Pit. Mr. Beef Stew was here. An elk, Gossamer, had the honor to bring him. Mr. Mizithra had his minor injuries tended. The hug between he, Molly, and Graedy was moving.

To survive this difficulty caused deep emotions to rise. Their love increased getting to know Mrs. Saffron, Miss Paprika and Mr. Murray. Family life agreed with the marmot. It was time to stop wandering, a motherly Molly chided. Good for him to get married. She smothered Mrs. Saffron in hugs and kisses most wondrous, other than the fish breath.

The injured needing care quickly reduced. They wanted to get well to enjoy the show. Very few would soon be in hospital. The old infirmary at the pond was vacated after the intense violence. The dam's damage was too severe. The region nearest the pond reached by fire, laid soot on the entire previous medical practice.

They relocated. The new, permanent home of the otter's hospital was selected. Once celebrations concluded, they would begin. The spot was to the right from the lake shore. A rippling stream bumbled down the hills to a large plateau. Caves existed where they could build a fine hospital. They hired otters, Keen's mice, and Douglas squirrels to set up the place.

Beavers and muskrat hewed and hauled logs for massive structures. Soon, all forest denizens would have an exquisite medical facility to help the needy: Graedy & Molly's Otter hospital (known as the GMO). Water flowed past the cave. They hired Bette and Tobias to dam up a small pond. Paradise would replace hell on earth. The cave had room for many beds.

The O'Grady's had other news. Back down south, the attacks came to porcupines and Winston and Beverly's crows. Raccoon and raven hit again. It was not Carnival island, but the peninsula and areas surrounding the Naselle River. They, too, employed bears as they are plentiful on Long Beach Peninsula and hills encompassing Carnival Island. Results were the same, only Paula and her pals made pin cushions out of raccoons.

Ravens had much smaller numbers. There was no air force factor. It seems Prince Bjorn turned out to be a wimp. Bears were another matter. Here again, the leader of the southern wapiti, Quintus, the fifth and most noble of his father Artemis' sons, stepped in and fought a heated battle with sleuth of bears a dozen strong. They were led by a foul beast of legend, Ursa. Eight bears met death in the fury, nineteen wapiti died. Ursa fled, yet his cave was found. Three with him were tracked.

With deft precision, a prickle of porcupines, fifty strong led by the brave warrior, Clancy, surrounded his hideout. Ursa flung himself from shadows, only to be regaled with a tsunami of quills. Many hit, though he felled three before going down. The bear succumbed to darts, dying alone in a heap outside his hollow.

The other three met a similar fate. The porcupine losses: a dozen. Thankfully, Paula was safe, ditto Murray's band mates. Raccoon, Eastern gray squirrel, and vole populations thinned. The southern uprising fell due to the solidarity and love of the defenders.

Roje and Agwe resigned the job of talent agents. They were not very good, and returned to help their parents. The stress of the shake up was too much. They were needed at home. The festival was off this year due to war. They would look to next season. Their parents could take a back seat and relax. They raised good jacks.

Boston Charlie confirmed the information. He introduced Murray and Miss Bluesberry to new agents: Darby and O'Gill. Darby was Simone's son, O'Gill was Vivenne's daughter. They were classically trained, bringing order to disorder. Tutti Frutti and Mon were loose, but tightly organized. Roje and Agwe never really caught on. The move home was perfect.

Darby and O'Gill slid in place as if Tutti Frutti and Mon were still at work. Meeting Miss Bluesberry and Murray was a joy. This year was shot. The skunk would soon enter her long winter's nap. Concert season was obliterated. Next year, they would have a line-up fit for a star. Each would have a curtailed, well planned itinerary per their schedules. Right now, though they better get ready to play.

First, a little treat was planned by the powers that be.

The Belted Kingfisher hopped up front of the stage, waving dusky blue wings above his head to quiet. A complaisant hush befell the crowd: elk and deer (they did the Switzerland bit, usual for deer), to rodents on water's edge, to floating fowl bobbing up and down on sedate waves.

The audience was immense. Every eye was on him. The only thing anybody knew was, Miss Bluesberry and Sugar Murray and a band were to honor them. As usual, the little birdie had a trick up his clever sleeve.

"In a moment, dear allies, we will be treated with the dulcet tones of our Darling Doe, Miss Bluesberry, heroine of this unpleasantness. Yes, yes, shout as loud as you can. Rattle that rack, y'all at the back. Splash water you, fowl. She is the very best. By her side, you know him, you love him, you want more of him, Sugar Murray, king of the saxophone."

"Shout loud for your hero. With him, his partner in war and peace, Momo. Yes, yes, give her shouts of love. The band, your friends: Deirdre lovely lady beaver on guitar, Ronaldo manly muskrat on piano, and Icy Ishmael thumping on bass. New to the band, but a short tailed weasel drummer of renown, the Whiskey Bend Kid."

"Oh yes, rain shouts of praise for them. Before they are to entertain you, I have a special announcement. Here to address us, all the way from his hallowed home at Crescent Cedar, I give you our Elysian leader, General Lloyd Joseph.", shouts and howls of joy exploded across the pond in a wave for this black, feathered genius. He came forward, standing before them, welcoming the approval of the entire Olympic range.

"Friends. I have few words. Victory against evil King Olaf and rancid Ralston succeeded due to the efforts of those like Major Murray and Major Mizithra. It was beavers under Lieutenant Bette, skunks under Captain Loganberry, Keen's mice led by Lieutenants Zor and Zax. It was medical attention, Graedy and Molly. It was the dual duty of the Douglas squirrels under aged General Douglas MacArthur squirrel. And it was the crows, who did me proud."

"It was my old friend, Boston Charlie, take a bow you old so and so. Lieutenant Ripple and Sergeant Twigs did more than any can tell. Finally, those most secretive, but who did a great deed saving Major Murray and Momo. Soaring overhead, pay witness the aerial finesse of Lieutenants Nicolo and Allessandra, the Gliding Ghosts, the Night Ninjas!"

Humboldt Flying Squirrels in formation, lit on stage. Odd square shapes guided by little front paws and fuzzy tails, hit a perfect landing, surrounding the cawing and bobbing crow leader. Murray hugged both. Now a unit of seventeen flying squirrels, they became a formidable force, guarding those wise enough to negotiate for their services.

After this recent violent incident, some wanted security available by these cryptic rodents. As mysteriously as they appeared, poof, vanished. Humboldt Flying Squirrels are no social butterflies, enjoying life as shrinking violets. Lloyd Joseph stood alone on stage.

"Finally, great thanks to K'wati and incredible wapiti gang. Thanks to the Exception, who warned of the evil Igor, saving us. Now, friends, before Miss Bluesberry and Sugar Murray take you on a magical musical journey, I have a surprise".

The final words departed his beak. He dove off to the ground below. The audience's attention was instantly drawn back onstage. A flash of brilliant yellow and rushing sound of wings was the cause. He had one thousand Townsend's Yellow Warblers explode full speed over the crowd. The intense yellow of their heads and feathers gave a splendid fireworks display, causing oohs and aahs.

Young marmots gave whistles of delight, bellows from wapiti, and chirps from chipmunks and birds. Instantly, gone. In their place stood a miracle. Nobody, not one of them would ever expect what came next. Good old Lloyd Joseph had the ultimate trick up his sleeve. It was all for the troops. They deserved a chance to unwind. Coordinating with Boston Charlie, he was able to contact the surprise act.

It was the incredible Covey of Quail Quartet. Angelic voices rose above the din. The warblers vanished to enjoy the diversion. Every ear turned their way. For a solid hour, Mrs. Queisha Q. Quail had her ladies in fine voice. Her little bevy had their way with them, singing so beautifully tears fell from many an eye.

Mrs. Querida Q. Quail, Mrs. Queisha Q. Quail, and newly wedded, Mrs. Qubilah Q. Quail sang lungs out. Bringing it up with a powerful soprano solo, Mrs. Qitarah Q. Q. Q. Quail (Yes friends, sadly twice widowed, I'm afraid. This one: Tandoori marinated quail on a bed of saffron basmati rice, garlic roasted artichoke hearts, curry butternut squash soup, and mango gelato for dessert) brought the house down. It's tough being a tasty quail.

For a full fifteen minutes after they ceased, the forest exploded in calls of approval and love. Bowing individually and finally as one, a young spotted fawn shyly neared the divas, mouth full of mountain blooms. Queisha took the bouquet, bowing, and wept. It was an overwhelming response. She was proud to participate. Appreciation for music of this caliber swelled hearts with joy.

Horace and Boris held Miss Bluesberry, Murray, and the band until Boston Charlie's signal. Profound bass notes from the very back supplied by the subtle paws of Ishmael echoing off the black rock backdrop, set the tone. The Darling Doe began to nod her head. She sported a red bow-tie around her black and white neck, a continental look. Another was trussed to the side behind, flowing down her back, a bit of whimsy.

The Whiskey Bend Kid set drums to tapping as Deirdre zapped forth a shower of blistering blues notes. Sugar Murray had enough, whipping out a saxophone blast, emitting shock waves across the water. Ronaldo picked up the tempo and clobbered his eighty-eights with a blues layer, engulfing the crowd in musical love. They jammed on stage for a full five minutes, Miss Bluesberry all the while dancing right along with them, freedom her only thought.

Then, oh yes, then she began to sing.

"We done fought the beast, who came from the east. He fought the battle way too soon, so we killed that nasty fool raccoon!", she hit this last word to an explosion from the crowd, "These is our hills, our valleys, our mountains, too. You come mess with the Olympics, you gonna mess with you know who.", she said, pointing a paw to her sax wailing, kippah wearing, scarred up old woodrat partner.

"Come up here, learn the rules, cuz the rest of them is fools. Up here we is full of love, but come to fight us and we'll give you a shove. Off the Valley of

Stones, follow the bear down all alone. That's right, we the ones we the ones, we the ones, who tossed a bear. SO THERE!", she shouted to skyward, releasing fears and frustrations the entire Olympic Mountain Range endured. It was her way to take the tension and make it fizzle.

The song lasted twenty minutes. Miss Bluesberry stood on stage, rocking and singing. They continued, playing old songs and a few new ones. The sole reason she was there was due to the love of Mr. Mizithra. What she did after he got her in the front door was her doing.

They played and played, the audience appreciative for every note. The Whiskey Bend Kid, fit perfectly. He could really bang out a solid rhythm. Other players, from Grand Meadow, begged him to join. He did gladly. They would be tight in time for the end of summer festival.

"Before I get off stage and run to my boar, my love, Mr. Loganberry of the Hills, well, I just have to thank one certain fellow out there. Now, he isn't the type to seek the limelight, like we crazies up here. No, he is happy to sit in the background. That is, of course, until it is time to tell a tall tale, then he is front stage center with stories to intrigue".

"To all of you, I thank you, we, the band, thank you. But most of all, from the bottom of my skunk's heart, I want to thank the one who helped me from the very beginning. Dear Mr. Mizithra, will you please put those lupines down for a moment.", the eyes of everybody in attendance turned to this humble marmot and began to laugh. His mouth was crammed full of lovely purple blossoms.

He came up on stage and bowed to a wave of applause and calls of joyous approval. Mr. Mizithra stood alone, uncomfortable being singled out for doing his duty. He was not heroic. He hugged Miss Bluesberry and Murray, then went off to the side of the stage. How his friends did this was beyond him. The poor rodent felt a bit nauseous from the jitters.

They exited the stage as well. A single stool was set in the middle of the dais, up front. A somber move, the place quieted. Birds, noisiest of all, ceased chattering and turned to witness the close of the show. Darkness approached, the sun already surrendering it's lovely rays of light to give night a shot. Soon, it would be time to sleep.

As a lone figure sat quietly on stage, silence reigned. Momo lifted the Shakuhachi to her woodrat lips and began. Soft, haunting, woody, notes wafted across the pond to ears of ducks back to the wapiti. Each note felt personally selected, building layer upon layer of emotions. A hush fell over the veterans. The music compelled them to think of the lost, Mr. Guar Gum, Buford and Beulah, Momo's parents, and the rest.

War is felt longest by witnesses who survive. This is more difficult than thought. Images beholding the horrors of war do not run gently to the back recesses of one's mind. Forever, you remember faces. Memories stalk, torture, leave the living to tell the tale, a sense of guilt for surviving. Many beloved brothers in arms do not. Her song went forth, then silence. Without a word, the audience went to their respective abodes. The show was amazing. Everybody enjoyed the Covey of Quail Quartet. The yellow warblers. Miss Bluesberry and Murray were exceptional. The band was fantastic, the Whiskey Bend Kid a real find.

Boston Charlie bid the performers a good night. He hugged Momo, a tear in his eye. Hers was the most moving. Without a word, her song said it all. It is incredible how a sweet piece of music such as her selection could so effect the hearers, but it did. Each went home happy to be alive, to have family, love, a teardrop, and to be a part of something so noble.

Chapter 21 Back Home Again

Summertime passed quickly. The weight gaining process was in full bloom. All over Grand Meadow marmots ate, faces buried in greens. The only goal was to increase. Winter came soon. War cost precious time shoving mouths for hibernation. Wrestling matches were at a minimum.

Towards the end of the season, Mrs. Lip Gloss had a meeting with the other surviving elder, Mr. Root Beer, and Mr. Milk Chocolate. Due to losses, the best solution this winter was combine tribes. They would dig one set of tunnels and chambers instead of two. This simply made sense. It was not to deny Mr. Milk Chocolate his place. Grateful, he concurred.

"Well, Murray, old friend, this was not the reunion either of us envisioned when last we parted. What a wild year above ground. The elk are ready. We are packed and fat. We should be alright. There is time to dig good homes for Sleepy Time. It is difficult to put in words, isn't it? Ralston tossed a curve ball! The day we met, denying the bandit a Pacific Northwest banana slug meal, set this in motion. Think of poor Ernie".

"Next year, let us not have so much excitement. None of our little walks about the Olympics ever had so much intensity. Not even when we faced Borya, which was the closest of all time. My thought is to have you come up with Momo and spend part of the summer with us. She and my lovely wife already made plans. You have to play sax here and there, but you must take some time for us".

"Who knows why we survived this ordeal, but we did. It is with the utmost of humility and love I bid you farewell for this year, Murray. You are the closest friend I will ever have. I am very grateful for this truth. God bless, see you next year".

Murray, unable to speak, held his kippah in his bowed hands, tears streaming down his face. Mr. Mizithra leaped aboard his wapiti taxi. He waved adios to the crowd gathered to say, bon voyage. Within an hour, they were up a thousand meters on the rim of the Valley of the Stones. Other than some bones piled up hither and thither, all clear. A few crows adorned the trees, otherwise they were home alone.

Elk were in a hurry. Once snows begin, they move to the lowlands to feed. The wapiti nodded good-bye, then vanished in a breath over the ramparts. K'wati's word was good. His troops gladly aided the marmots. This was the only way they could survive harsh winter in the Olympics.

By riding elk style, no one lost weight getting home. This blessing was the last piece to the puzzle of war's aftermath. The elk were the final domino falling perfectly in place. The marmot population had a chance to make it through dormancy to repopulate.

Per K'wati's instruction, they ran past the site of the fire. After a fast assessment, they headed towards their king. He was with Lloyd Joseph at Crescent Cedar. The damage to the forest was minor thanks to the fire. It could have been a disaster. No real leadership now existed for raccoons. Lady Satin and her few ravens were spotted east, licking wounds as it were. They would present no trouble for some time. Once he knew the extent of Ralston's atrocities, he became overcome with anger.

He set up a gathering with the head of bears in the Olympics, Gunner (warrior), an old, somewhat decent black bear. He lived up high, seldom contacting anyone. He acquiesced to K'wati's request. Faced with solid evidence, he went back to his sleuth. His message: were any of their species to ever do business with any other animal again, that bear loses his or her life, their family and friends, too. The indignity Ralston visited upon them was grave. Igor was a useless useful idiot.

Work began at once. Armed with hard-earned wisdom, burrows went straight back under the thickest parts of the slope. If a healthy sized rock lay in the path, two holes were dug with chambers back behind. No small rocks were considered. Engineers called the shots. Size mattered. Too big and a bobcat was inside, feasting on marmot tartare.

Escape passageways and booby traps abounded. Making the wise choice to build one colony, work progressed rapidly. Mr. Milk Chocolate and his few colonists were happy for the decision. A communal effort ensured more to make it until next year 265 days away. Oh wait, this was a Leap Year, 266 days. Sorry, I forgot.

One morning, it was underground time. A nip filled the air. Eating and digging finished on time. Wisely, a few empty chambers were dug to fill with dirt as marmots modified rooms. A wedding or two was held. Sadly, they lost Mr. Egg Flower Soup. He died in defense on the ramparts. Valiant to the end, he held off a raccoon alone until, beaten, he fell. He stopped the assailant long enough for others to fill the gap. The bandit died, never gaining entry. In his place, his son had the joyful task of uniting pairs in marmot matrimony.

His name, Mr. Crab Puffs. Like father, he was a fun loving rodent. He made couples feel at home. He saw daddy do it often. The ceremony was solemn, but need not be stuffy. They had such a rough year. He felt it ideal to perform wedding services on the cliff's edge. He donned a crown of lupines and lilies, causing well intended peals of laughter. The backdrop was sensational. At precisely the perfect time, a flight of doves soared by at the finale`. He was a chip off the old marmot. Pops would be proud.

Mr. Gilligan's Island and Mrs. Maryann were very healthy, having guarded kiddies underground against bobcats. Their sons were safe, too. Their wives helped, Mrs. Pebbles served on line in the trench alongside her husband.

Mrs. Wilma fought with distinction along the pathway. Mrs. Betty helped young ones, staying up front to be first to bite a bobcat. This core, with Mr. Milk Chocolate and Mrs. Mascara would rebuild. They were pleased with how the colony held together.

No time to dwell on the past, plans were already formed for the next colony. Time outside passageways was used to select the proper place for a new home. They chose an area south, under a large stand of deep rooted trees. It had the best sunlight and a natural ridge. Enough rocks lay strewn in big old clumps to make stubborn blockades. Lessons learned in this near death experience had defensive measures initiated.

Sharpened spikes were a favorite: easy to make and set in place. Right before going in for the next 266 days, they dug a few minor tunnels, wide mouthed for central entries. They were over a dozen meters apart. Each was twenty centimeters in diameter, back ten meters. A large chamber filled each. Completed, they covered the entries with large stones and dirt to protect in deep snowfall.

Sadly, war took a heavy toll on some families. Mrs. Jazz never recovered from her daughter's murder. She passed away during the time at Grand Meadow. Mr. Perry Mason, a most devoted husband, did not last long after. Orphaned, Mr. Burrito had one more loss, Miss Coupe`, his intended, died in the trenches. She was an Olympic marmot warrior, first to taste the omnipotent power of the mighty Igor.

Mr. Burrito survived the conflict healthy and orphaned. The hero of Fort Ballerina could not handle ending battle. He brooded throughout the celebration. When Murray was up playing, he was sulking in his transient home under a few knotted pine roots. It was a comfy cottage for one. Burned on every relationship, his attitude changed. He began to wander off, to walk alone in the woods south of Grand Meadow.

When it came time to load elk and ride up to the Valley of the Stones, he was absent. Rather than seek love and support available to him in large heaps, he drew within himself and hid his love away. So hurt was Mr. Burrito, grief overtook him. He had to be alone. It was rumored he spent winter in the safe keeping of Graedy and Molly.

They could not know. By the time they reached home, dragonflies ceased delivery for the year. They could only hope he did, waiting until spring to know. He did not avail himself of the tools at hand and lost his way. If you endure hardship, there will always be somebody willing to help. Bottling up or running away from pain are not good ways to deal with damage. Finding help is hard to do, understood, but it is better than any alternative ever tried.

Mr. Opie and Mrs. Perfume were two others who died in battle. They were amongst those on the pathway. Mr. Mizithra saw them go up as raccoons came down. They were never seen again, bodies vanished. These were two of the very best. The foolishness of war is the cost in life.

Many on both sides did not see another year come and go. Why? It was for the egos of allies whose entire purpose was to disrupt the entire mountain range for their own desires. The price was steep, far too steep. Both lay dead. Many of their followers here and down in Willapa Bay were history. War, what is it good for? Absolutely nothing!

Days darkened, sun less warming. In early hours, frost blanketed green plants like powdered sugar. Every moment above ground, they fed. Young were chided to keep eating lest winter take them. It would be the first Sleepy Time for Mr. Murray and Miss Paprika, both rollypoly. Silly siblings, chubby, laughter erupted looking at each other's swollen cheeks. Mrs. Lip Gloss said her rheumatism was acting up. This meant a cold winter according to her bones. No one thought this ridiculous notion a true signal of prediction, but time was at hand.

Everything was in preparation. The first flakes fell around noon. The unique, fascinating white crystals fluttered about in the gentle breeze. Those falling nearest the ledge were whisked back by the current, carried across the floor. It was not much. A few scout flakes sent as a preamble to the Olympics' nippy season. Once snow began to fall in earnest, the mountain range radiated a heavy alabaster coat.

Wapiti made for lowlands. Woodrat, muskrat, and beaver worked through winter. Bette and Tobias finished on the old dam then came up to Grand Meadow. There were plenty of places to build a lodge. They had one erected with a fine dome in no time with the help of Twigs and Ripple.

Murray, Momo, Nediva, and Frema stayed with them safe against invaders or the weather.

With snow falling, Mrs. Lip Gloss begged Mr. Mizithra to step out for a brief conversation. It was hibernation time. Before going in for the last time this year, she needed a little chat. Her face was very serious. He obliged, knowing this lady to be a marmot of grace and dignity.

No other elder was so revered. She was an excellent example, not to female marmots alone, of what was doable working hard to achieve goals. Everybody was inside, his wife and kiddies awaiting his big, warm, fuzzy body to waddle in and get them warm.

"Mr. Mizithra, once again you amaze me. I have known you since you were young. When you married my daughter, it was a pleasing event. Your girls, my granddaughters, were my loves. After your sojourn to recovery, it was joyous to see you marry again. She is a fine companion. Your young are well trained. They respect the old ways. This is good. I am very pleased with how you behave. You are a role model."

"The war, stupid Ralston. What an idiot. Did he honestly think he and a flea bitten raven could beat us? Please. They had the trick of that nasty bear, but we had right and unity on our side. What surprised me, what shocked others, was your leadership in battle. The stories you tell about heroics in your fables, well, I thought you embellished a bit. Now, I see you are brave, decisive, strong. At every turn you proved a marmot of preeminence. As an elder of this colony, I offer you this sincere hug of love and the admiration of every member."

Mr. Mizithra nodded, tear in his eye, hung his head and smiled.

BOOK 4: PEACE

Chapters

1. Rise And Shine, Marmot 378
2. Murray's Update 384
3. The Patient 391
4. The Lodge 398
5. Murray's Place 403
6. Clancy 410
7. Miss Bluesberry 416
8. Spring Springs 422
9. Dublin 427
10. Bette's Beau 434
11. Dublin Fits In 437
12. The Valley News 443
13. The Saga of Mr. Beef Stew 447
14. Mrs Lip Gloss's Charge 451
15. Old Friends Talk 456
16. Murray and Momo's Sacrifice 461
17. Mizzy & Mr. Beef Stew 465
18. Erasmus 472
19. It's Time 475
20. The Lesson Continues 479
21. Tell Me About Love 483
22. The Unavoidable Truth 488
23. Fade To Black 491

BOOK 4: PEACE

Chapter 1 Rise And Shine, Marmot

Each hibernation of his life, Mr. Mizithra spun tall tales to anxious ears, eager to absorb what the lessons taught. This specific time he had a massive audience with the combined colonies. War ravaged both families. It became prudent, common sense, to unite. When spring sprung, Mr. Milk Chocolate and his marmots would build a new, separated home.

Situated at the very back of the slope, ten meters underground, one large tunnel was dug between colonies, just in case. This excellent idea met with approval by both sides. Excavating soil would take a long time, but they had the whole season. Most were content to rebuild the bulwarks, repair damage to the pond, and modify the existing interior.

Within a few minutes outside after a long, long nap, Mr. Mizithra was attacked in his fuzzy face by a flurry of dragonflies. He had messages from all over the Olympics. K'wati, Lloyd Joseph, and Murray, to name the top three, wanted to confer. His first inclination was to get together with Murray, but when King of the Wapiti calls, you go. He emitted the buzziest red dragonfly to answer the plea.

Since he could not come to the elk, the elk would come to him. One thing to understand, kings, animal or man, travel with an entourage. K'wati's posse was two hundred strong. The sight of them storming the ramparts after waking up was frightening to the kiddies. Everybody else loved it. Not one mite was disturbed on the damaged fortifications, save a handful of dust whipped up by passing elk.

"You know the war was lost? You know the price we wapiti paid for you? You made it rain bears and marmots on our friend. In Willapa Bay, elk died. Here, we were injured, called unwillingly to action. The story is told. You were trying to save a what, a Pacific Northwest banana slug? This entire episode started over a slimy, ground crawler?"

"Well, it is true then, Ralston was insane. I am told it was you and Murray who saved the bug and drove the raccoon onto the road. It was there the Thumb's machine struck him. Stupid animals these "people", save the Exception, of course. He is not like them. He is not a man, really. Odd trying to explain what I do not fully understand."

"You stay here in the Valley of the Stones. This is home. It is under our protection and watchful gaze. From now on, wapiti will patrol your settlement to keep the nastiest carnivores away. Here you grow, flourish, and profit in this valley. No more adventures for you, my friend. Stay home, be a husband and father."

"Visit Murray and your friends down at Grand Meadow, but stay away from the rest of the mountains! I have never had so much trouble with a colony of marmots in all my born days! You, sir, are a menace!", said K'wati with a grin from furry ear to ear, antlers included.

K'wati and his companions leaped up the sides of the slope in a loud crowd. Noisy beasts, these elk, thought the marmots. They were gone. The marmots were aware of their blessings from their antler clad friends. The whole time underground they told and retold the tales of war. Some were here, others there. Some saw little, many too much.

Scores were damaged not by physical injuries, but hurt hearts. In the heat of battle, they saw loved ones fall, beheld horrific death scenes, or killed the enemy. Marmots are not hunters, they eat flowers. To put any other creature to death is detestable. Yet, defending the colony was done with resolute spirit.

True to his word, K'wati's wapiti kept an eye on the goings on in the Valley of the Stones. Elk actually liked marmots. They were not inclined towards preferences, but the fuzzy little guys got to them. Much larger mountain monsters befriended rare, rotund rodents, visiting often to enjoy their company. Elk travel in herds or gangs, as we have seen. Deer, cousins if you will, do not.

Wapiti are called a gang or herd. Yet, amounts stagger in difference. K'wati could muster several thousand in one massive group were he to ask. Deer had no such numbers, leaders, or way to carry out such a thing. Wapiti rule the Olympics because they were meant to do so. It is the way nature laid it out, not ours to question. Their numbers are more like the style of Jon Deer, just sort of hanging out enjoying life in the Olympics.

Their governing method served all evenly, in balance. From bugs to bears, they had to consider every species. Harmony demanded harsh acts. For example, cougars, coyotes, bears, hunt and eat elk meat. Yummy. If the king excluded his species from the menu, it would alter the balance of nature in these mountains and valleys. The end would be disastrous.

This is the main reason man needs to take a hands off approach, if possible. Too many cooks spoil the broth, so to speak. Lest any think this an easy decision, the King's own daughter fell to the deft pounce of a well hidden cougar. She died after a horrendous battle. The mountain lion won. She was eaten.

The king did nothing. Had he, his followers would dismiss him. Leadership is never easy. It was not for him. K'wati's reign was a long one. His son carried the legacy with pride and honor. They lived in union even with those who killed and ate them. Such, my friends, is the true way of nature. We may not understand, but it is their world. We visit and learn.

The method K'wati employed to personally keep tabs on our two amigos was attending the Grand Meadow Festival each year. A true music aficionado, he always rattled his rack to the racket. He found Murray exquisite. Miss Bluesberry always impressed. As the music scene began to revolve around this most picturesque place in the Olympics, he sought more reasons to hang around.

Unable to play, of course, he appreciated the talent of others. This is a true sign of the well-rounded. Nobody does everything and certainly no one can do everything perfectly. Living life, you find it best to let experts do their thing. You would not, for example, perform surgery on yourself simply because you know how to cut with a knife.

Our two best friends held the noble elk leader in high esteem. Were he not to have intervened, their families would have died. Forever in his debt, they always played any request. Mr. Mizithra made certain to tell his very best tales when K'wati visited. Brushes with royalty became as much a friendship as one can have with kings and such. He respected the valor of the marmot and woodrat. Both were worthy of salvation.

Besides, truth be known, raccoons made his fur crawl. Nasty, foul beasts. Ugly omnivores. They were not to be trusted. He was a learned elk, but it did not take a genius to realize raccoons are not reliant. After the sins Ralston committed, the entire species was held on a short leash in the Olympic mountains. Ralston's transgressions left a lifetime residue on the wapiti's heart.

Next on the agenda, flying in to visit, as marmots do not travel well, Lloyd Joseph came a calling with a five hundred crow escort. He bade Mr. Mizithra meet him in a small hollow above the ridge. This was done for privacy and security. The aged crow was wise to use caution. Victory did not dissuade every bad guy. Assassins lurked in the shadows, waiting the chance to pounce. Mr. Mizithra knew it, telling the crow messengers he would meet at the agreed upon time. They did.

Trees became blackened apparitions, sagging under the weight of so many guards. Before war, when Murderers Row came to visit the Beaver Blues Bungalow, Lloyd Joseph was ancient by any crow standards. He had outlived all contemporaries. Sister Beverly was several years younger than he. Now, this burden unloaded, he was clearly worn to a frazzle.

Post-war, he never traveled with less than this many. Paranoia was his price for leading the animal kingdom in a reluctant war. The flight up now took two days each way. He was old before war, now some feathers were white. Around his beak, on his chest, plus a few on his tail bore evidence to internal stress. Worry dulled his eyes.

No one leads for free. Being in charge means the buck stops with you, as the saying goes. This is why so few people choose to do so, and so very few are good leaders. This is true whether in battle or some other endeavor. Power is far too easy to abuse once acquired. History proves to us, corporals do not make good leaders, for instance.

The truth: his days were dwindling to a precious few. Knowing the inevitable, Lloyd Joseph made the proper arrangements. He honored his people as they did their vaunted leader. In his place, the lady crow who in times past delivered private messages, Neferteri (beloved companion), would be empress of the Olympics when he passed.

She was still too young, his great-granddaughter. Of all young ones in his line, she shone above the rest. She had quality one cannot measure, only possess. Lloyd Joseph saw. Training began early. Until deemed ready, Simone, Vivenne, and Darla would act as queen regents. They would train and prepare her for ascension to the throne. As empress, she would rule the vast mountain range. This is unlike the audacity of ravens.

BOOK 4: PEACE

Leaving a home known all her life, Neferteri would go to Lake Crescent for finishing school. To rule this important kingdom was a huge responsibility and undertaking. A sovereign could not go in blind and unprepared. Lloyd Joseph knew it was time. She was the best of his line to succeed him. While young, she was wise, observant, patient, loving, and kind. Her mother was his favorite granddaughter, Chloe. Ravens took her sweet life when Neferteri was young. Lloyd Joseph raised her as his own.

He thanked Mr. Mizithra for his leadership and heroics in war. He presented marmot to future empress. Lloyd Joseph told her this was the best marmot he ever knew. Other rodents came and went but this chubby fellow and humble old Murray were the best of the best. Were she to need help, this was the marmot. All she had to do was send a message to the Valley of the Stones and this fellow would run to answer the call.

To his credit, Mr. Mizithra, stunned by the boast, humbly bowed as an offer of proud service. The festival at Grand Meadow would be the biggest ever this year. As a gesture of appreciation, the crow booked the very best accommodations for the marmot and his family. Murray would be too busy to help, so Lloyd Joseph did so for him. That old woodrat was too much, beloved by the aged crow.

"Soon, dear friend, soon I must go off by myself. My days are down to a wing full. During my reign, I tried to keep peace. That damnable raven and crazy raccoon shortened my life with war. This was not my desire, Mr. Mizithra. In coming clean, you must know bravery is why you were selected to help defeat Ralston."

"When you took on a bear for eating your family, we learned of you. When you began to travel the walk of grief, we followed you. We saw you kill the kestrel. Nasty little things are they not? Time after time, you found out ways to keep from being eaten. This was impressive.", Lloyd Joseph's knowledgeable grinning did not help lessen the memory of his various aches and pains.

"You and Murray are legendary within the confines of Crescent Cedar. We sent the sow skunk your way. Later, we sent her a boar to help them and for our war efforts. Boston Charlie worked with me for many moons. He guided you to the Beaver Blues Bungalow so Murray could play and Miss Bluesberry could sing. With the two of them you went to the Valley of the

Stones, everything was in place. The only enemy we could not foresee was that bear. Thankfully, King K'wati came to the rescue."

"I hope to see you once more, at the festival at Grand Meadow. You will have the best place from which to see the show. My health weakens me, but I feel time will permit me one last time to enjoy this event. This year we have victory over evil to celebrate. This is going to be the theme of the festival, too. Songs, dance numbers, and comedy routines will revolve around the whole dramatic episode."

"Mr. Mizithra, this triumph does not happen without you. Never in my days on earth have I come across a marmot as madcap, as marvelous as you, sir. Your home is blessed. Defenses you erected and fought for are worthy of rebuilding. My very best wishes for you and your animals. I am humbled in your presence, brave knight." A tear in his eye, Lloyd Joseph curtly bowed, then flapped away followed by a shadow of crows which blackened the sky.

Considering his weakened condition, Mr. Mizithra felt honored by the visit. Most came to the crow, the crow seldom left his castle grounds to see anybody. It took two days to fly back down to Crescent Cedar, two more to recover. In this late spring, he knew his last summer would soon come and go. Accepting time was limited, he made this effort because of the sacrifice and bravery the marmot displayed. It deserved recognition.

By the fall of autumn, he would go to an outcropping alone in his private time to pass. No entourage accompanying to this secret place. Flying past long ago, he found bones of a cougar, part of his massive lower, right, front tooth chipped. If good enough for a puma, it was good enough for an old crow.

His final plans were set. He felt positive about the Ladies of the Lake taking charge with adolescent Neferteri. She would make an excellent empress. Until she took full care of the reins, crows would manage. They loved the aged leader. Each gladly helped. Once Neferteri became queen of the Olympics, crows would dominant. Lessons from great-grandfather and from queen regents prepared her nicely. Hers was a reign of love, peace, and prosperity.

After the meeting, Mr. Mizithra asked his loving wife when Murray and Momo were coming to visit. Two months. He pleaded with her to either come with him or let him go visit. He had an urgent request to go see Murray, for what he did not know. With a look, you know that look?, he blushed and headed for Grand Meadow.

It did not take a rocket scientist to figure her husband's desires. Mrs. Saffron was happy to let him go. He needed to burn off nerves developed not being able to tromp all over the Olympics on some crazy quest. He was on the fringes of the most beautiful place any knew. Grand Meadow was in full bloom. Every animal, bird, and bug availed itself of warmth and food abundant in every corner.

Murray and Momo had a lovely hovel not far from the stage. Mr. Mizithra was guided there by Horace. Boris was busy elsewhere. If he ever needed some help, he was to call Horace and Boris. Anything, he was assured, anything at all for the Hero of the Great Wall of China. He was a war champion. It was not a reason as far as Murray was concerned.

Neither Horace or Boris fought invaders, volunteering for medical duty. Sometimes, those who do not fight witness the worst of battle's aftermath. Every animal did it's best with talents owned. You cannot judge those who did not fight by the standard you did. Medics, with no weapons, go right to the front lines, too. Unity, loyalty, and love, those qualities spell victory. Jealousy and self-righteousness are right out.

Chapter 2 Murray's Update

Reuniting, Murray waved Mr. Mizithra down the tunnel to home. Momo hugged the marmot once he entered the major chamber. They had two rooms off to the side, one for them and one for Nediva and Frema. A steaming sauterne of soup entered, carried by a woodrat missing the ring finger on her right hand.

It could only be his most wonderful substitute mother. Nediva came in and set the bowl on the table. A warm, glowing grin planted itself on the marmot's happy mug. She waddled slowly over, very close to Mr. Mizithra. Looking up, she set her eyes to make him out as her eyesight was fading. She squinted and spoke.

"Sit, eat, you big galoot. Ve have been vaiting for you for hours. You are schtill so slow. Sit, have a bowl of schoup. It is hot und gut. How is your frau und kinder?", she rattled away for hours never pausing to take a breath. He never answered, just let her chatter while he beamed.

They had a lovely guest room prepared, a bowl stuffed with fresh lupines and lilies. Bed was soft leaves and pine fronds. He slept well, rising early. He left the warmth of the burrow, stretching in the first rays of dawn. He trundled down to water's edge for a sip of water and wash a furry face. It was a wonderful way to greet the day. Across the pond, ducks quacked about, feasting on waterlogged greens.

Duck butts would flip up, then back over constantly. It was rather humorous to watch, which he did. Bugs flitted a few centimeters above the water, too much temptation for eager fish. Once a gnat got close to liquid, a fish face would jet out of the calm water splitting the surface. An open mouth would shut and the bug would disappear. The action was so quick the insect had no chance. Yummy for the fish.

Mr. Mizithra spotted Boston Charlie and Matilda fishing on the far side. Each perched on a selected bare branch. They faced away from each other, providing a much larger circle of fishing. Two is better than one as they proved. In moments, each successfully snagged their victim. They flickered to their hidden nest. Another young one this year. Matilda was fertile. They were a fruitful couple. She was an excellent mother. Boston Charlie quickly returned, landing next to his dear marmot friend.

"Well, now look who's here. Mr. Mizithra, I am so glad to see you. It has been a while since war ended. There has been no time to talk over those days and deeds against that foul raccoon. General Lloyd Joseph had me flying all over the Olympics. Murray shared his thoughts on the battle at Fort Ballerina and the Great Wall of China. Such a terrible time. Still, everybody says you, sir, were very brave. That does my heart well."

"Yes, Lloyd Joseph told me everything. He told me how you obeyed his orders, sending Murray and I to the Beaver Blues Bungalow that first time. This was long ago. It is a hard nut to crack with me to be deceived. Trust is an Olympic marmot's most coveted attribute. To have others have confidence relying on me is one of the pillars of my character."

"So, for this specific reason, I have something serious to discuss. Boston Charlie, we are friends. I will tell you, deceiving a marmot is not wise. We are not a species who forgives betrayal easily.", the visage on the marmot's face was grave.

"Yes, sir, yes, yes, I do understand and agree with you. However, I will offer two indisputable facts in my favor: Lloyd Joseph had a plan the whole while. I was merely a cog in the wheel. In fact, were it not for the bear", here Mr. Mizithra visibly winced at the memory of the foul beast who ate his daughter, "we would have won."

"Second, it is my firm belief both you and Murray are better off for my having directed you to the club. Since then, sure you have had some troubles, but over all getting to know beavers, muskrats, and the rest has been a blessing. Murray found a home and you as well."

"That much I agree with heartily. War is evil. Forces aligned against us were formidable. Hear this then, I will forgive you, but do it again and our friendship will end. You cannot break a promise with a marmot and get anything but a cold shoulder."

"Yes, war is brutal. We fought a much larger enemy. Crows were spectacular against ravens. We faced raccoons, rebellious beavers, large squirrels, voles, bats, and that bear. I know many died, friends passed in intense fighting. The path littered with dead on both sides. We cornered Ralston and his gaze with the help of Mr. Loganberry of the Hills. His surfeit, tails raised, guided that evil animal uphill. We trapped him and his garbage up on the Great Wall of China."

"Bette, Tobias, Twigs, and Ripple built a dam. Tobias took care of the diversionary canal. Muskrat worked toting and lifting. They cut off water long enough to ford the dry bed and kill the enemy. We are not a violent species, Boston Charlie, but all saw red that day. He was killed and tossed in the riverbed."

"Bette and others pulled away large logs of the dam holding back water. In a short time, the creek rose enough to cart his lifeless carcass downstream. His horrors were at an end. We waved good-bye to that rancid raccoon."

When Mr. Mizithra spoke earnestly with his friend, it was not to hurt him or make him feel guilty. In being honest and sincere, he made Boston Charlie understand the value trust had to him and marmots in general. If someone wrongs you, show love and maturity. Let them know in a kind, but straight forward way your feelings. No one likes to find out problems between two from gossip. That is worse by far.

If you are hurt by one who is a friend, it may be they do not know you suffer. It may be they do not expect you to say anything. So, rather than assume you know someone's heart, it is better to speak in love. Then you know truth. It shows respect for them and the relationship. Raising concerns in a caring manner versus accusative permits friendships to grow. Silence leads to thoughts which quickly distort the truth.

He told his friend the price trust holds. Boston Charlie now knew. It never crossed his mind to do a thing to hurt Mr. Mizithra. Only by being told did he realize the impact on his friend. The marmot loved and respected the Belted Kingfisher. He spoke truth. Ironed out, friendship grew. The one offended never brought the incident up again.

The point was made. Just between the two of them, told a friend the pain deceit brought and effects. To harp on it once the problem resolved would make Mr. Mizithra petty. He would not truly accept the apology. If he had, he would not feel the need to resurrect the dead. It is simple. Be honest, not cruel. Cure schisms, move on, remain friends.

After a bit more chatting, they went to the area Graedy and Molly selected for a lower medical center. Their hospital was up the little brook directly behind their smaller place. Here, they could service easy cases during the day. The upper hospital held patients who had to stay down.

Sadly, Mr. Burrito did not come to them. He was never found, mournful reminder of the price of war, it's aftermath bearing evil.

Otters in training to become doctors took larger patients. Scurries of Douglas squirrels cared for smaller ones, the anxious chirping making rest impossible. Still, they meant well. Molly constantly had to chastise them to keep quiet. If you know Douglas squirrels, this is not achievable. This made the hospital a comedy center, too, aiding in recovery.

The goal was to create doctors, acupuncturists, and chiropractors to spread throughout the Olympics. To this end, school took a larger focus. Once adequate doctors, nurses, acupuncturists, and chiropractors were ready, Molly and Graedy spent the bulk of time teaching students what they knew. This medical center became famed in the hills and valleys. Anyone who made it to

Grand Meadow would find Graedy and Molly's Otter medical facility willing to accept any critter.

Working on centipedes to wapiti, they were known for acceptance of other species, even meat eaters. Weasels came to her regularly, tending to get injured trying to kill other animals. Ignoring images of dead creatures consumed, she would repair injuries. After all, she surmised, I eat fish. We are from the same family. Dedication to saving others made her a prize in the Olympics. Skunks took a special touch, of course.

After a hug from both, Mr. Mizithra went to the stage, finding a gathering of musicians. Not one himself, the marmot loved music. Hey, at least he could tap a toe or two. From days back when Murray took the stage, he had extra pride in his woodrat chum. Talent came from practice and playing back home before they moved.

When Murray played at the club the first time, talent was there to be a professional. He did. The initial foray on stage proved he had chops. The addition of Momo and her flute sent it over the top. Bette, Twigs, Tobias, Deirdre, the Whiskey Bend Kid, Ishmael, and Ronaldo joined Murray and Momo.

The band was intense. With Tobias and Ishmael they had two basses. The dual thump was outstanding. Tobias taught him the tail combination. They made an incredible back beat sound. Two lady lead guitarists was sweet sounding. Bette now employed the bottle Murray brought from Bear River Ridge. Lead and slide guitar at the same time, wicked.

Ronaldo made piano a focus as the talent he possessed arose. Twigs had a massive band to bang with. He was on Cloud 9 because of it. He was so happy. The Whiskey Bend Kid gave them two drummers. They banged kits with love. It was most amazing to hear as sound gelled to perfection.

Enter, the lady of the lake. Moving digs up to Grand Meadow, Miss Bluesberry was so happy, she was about to burst. Exploding on stage, she ran screaming towards her wonder friend, arms waving madly. Tossing a massive hug on poor Mr. Mizithra, she squished with love unbounded. He was her favorite animal in the world, save her husband, of course. This Darling Doe was now a diva extraordinaire.

Her entourage were Horace and Boris. Two is not enough, so they added cousins: Cloris and Delores. These lovely ladies, fans anyway, were devoted to *their* skunk. Her vocalizations were mesmerizing. Her song: "Lovin' Every Critter On Earth", was inviting to the forest denizens. Every animal who heard it, loved the song. Soon, animals, even classical music aficionados such as Trumpeter Swans, were heard humming the tune.

Music, songs, lyrics, why? Well, it is simple, notes emitted by piano, guitar, violin, kazoo, instruments touch your essence. Music is innocent. It has no religion, political party, or side. Lyrics are what they do to heart, mind, and spirit. Words added to music inspire school songs for sporting events. Opera to punk rock have stories in words. When you combine music, moving the soul and thoughts sung in a certain way, the results are astounding.

Poetry moves your mind. Put poems to a catchy beat, a glowing rhythm, and you have a certain something no one can fully describe. Spiritual songs stir religious people. Love songs, well, you know what those do. Operas, like rock have huge sounds which stimulate our hearts. Stories from past, musical tales: Carmen, the Marriage of Figaro, South Pacific, or the rock opera Tommy use many songs to convey a message.

With talent, her ability to perfectly phrase lyrics, this sow held them in her paw. With such power, her only goal was to inspire a crowd to revel in music, lyrics, vibe. Others use music and lyrics to actually harm the listener, inspiring hate with a goal of inciting physical maltreatment of others. Music abuse is a sin against not only music, but the world.

If you have a talent for singing, playing, or writing songs, please use it for good, for pleasure, not evil, injury and insult. Miss Bluesberry sang strong songs, like the one against Ralston, but for a purpose. After the war, she never sang it again. Gone, his vile activities need no longer be spoken of by her. She gladly put that song in her "never sing again" stack.

Once bands got to playing again, Grand Meadow became the center of musical appreciation for the mountains. Due to it's perfect location and copious food, animals with talent came to learn, play, and grow. Mr. Loganberry of the Hills settled in domestic life with his bride. They had a marvelous cave up behind the stage. She could go home at night and rest, her husband caring for her needs. He was a cave husband.

He spent time with his three brothers roaming the area and having fun. Yes, he had four brothers and none were slain in the war. They had to take care back home because the promise to big brother was to fight, then return to the loved ones over the way. Three brothers opted instead to remain because it was perfect. Home was not, so they stayed.

Mr. Salalberry, who watched over Miss Bluesberry, strongest of the brothers, went back. He became: Mr. Salalberry of the Hills. His older brother gladly surrendered the title. This "of the Hills" moniker denoted lordship over the region. Skunks, like most animals, have a territory. The one he took over from his brother was in their family tree for a great many generations. It is here they made surfeits and remained.

For one to move, usually meant too much competition from other males. Here, it meant brothers moved along, leaving a true, natural leader to take over. No sibling jealousy. On the contrary, they were happy to see him rise to a position of such responsibility. He was bent this direction. Someone had to do it; he the perfect candidate. Good choice.

If someone you know achieves success, what do you do? Is jealousy the result? Hatred, anger, spite? Oh friend, no, please. It is natural for a bit of envy. Letting emotions rule, allow it to harm you and the one you aim evil toward, is irrational. The best actor gets the part, best athletes go to playoffs, the finest musician plays in the symphony. That is the way of life. If you are not good enough, try harder. Work for the goal, do not whine at others' success. Appreciating your own talents, achievements, and abilities allows less time to dwell on those of others.

Mr. Salalberry of the Hills became a leader. It was deserved, earned on the field of battle and treatment of family, friends, and strangers. Enjoy the success of those you know. Do and they will do the same, so get off your couch and do something worthy of notice. He did, returning to their old homeland and becoming a famous polecat. His family remained proud of him all his days. His surfeit grew, territory secured, respected by all who met him.

After meeting everybody, he prepared to leave. Murray and he had a talk, per usual. Due to the nearness of the Valley of the Stones to Grand Meadow, it was wagered visits could increase. The only concern he had was fatherhood. Murray and Momo wanted to have a litter, but it would have to wait for winter, when they could care for young ones and not worry about any tour or show.

They were going to come down for a show later this summer, before hibernation. They still had a great deal to discuss. He and Momo could stay there a while longer than scheduled. Murray agreed. Mr. Mizithra gave a hug to everybody. That certainly took time.

Heading home, he saw an odd sight, a bird of prey resting up against a tree, on a limb. He leaned on the trunk, weak. It was a mature male Cooper's hawk. He could tell by the red eyes and small size. Due to the ways of nature, females of this species are larger than males. Their dimorphism is reversed from Olympic marmots. As we found, beaver are about the same size, no difference in gender. Nature does this. The hawk was clearly in some form of pain.

"You there, hawk. What is your trouble?"

"Oh, oh, oh. I am injured, marmot. The pain is overwhelming. Do not stand there whilst I die and mock my pain. Let me die in peace."

"Can you fly? I say, can you fly, sir?"

"Barely, why?"

"If you can make it to Grand Meadow, find Molly O'Shea O'Grady. Tell her Mr. Mizithra sent you. They will even treat a raptor."

"Do not jest with me, marmot. I did not think your species to be so flippant with the wounded. I was told you had compassion. Instead, I get derision in the face of my death."

"No sir, no taunting, I am helping. Flap your wings over to Grand Meadow. Ask for the otter doctor. She will help your wound. Quit being a cantankerous fool. You are injured, I am attempting to help. Your pain is directing this anger. What is your name, hawk?"

"Kelly. I am Kelly the Cooper's hawk. You are not lying? Very well, I will go. AHHH. Hard to breath. I must hop tree to tree. My chest is crushed. If they help me, I will return to thank you."

"If you recover, find me in the Valley of the Stones, you mean, crotchety cuss.", shaking his head he headed home.

Chapter 3 The Patient

Pain does weird things to animals and man alike. A Cooper's hawks' cry sounds like: *hick, hick, hick, hick, cheep*. This Kelly shrieked, as he slowly flew to Grand Meadow, following the marmot's advice. This took time to find. Calling to a crow on a nearby limb, she told him how to get to Molly, flitting away. A bird of prey is feared, dreaded by the rest of the animals.

He had to flap to Molly, but could not fly in a hurry due to pain. It was branch to branch. Heading north along the water's edge, he reached the infirmary. Fluttering to the ground, he landed awkwardly screeching out in pain. Animal and bird patients fled despite dressings and braces. They were evading death. He could eat them.

Molly emerged from the back. Her eyes blew up at the sight of this vicious killer running amok in her hospital. An otter is much larger than a Cooper's hawk, especially a male, so she grabbed a stick to beat it out of their aesculapian facility. Nearing Kelly, she saw him fighting to breathe. She saw his chest wheezing. He was clearly in pain, from his eyes.

This was no attack, it was someone looking for medical attention. With others fleeing, she went at the problem. Medical people, police, fire, military act opposite of common sense. They run into fire fights, burning buildings, machine gun nests armed only with bandages, facing danger to save the lives of others. This is a unique class, who willingly sacrifice to benefit those they do not even know. Yes, extraordinary people are these.

"Mr…Mr. Mizithra…sent…", all he could spill before collapsing to the ground, out cold.

Molly held him gently, carrying Kelly to a table for a good look over. Having never seen a living hawk up close, she was timid at first. He could rise and strike with his extremely sharp, hooked beak. Two otters came in the surgical theater to assist. No way was a Douglas squirrel going to venture this close to a killer.

Cooper's hawks eat birds for the most part, but a fat squirrel would make a fine meal in a pinch. It was understandable to fear this mighty hunter. They had no defense against such deadly weapons as beaks and talons. Any human being can see the sense here; you would not willingly go work on a lion or tiger were it not sedated, claws and jaws secured.

The patient was unconscious, supine on the slab, resting. Molly felt the body for signs of injury. He unable to blurt out what was wrong, merely experiencing extreme pain. After a tender probe, she found his chest concave. He slammed full speed into a tree chasing a Mountain Chickadee.

A little chickadee would be a nice repast for this feathered hunter. It is their style to either burst from some hidden nook in shadows up in a tree or strike from above. This latter practice is hazardous because larger raptors will take out the Cooper's hawk. It is vicious up in the blue skies. Think of aerial combat with human airplanes, that sort of the same thing.

This is a common problem for this particular bird species. They tend to fly so fast after prey, caution is thrown to the wind. It leads to this situation Kelly befell. He was after breakfast. It turned as he reached for the bird with his left talon. Focusing on his task at hand, claws at the ready, Kelly failed to see a broken branch. The chickadee did and darted off in time to avoid contact.

Kelly was not so lucky. He hit the branch with his beak, throwing him off balance, slamming his unguided body at the trunk of a cedar tree. It took his breath away. The chickadee dashed. He made a move for the meal. Pain stopped him. This impact of body to tree centered on his chest. It was clear, he was near death's door.

Graedy, hearing commotions, made for his wife's side. He could see the patient's situation. He recommended first adjusting bones to set them properly in place. This he did very gingerly. Next, they got the infirmed moved to a different room, carefully, gently wrapping his chest. A hood went over his eyes, keeping him from striking hospital staff.

Their fear was genuine. His feet were wrapped to keep talons from grabbing anybody close. His reactions to whatever they had to do to him was new territory. No other raptors came for treatment to Molly, so she was cautious. In dire need, it was a good chance to learn another species. He needed rest. All they could do was wait and hope for the best.

He slept a long time, required. Impact to his chest was so strong, death was a real possibility. This is peculiar to this type of hawk. Speed is their key weapon. Once in a while, something goes awry. The bird will slam into a tree or some other object. Finding a dead Cooper's hawk due to crashing into stuff is sadly common. Kelly's quest and pursuit of food almost killed him. The need was recuperation.

She had a room. He moved in the most private area of the hospital. They built rooms going up, so he was quite high. It was felt once he recovered this would be an excellent position to begin baby flights to regain strength. To feed him, they asked weasels to help. The Whiskey Bend Kid had a friend who was willing to share his skill with the bird.

Normally, he would die from injury in the wild. Fortunately, Mr. Mizithra happened by to lend a paw, so to speak. With the skilled hands of Graedy and assistance with exterior injuries from his loving wife, Kelly got the attention necessary. Alone, he was a goner. He was blessed the marmot was passing by and saw him. Mr. Mizithra is sure one cool guy.

Given time, some food, water, and herbal tea, he would have a good chance at recovery. He would endure pain in the body mending itself. Most difficult for him was the hood over his head and the pads on his talons to dull them. These were uncomfortable. He understood, but did not like it. They could not risk hurting hospital personnel. He was in isolation for everybody's protection.

When you think of medical practices, there are many factors to consider. Some fear doctors so much they never go. On the other hand, some run with every imaginary ailments found. Medical personnel are just people. Some are awesome, some not. It is truth, not the unrealistic practice of honoring someone simply for being a certain profession. One thinks of Albert Schweitzer or Father Damien and Mother Marianne Cope on Molokai, Mother Teresa in India, or the work of Jonas Salk.

Conversely, think of doctors who perform surgery or procedures simply to bilk people out of money. Or those who give a plethora of pills to quell complaints. That is hardly medicine. Here, we have two of the former: Molly began healing arts as a whelp, at the feet of her parents. They were healers for generations way back for many years.

Graedy stumbled into becoming a chiropractor. He whacked his neck against a log jammed oddly in the river bottom. A friend popped it in, saying he learned from Chinese chiropractors and acupuncturists.

This led Graedy to seek the couple for instruction. The blessing of being adjusted made him want to learn. For days, he journeyed down the Elwha, swimming past Crescent Cedar, to the lowlands. Finally reaching a swollen pool, he found them. Smuggled illegally to the United States, they were extremely rare Golden Coin Turtles from the mainland. Humans black-marketing in danger of being caught, released them in the Elwha.

They swam to this fat, calm pool and set up shop. Soon they were popping backs and necks back in place left and right. Word spread and students appeared. A school was created. With good luck, Graedy found them. They only trained a few students at a time, so he had to linger in the pond until it was his turn. Filling his time, he met Molly. They fell madly in love and soon married. She had medicine, he chiropractic labors.

The Golden Coin Turtle is a beautiful, medium sized freshwater turtle found in parts of China and northern Vietnam. Used for medicine by they who so subscribe, the species is endangered. This means human beings in a lust for money, hunted them to near extinction. This is not the only species of turtle or animal on the list. It is long, way too long for anyone to be proud thanks to the Thumbs.

It is, instead, the abuse of the planet which is of concern. There is no reason for Chinese turtles to end up in the United States, but it happens daily with people sneaking in critters: birds, snakes, lizards, turtles, are common. This is what some people do for money, as animals suffer. For our story's sake, they are merely characters in the play, so please, enjoy.

Her name: Ming Ye. His name: Ming Yen. Trained in southern China near Lake Fuxian to become acupuncturists and chiropractors, they were captured in traps set in their mountain home. Taken from all they knew, family, friends, and environment, they were slipped on a cargo plane bound for Seattle. Once on the docks, smugglers arranged to drive them to a small airport outside Port Angles, then to the client. Learned eyes spotted the criminals. Authorities chased them along Highway 101.

Evading police, the bad guys made it to the Elwha and released the pair in the wild. Later arrested, evidence was never found. The turtles spent their days in an American paradise. They had no way to return, so made the best of what they had. They became American turtles from China. They did not shake it all, enjoying being from one culture and embracing a new. They could enjoy both dragonboat racing and baseball.

Graedy stayed long, learning both, but specializing in his first love: chiropractic. Eventually, he and Molly moved up stream from the Beaver Blues Bungalow, to set up practice and enjoy life together. It was a quiet oasis for two.

Leaving his brilliant instructors was difficult. They wished him well, giving gifts, humble bows, and a certificate for passing all courses. He graduated.

He and Molly were grateful. They took what they knew to high alpine reaches. This was good country. It was a lovely brook with a nice pool for swimming and food, an otter must. Home was pleasant, attached directly to their medical facility. For many years they tended to wounded. It was only much later, when coyotes hit and destroyed everything, they were forced to come down to the pond.

It was here, of course, they again saw former patients so far back in this story. Mr. Mizithra and Murray were greeted with a great deal of love and affection. Another home was demolished, this time by a stinking bear and rotten raccoons. Due to war, they had to go where they were needed. This brought them to Grand Meadow. After war, they agreed this was ideal. The spot west, up the little brook, was perfect for their needs.

They built a huge hospital, recovery area, and school. It was odd how life tossed them here and there, but with love and strong will to live, they survived and thrived. It was here they stayed for the rest of their days, a tribute to the otter species. Their legacy lived on in the many students who went on to fill the Olympics with animal doctors, nurses, chiropractors, and acupuncturists.

Kelly the Cooper's Hawk was in rough shape after whacking a tree. Cantankerous by nature, this did not help. He did not appreciate the hood and talon coverings. Mr. Mizithra, nice guy we know him to be, helped the wounded wing master. Good fortune was on his side. He found Molly and Graedy with the help of a crow. Thanks to the Sergio, the Whiskey Bend Kid's friend, food was constantly supplied. In fact, Sergio and Kelly had some long talks.

To help his new, feathered friend, Sergio hunted up a bird or two as he was able. A short tailed weasel is not the best at this sport, but he was willing to help a friend. This hearty species, a subspecies of the European Stoat which is much larger and turns white in winter, feeds on voles for the most part. Their bite is the worst or best depending on your end of the teeth in the animal world. Their bite is stronger than a tiger. It is a size deal is all. They do not turn white in winter, as do other weasel species. They are so strong, one can slay a rabbit five times larger.

With this, and physical therapy, Kelly was soon able to make short flights, hitting the lake only once. He flapped in the water to shore. It was not far, but showed how strong he was. Within a few weeks after this, he was ready to leave. He thanked Molly, Graedy, the entire staff, and Sergio. He promised Douglas squirrels he would not hunt their kind for the help they gave in helping him mend. Sergio would see if the wapiti would allow Kelly to come to the Grand Meadow Festival.

Raptors were strictly forbidden, which makes perfect sense. You would not want the audience eating performers. Bad taste that. Turning to his healers, he hugged with wings around them. They could keep the hood and talon protectors. Kelly wanted freedom. He had to hunt!

Flapping with real strength, he rose from his lofty perch, circled the hospital twice, waved a wing, then off to the Valley of the Stones to pay a visit to a certain, kind marmot. Molly and Graedy waved to the most interesting patient either ever had, hugged for a bit, then turned to tend to the next case. She had another Humboldt flying squirrel injured in training. Those gliding nut jobs were too much. Still, no one could ever doubt their effectiveness.

Graedy had Mrs. Holstein, a well-known hypochondriac otter, who wrenched her back yet again. Were she to listen and lose some tonnage, problems would dissolve. Too many fish cakes. Ah well, such was life. They were happy and the Olympics benefited as a result. Not far from the pond, Kelly spied a passing thrush and quickly had a fine meal.

"You. Marmot. Yes, you. Come here. You remember me? You said to come see you if I lived. Well, I did. This, sir, is difficult, for I am a most reclusive fellow. Thank you, sorry to choke up, thank you for what you did. Most other animals would run away and be thankful another hunter was dead. You took sympathy and had compassion on me. For you to find an enemy and not pass by, but help is magnanimous. You are unlike any other animal I ever met."

"You and Molly and Graedy are special. I am in your debt, sir. If you ever need me, send word, and I will come. I can muster a cast of Cooper's Hawks if you need help. Thank you, Mr. Mizithra. You are something else. Thank you.", with this he flapped over the side of the ledge, the stream's mist masking his trajectory. In a flash, he was gone. Mr. Mizithra smiled. He had another tale to tell, yup, he sure did.

The 100 days over, marmots fattened to bulging, they covered the tunnel entrance, preparing for a long winter's nap. Mr. Mizithra reflected on the year and smiled knowing his family and colony were safe. It was one for the books. Stories he could now tell would be echoed by those who witnessed and participated in war. Key was they were alive, survived violence, and had a colony which was going to thrive come spring.

Chapter 4 The Lodge

Life has it's twists and turns. It is a river complete with a beginning and an end. How you fill the space in between is up to you. You may have a stream as short as the D River in Oregon, or as long as the Nile, one never knows. It determines the individual waterway you create. Life, like rivers, has waterfalls, rapids, curves, pools, and a final delta which flows to a lake, another river, or sea. Such was the case of Bette and Tobias, orphaned siblings, and their sidekick, Twigs.

These three began at the Beaver Blues Bungalow long ago. Beavers were born there, Twigs another part of the pond. Playing together was a habit none wished to break. With war, death of parents and brothers, it was impossible to live there any longer. The club was a shambles. They had no desire to rebuild anyway due to memories. It was too much to expect. They were not going to go back to the scene of such misery and death.

Enter the solution: once hostilities ceased, they finished repairs to the dam. A young family of beavers, Festus and Clementine, with a brood of four wee pups, were given charge of the pond. For generations, they dwelt in the same place, never rebuilding the club. It was hallowed ground. A plaque was erected at the site. The Beaver Blues Bungalow Massacre was commemorated for all time. Bette and Tobias' reward was the lake at Grand Meadow. This was in gratitude for combined efforts during and after war.

The ingenious blockade and diversionary dam saved the day and helped defeat Ralston. They had the whole lake. It did not need a dam. The water flowed from the source. A dozen small streams collected here, the depth over twenty meters. The lake had a long shoreline. The river left the tarn, ending in their former home before eventually arriving in the Strait of Juan De Fuca.

During winter they planned, knowing they had time at hand. It would simply be called: The Lodge. This would not be just a plain, old beaver home. It was going to be over the top. This was going to be The Place to go. Bette took her wild imagination and blew it up. Her brother and best friend, Twigs, were on her side. The club would be much more secure, too. A team of otters, always at least six, would have a vine rope across the entrance at the shoreline to keep out the riffraff.

Past security, one walked a guarded plank to a large door opened by muskrats on either side. Inside, a large foyer greeted you to shake off the outside before going either left or right. The left side was: Beulah's Boudoir. The right side: Buford's Billiards. Left was a restaurant and guest accommodations. Right, a club with pool tables, a huge stage, dance floor, and well placed seats.

Both areas had ample escape hatches directly to water for muskrats, beavers, and otters. Due to increases in music festivals, additional sleep chambers came at the far southern side. Water lapped up nicely to a small ridge, a stand of trees behind. Here they had teams busy digging a dozen extra areas for camping. These they sectioned off from the rest of the audience zone. Pushing up rocks and packing soil, they created a stadium effect with rising rows of seats. This way, parents could let young ones play and listen to the music at the same time.

The restaurant was the front of Beulah's Boudoir. The rest chambers were to the far left, so neither area was disturbed. There were three areas. You could eat in front on single seats cafe style. You could go to a large portion right with tables and chairs. This is where most came. Then there was the private area around the side. This was fairly dark. It was for those who did not like mixing with others or had a cloistered meeting and desired no disturbances.

A new bark display from the bungalow was larger, varieties of bark coming from all over added to the list. Tobias worked a deal with crows to import it from the lowlands. There, trees were more diverse. By bringing it to Tobias, he could increase differing bark species and make outer tree casings something more elite.

The menu was enhanced, bringing in connoisseurs from all over the Olympics. Red Cedar, Pacific Yew, Japanese Maple, and Ponderosa Pine were new bark varieties on the bill of fare along with standards. Flights of bark became very popular, a sampling of several types to create a flavor flow. These were washed down with special Lodge Cattail Tea.

Lincoln, Cain, Abel, and Seth were regulars. They moved to Grand Meadow in the postwar world. Their lodge was destroyed in the raccoon uprising. As elderly beaver, they were not capable of fighting. Instead, they assisted the rebuild with Bette and Tobias. Tobias urged them to resettle. They did, taking up residence in four lovely chambers way off, deep inside the pond. This was bachelor's paradise. Here, they could dine to their heart's content.

The food was like before. The chef was Ripple, whom they found to be a perfect choice. He was private. Staff out front cared for customers whilst he whipped up meals fit for royalty. Wait staff were otter, beaver, woodrat, and Douglas squirrel. Muskrats, well they just were not built that way. Muskrat did security, played music, but would not work in Beulah's Boudoir.

Bette took particular pride. The restaurant was her mother's crown and queendom. This was an homage. Her old recipes were unchanged. To do so would be a travesty, a sin. As time passed, the eatery at The Lodge became a swanky joint, still dishing out Beulah's honest eats, mixed with daintier fare. In season, it was a good 10-15 minute wait to get a table. They made the best cattail tea and other drinks. Mineral water was tasty, too. The Lodge was a hit.

Buford's Billiards had two billiard tables, a huge stage, dance area, and chairs. A small dressing room at the back was the only way to hide before a show. They called the stage: Homer's Hayride, backroom: Henry's Hangout to honor their late brothers. Over time, every musician, singer, or performer put a paw print on the walls or ceiling to prove they played this magical place.

The stage outside for festivals was legend, but this underground dais was just plain wicked cool. The reverberating volume rippled the lake. Fish avoided the area at all costs. Jam sessions were revered by every musician in the mountains. They would come to make their mark, pass the test, and be allowed to play with the best.

Deirdre left early, getting a call from a male beaver who played bass. His dam and lodge were many kilometers away. She quit the band, and made her way to him. She sent word back and the Whiskey Bend Kid left to join her. The band she and her husband, Hightail Hank, with the Kid rocked their lake with gusto all her days. Bette missed her jamming sister. Six-string gunslingers kept her fingers busy.

Ishmael left to join another group over the hills, closer to the Valley of the Stones. There, a small lodge and dam existed. His sister needed him. Her husband was overwhelmed. She had pups to whelp. They had the family dam to consider. Music was important, family more so. Ishmael came home, found love with a lady beaver. Sausalito Sue could flat out sing. Once the dam was repaired, Ishmael proposed, Sue accepted, and they built a dam family band.

Bette had plenty from which to build. Tobias, Twigs, Murray, Ronaldo kept music flowing. The stage had piano, standup bass, a row of guitars, and two full drum kits. Various percussion instruments were available. Branch and twig stands for saxophone, flutes, etc. laid to the far side, dubbed 'Sugar Murray's Spot'. This was a 'Do not touch' zone. Do not go near instruments, unless you are a player going on stage. Muskrat guards stood watch over the platform. Homer's Hayride was a big deal for locals. Animals were known to travel up to fifty kilometers, just to get a chance to play with the best the Olympics had to offer.

Murray and Bette lived against each other on stage. Tobias thumped bass and tail in his wild style. Ronaldo took to the band like a regular and was soon the talk of the town. His hard blues style made piano a driving force for their sound. You had to keep up with Bette, Murray, Twigs, and Ronaldo to measure up to the big show: Homer's Hayride.

Bette and Tobias had rooms sequestered behind Henry's Hangout. Special underwater passageways gave secrecy. Home was off one of these. They entered from beneath, rising to a drying cavern. This led to a big central room. They entertained others who could swim. Each had a large chamber unto themselves. Twigs had a place with Ripple.

They lived in a cozy spot behind the restaurant. While visitors slept on one side, they were alone in a private area. From the surface, all you saw was a small rise of sticks, branches, and grasses held in place with muck and mud. Their underwater escape was ideal, too. Under this facade was a bachelor pad, fit for two wild and crazy muskrat guys.

This they were not. They mostly went back after work and playing music to collapse. They loved Grand Meadow. The old pond was nice, but this was epic. On days off, the two would swim all over the lake, discovering hidden underwater caves. It was Paradise. For Tobias, music was everything, was happier than he could recall. Ripple was in seventh heaven going in sub-aquatic caves. He found all kinds of stuff.

The Lodge held contests to see who was best. Every once in a while a really good player would be found. These were usually nabbed by Darby and O'Gill. Guitar god wannabes would come challenge Bette to playoffs. Whoever the crowd liked won. Otter, weasel, muskrat, or beaver, no one unseated her top spot. Taught by Buford as a whelp, she took to guitar picking like a duck to water.

She lived blues, could strafe the stage with sick rock riffs, or pluck the sweetest slow ballad leads ever heard. She could change tempo mid-song and play slide. Bette was a guitar goddess. She enjoyed playing with other guitarist, missing dear friend Deirdre deeply.

Miss Bluesberry's voice ruled The Lodge. Mr. Loganberry of Grand Meadow (his new moniker) beamed with pride at her voice. Other clubs wanted her to play shows. Darby and O'Gill kept her schedule full. Any and every day off road was spent singing at Homer's Hayride. She was the first performer to put a paw print in Henry's Hangout. From then on, only those playing were permitted to leave a mark. She was the Darling Doe, star and diva, as meant to be.

In her humble glory, she starred in shows at Hart's Den, Squirrel's Center, Crow's Nest, and constant repeat performances at old Muskrat Cavern. Mr. Loganberry of Grand Meadow accompanied her to every show, always glorying in her lilting tones. She calmed his spirit and filled his soul in ways no other animal could. These two were so deeply in love with each other it was disgusting.

The Lodge gained an earned reputation. No other club could offer the ambiance, food, or music. Once on top, their goal was to maintain and grow, changing where necessary. If you are a part of a business, band, squad, getting to the top is a struggle. It takes teamwork and great effort.

If you are the acme of your selected field, let's say the best baseball team, you see vistas thus far unknown. To remain the pinnacle, you must stay ahead of the rest. The ball team will add a new left-handed pitcher, try to improve batting, and cut down on errors to stay in rarefied air. It is said getting to the top is difficult, staying is harder still. It is the true test.

The Lodge house band had excellent musicians. Momo would come play. She loved Shakuhachi. Learning percussion added to songs. Players would show and tryout under the talented ear of Boston Charlie. Horace and Boris were usually a part of selecting. Cloris and Delores were called strictly for singers. They had "ears" to hear perfect pitch. Darby and O'Gill had a table up front for frequent visits.

Experts gathering at The Lodge made it easier. The pool of talent on this pond had a habit of spilling out to the big stage. Canadian geese kept pretenders away. Loving metal rock, those playing tended to destroy the stage while performing. Nonetheless, the new digs became famous, Bette, Tobias, and Twigs happy at home, glad to survive what the rest of their family did not. These were grateful, happy critters. Life was good. Music flowed like fine wine.

Chapter 5 Murray's Place

Murray and Momo had a lovely home. They did not want to live in a place they played. This was Momo's decision and proved wise. Nediva and Frema were aged, especially Frema. Given any place, they selected a little knoll north of the stream, towards the hospital Molly and Graedy built.

Everybody pitched in. Soon, they had a small pool filled with water bulbs, and a cattail lined walkway to a tunnel. It led down, opening to a large chamber. Off of this left, a kitchen, always something good on the stove. Tea was a true joy, served with humble pleasure.

Murray and Momo had a large bed chamber at one end, shaded with a dried lichen curtain for privacy. Ladies had rooms in another section. Instruments lay strewn here and there as Murray liked to play at all times. He was indulged as any other woodrat musical genius. They had a lovely home, made better with a healthy dose of love.

He had a nice swimming hole out front where he could frolic and snag a fat water bulb for a snack. After wild traveling adventures, he was a happy home husband. Momo made him so halcyon, all he wanted was to love her and play music. He felt this a great, undeserved blessing, but was glad to have.

BOOK 4: PEACE

Recovery from war, considering these elderly ladies were assaulted, kidnapped to V-Island, and rescued, was understandably slow. The new home was built while the lady woodrats rested at temporary digs Mr. Mizithra visited. Frema was old and frail.

Molly came daily to check her health. They formed a strong bond, the physician's bedside manner a tonic to the old lady's soul. After the rescue, they were flown by trumpeter swan, not something Frema did willingly, to Crescent Cedar to wait out the war. It was woodrat paradise. They were spoiled rotten by the crows.

Murray and Momo were vital to war's success. Caring for loved ones was no chore. It was an honor. They were flown to Grand Meadow at the end, setting down, eight trumpeter swans, a tight, solid plump landing in perfect unison, becoming a gaggle. Vowing never to fly again, she ran and kissed the shore. Now with this lovely home, she was a very happy old woodrat. No Cossacks could get her here.

Nediva wanted only one thing. Murray assured her this winter the young couple would do their best to make babies. "A litter, kinder, a litter, all I vant is a litter!" words which roamed his mind day and night. She wanted to be a grandmother in the worst way. She saw herself doting on young ones.

Momo eagerly desired becoming a mother. Like her husband, she wanted to fulfill a promise: back to Carnival Island, the festival, scene of her parent's death. She needed to complete the healing circle. It was fortunate to do now. Once parents, traveling days were over for good. Raising little rugrat woodrats was more important than Waadah Island or Willapa Bay.

Darby and O'Gill set the same schedule, local stuff, Waadah Island, six spots south, then the last show at Carnival Island for three weeks. Agreed, Momo would join only for the festival. The rest of it held no interest. She wanted him to have a fun time with musician pals. He loved to play. Long a bachelor, it was hard to shed all characteristics which made him who he once was.

To her credit, she recognized this and saw the need to let him fulfill dreams. Getting it out of his system would leave nothing to complete. He could have a blast, spend three weeks playing with his favorites with Miss Bang

Boom Bang, The Sir Franklin, Esquire, Dylan, Giovanni, Corinna. Agreed, he prepared for upcoming events with joy.

He hit the Hart's Den, The Muskrat Club, Crow's Nest, then Waadah Island. They knew it would be his last time, so played their very best. Tears flowed from Ness and Paisley at the final show. Hammer, Chicky Baby, Mamiko, and Graham wished him the best. The shows they held were incredible. Haggis begged him to reconsider.

Any time he wanted to come back, he was welcome. It was a no-brainer. With the addition of mouth organ, the sound grew. Like Bette and Tobias at The Lodge, Club Club had high standards, constantly raising the bar to keep number one status along the northern coast.

Murray enjoyed playing with them and grew each of his three visits. Hugs all around, then on Trumpeter Swan Airlines, south to the next gig. Murray had instruments along, and reveled in soft eider down Theodor's feathers provided. Flying with them was wonderful.

They played six venues as before. Sometimes it rained, but for the most part, it was lovely. Summer was in full swing. Travelers enjoyed the songs and the way it made them feel. Audiences were bigger in venues due to previous shows. Billed as 'Sugar' Murray, he had a following.

Each site, performers saddened to know he would not return. Such is the way in the animal world, where lives are briefer than humans. The times were fun, save one upsetting issue. It got so bad, in fact, Murray almost canceled the tour. He sent for Darby and O'Gill to explain.

He enjoyed playing with animals: Plinky, DeMond, Zigzag, Sweetie Pie, Roddy MacNamara, Jacinto, and Consuelo. At each venue, the band and audience seemed interested in talking about war, not music. The only reason he left home was to get away from what happened, not rehash the struggle He was touted as a war hero, not a musician.

In a fit of rage, Murray informed them the only reason he agreed to come on this last tour: fulfill his obligation to Murderers Row and visit old friends. He wanted to play with them one more time because his days on the road were at an end. Announcements sent ahead to various venues instead of championing his saxophone prowess, focused solely on his war record as Major Murray.

Murray fought to defend his family, friends, and way of life. He did not seek to fight. If war were unavoidable, he would battle so hard the enemy would never fight him again. He lived. His world was razed. Families suffered. Homes vanished. He killed animals to defend wife, mother, grandmother, woodrats as a species. Stress, fear, injuries took a toll. He aged from the burden borne combating evil.

Most enemies were larger in size. No woodrat takes on a raccoon on a whim. He saw others fall, woodrat or marmot, it did not matter. He saw death and injury. The last thing wanted was constant reminders. If you cut your finger, you do not remove the bandage and flick it so it will hurt, do you? It was the same for him, reminders meant pain.

Those who survive combat, soldier or civilian, do not usually wish to recount pain over and over again. It is identical for events of serious trauma. If you go through a bad time, a sad chapter, other than for purposes of recovery, it is not a subject you broach with joy. Bad stuff needs to be addressed, then put away. The hurt is too much, stays fresh for some reason. Keeping it in the dark serves a good purpose.

Murray had no desire to be a hero. He wanted to be known as a sax and harmonic player, not combat fighter. Darby and O'Gill flew at once, realizing the mistake. They went to each venue, removing any evidence of his war record. The only thing anybody was to talk about was music. If it was lousy, fine, but let music stand on it's own. No help was needed.

When Momo reunited with him at Carnival Island, he was his old self. Leaving bygone days behind took time, self-reflection. He and Momo went to the far northern part of the island once Ivan got them settled.

She took out her Shakuhachi, sat on the same branch, and played soulful music while her husband sat weeping at her feet, baptized in love. Fog lay heavy on the waters, kissing full trees with fluffy, wet whiteness. They stayed alone at this romantic oasis for a long while.

Walking back through the maze of animals, a voice cried out. It was Paula the porcupine. The hug given Murray and Momo was one for the books. Everyone feared quills would be stuck in them. It was a squeeze play if ever there was one. She missed this little woodrat. She knew Momo from childhood and blessed them both for their marriage.

It was a boon and she was happy. Comedy still flowed. They simply had to meet Clancy. War and the fight against the bear was fresh in her mind. Fellow quill warriors died in the cause, but she was fierce. After they spoke of it briefly, she was good, old Paula. The lady could go from sweet and sassy to kicking rear ends in a flash.

Respect, grown by assistance of other animals, was admirable. She was more revered than before war. Forced to fight for life and others, she rose to the occasion, becoming a heroine. War veterans were allowed to sit up front. Her shows were always packed. She spent time helping those wounded recover. Those gigs were private. Vets loved her.

Once back at the main area, Dylan O'Tanner waved them over. Corinna burst forth, flapping over for a big hug. Miss Bang Boom Bang nodded, drumsticks in hand. The Sir Franklin, Esquire bestowed a hug squeeze of brotherly love. Giovanni, in true Italian style, kissed both cheeks of the woodrats. Then he gave him a tiny bite. Ivan had them in the same spot as last time. The old practice area was cordoned off, so they were alone. They went while Momo sought old friends.

Music picked up where they left it, Ridin' the Zephyr, harmonica juiced up and humming. They jammed and practiced for a bit longer than normal because it was so much fun to play with these special critters. By the time he got to them, he dealt with demons, casting them to the winds.

Prayers answered, he played his little rat heart out, stomping with authority. It felt like home performing with this group. He was soon wailing away to his heart's delight. Momo joined the band on stage with her flute. She and Corinna did a sweet, slow number, "Yesterday's So Long Ago", about a lost love.

Corinna's soulful, honey dripped vocalization drank lyrics. Momo's lilting notes carried across the crowd touching each heart. Words came from deep within. Corinna knew of lost love. She had truth in her utterances. The assemblage, held willing captives, were released from a musical, willing enslavement after a record three encores.

Knowing it was Murray and Momo's last time, they wanted to drink up as much as possible. She even performed some acrobatic stunts with others at the smaller stage. A fun filled exhibition of duck synchronized swimming was spectacular. This time, Coots took the crown away from the upstart Goldeneyes. Watching Swan Lake a second time was magical.

This time Fur was not offered. Instead, the musical was My Fair Sow, about a lady skunk emerging from a stinky life to one of perfume and love. Ivan had a cousin, Stanislavsky Jones, with whom he shared Miss Bluesberry's story. It became a rock opera, featuring ten new songs plus five standards from Miss Bluesberry's catalog.

This was it's debut. They caught the first performance. Afterward, Murray and Momo met backstage with the playwright. He had a number of facts wrong, which they corrected. If he was going to honor Miss Bluesberry, better do it right. Most writers have egos, he did, but not enough to ignore live witnesses, especially these two.

Major Murray and valiant wife, Momo were heralded for front line combat. He bore scars earned in battle. Stanislavsky took the information to heart, added a Russian flair, offering a fresh, better product next show.

He took two days to make changes. Errors were minor, but effected truth. Murray was pleased and made a speech after the finale`. The actor who played him was a young woodrat, Nehemiah. He was excellent, getting Murray's accent down perfectly. He did chide the fellow, 'I don't sing, ever!', wearing a big grin. It was marvelous. He had to tell Miss Bluesberry. She would faint.

One new type of music, if you will, emerged this season. It came from the rabbits. They called it, 'Hippity Hop'. Using musical tones as a mere background, these bunnies would utter strings of words in a lyrical fashion, weaving the words spoken instead of singing lyrics. Murray and Momo did not care for it, but some seemed to love the stuff.

Murray and Momo made a special side trip to Bear River Ridge to see Shadow and the rest. Swans were happy to help. Hot springs were Momo's favorite. Perhaps, she said aloud, they could find the same thing in their neck of the woods. Hot springs agreed with her. Soaking was relaxing. Shakuhachi playing soothed everybody in the pool. It was very good to unwind for once. They got dry, then got down.

Music bounced off hills, down Ellsworth Creek, bellowing across the Naselle, down the other side to Dismal Nitch on the mighty Columbia.

Jamming all night, swans flew weary woodrats to their final appearance on stage with the Willapa Bay Blues Band. Murray was psyched.

"Birds, beasts, and anything else hanging around," began Ivan with a smile, "today for the final act of this year's highly successful Festival, we have the greatest band Willapa Bay has ever known. Blessed with the best guitar player, our own Dylan O'Tanner, back ups Miss Bang Boom Bang and The Sir Franklin, Esquire, keyboard play of Giovanni, we have added sweet sounds of lovely, vivacious, sensuous, lady red Cardinal: Corinna."

"If that is not enough, my friends, we have all the way from the Olympics for his final performance of all time, the one, the only Sugar Murray and his sweet saxophone", applause hit as Ivan faded in shadows.

Dylan blasted a riff exploded to shock the crowd. The Sir Franklin, Esquire was on the bass line in a note, Miss Bang Boom Bang on his heels. Murray stepped up at once, joining the throng. Giovanni came over top, plink, plink, plink snagging lead. Not a word, only groans, Corinna stole the crowd, stuck it in her pocket, and would not let go. Guttural bird utterances shook the audience. It was animal, primeval, naughty.

Not relenting, she forced the issue. Dylan, wasting no time, lit a blistering lead. She slithered against him. They moved as one in slinky rhythm, audience mouths agape from the water to the top of trees. She sashayed over to the beaver on bass. Why, one could swear The Sir Lincoln, Esquire was blushing by the time she swished away.

Pausing to fuss with Giovanni's weasel fur, she perched on top of his piano and began belting out lyrics to Willapa Bay Blues. The slight change in tempo was instantly altered by drum and bass while sax blared. They jammed this first number for a long time. Each song had an extra special feel. Momo joined them for a couple of numbers playing percussion.

After three encores, it was dusk. Music had to stop. A party after the show ensued for hours. Murray and Momo finally made it to the stump. They collapsed, smiling in complete satisfaction. Next day, Ivan had a big bon voyage party. They boarded swans and flew home. This two days, but they were worn out and happy. The final flight set down right along the shoreline in front of the path to their place.

Murray gave most humble thanks to Theodor and the trumpeter swan gaggle. Winter was at hand. They had to skedaddle south quickly to avoid the snow. They headed off, a beautiful skein of white jetting across a blue, gray sky. He heard Ride of the Valkyries clearly as the departed. Storm clouds were on the move, no regard for anything in it's path.

Preparations needed to be carried out. Home, they got as ready as feasible, knowing Frema was in ill health. They strove to make her last days good ones, full of comfort, warmth, and love. This they did and she passed in dead of winter, a happy old woodrat surrounded by family, and Molly and Graedy. Such is life, such is a well lived life.

Chapter 6 Clancy

When battles south occurred, a star was born from dark shadows. Male porcupines tend to be solitary fellas. Clancy was. He wandered the peninsula, seeking others of his kind, eating, and spreading good cheer. He was lonesome, not unsociable. His philosophy was: life is a bowl of sweet berries and most porcupines are starving. Enjoying the world in which he lived, Clancy explored nooks and crannies. Being nongregarious was fine. His variety was unique in the rodent world.

As with other species, they are not native to the Americas. Somehow, they floated across the Atlantic from Africa eons ago to Brazil. Skeletal remains show a migration north. Soon, they spread all over, adapting to new environments rather well. Arriving in the Willapa Bay area just above the huge Columbia River ages ago, they thrived in the soggy region.

Their worst enemies were: fishers and cougars, though other predators were not so friendly either. This includes bald eagles and large owls. Fishers climb, so porcupines cannot escape predators in their most common fashion. Fishers, hunted to extinction, man's kindness brought the natives species back to these climes. They thrived as before. Cougars claw past quills with power. With the rest, defensive actions are adequate. Most of the time the porcupine wins these defensive battles.

Clancy was from a good family. They were part of a prickle in the hills near the south end of Willapa Bay. Here, land was void of Thumbs. It was too soggy to build buildings. It was considered wasteland. This suited animals well; human interference stinks. He struck out on his own as a young adult. He enjoyed trekking the whole region.

When the festival occurred on Carnival Island, he was kilometers away, oblivious. He never heard of the thing. Only when war struck did he appear. The buzz in the woods told him war was on the way. He knew the stakes, not the participants. He was a proud porcupine, vowing help where he might. Crows filled him in. He headed for action.

Arriving at the battle in the nick of time, he fell on the bear Ursa from above in total surprise. Hearing loud bear sounds, Clancy came upon the fray from above. He leaped on the perpetrator. He turned the tide of battle, stabbing the final quills in to finish Ursa's miserable life. His appearance shocked even Paula and her porcupine army.

Enemy vanquished, Clancy simply shuffled away not bothering to communicate with those he saved. The opportunity presented itself to fight. He did. He saw no reason to stay and dwell on the past. He was off finding something. If the crows returned, he would respond to requests. Otherwise, he had places to go and porcupines to meet.

Paula and her prickle were shocked when Clancy simply took off without so much as a conversation about help provided. Not like others, he supposed, violence was shunned. He only did what he did because the need presented itself. Heroes are not made from those who seek fanfare, but act in the heat of battle. Clancy saw no need to tarry. As quickly as he appeared, he was gone.

More problems came days later. Crows assailed him, as they did any others who could fight. They were still in open conflicts with the enemy. Ravens hit briefly, but were of no large quantity. Warfare came down to raccoon and bear. Others fought, but porcupines were in danger and could wage war. In the Olympics, no such species dwelt. It was too chilly for their needs. Carnival Island, the peninsula, Bear River Ridge, the Naselle River were home. No invaders could come take over. They fought.

Crows called yet again, Clancy answered the need. He holed up in a safe tree fort. He could see and sleep in peace. After fighting, he had to recover and reflect. Killing a bear was intense. He did not like the way it happened, but who gets to choose where they scrap? Clancy made for the newest combat scene. He would help where needed, feeling no special camaraderie with any in particular. They were merely other porcupines. If they wanted to talk, it would not be about skirmishes.

At the next to the last battle, Paula made a boo-boo. Seeing a bear's head, she moved alone, not telling anyone her intentions. She left the safety of her prickle behind. If she were to get in the perfect spot, she could drop on his head and quill the jerk. A badly placed paw caused leaves to rustle loudly. The bear turned, lunging. Clancy appeared out of nowhere to stand between Paula and bear. The carnivore was stunned.

Clancy shot off porcupine stink, raising quills at the enemy. It was angry, throwing caution to the wind. The bear moved for Clancy. His whistle called down judgment. The rest of the prickle suddenly showed on the ridge above where they stood. The big, black beast had no escape. He took out one trooper, but succumbed to pain each deadly arrow shot by rodent power delivered.

Paula's pals mopped up the mess. Clancy quietly escaped to his tree. He saw Paula twice, a lovely lady, a sweet sow. Her move to intercede the large, black beast was foolish. Without back up, doomed. He whistled low to the prickle. They recognized him. He motioned them to follow, then launched his shock attack to flank the enemy. Caught off guard, the predator died by a plethora of lethal quills. Death came quickly. Paula, as far as Clancy saw her, was not bright. Pretty, nice quills, but not bright.

As the final confrontation arrived, Clancy was again called from his sublime tree fort by a murder of crows. He grumbled at the irritation, but agreed to scrap. Led by flying streams of black feathers, he found the scene. The large male bear was in the entrance to his lair. A few dead porcupines lay about in a bloody painting of pain and death. A fallen smaller bear was to his right. Dead animals lay around the corpse.

Paula and her prickle surrounded the monster. No one was willing to throw caution to the wind. Clancy was. He dove from a high position once again. He hit pay dirt. The blow was a forward flip, hitting upside down, quills landing perfectly on the enemy's muzzle. Nose, eyes, and mouth were pinioned. The bear screeched in pain and terror.

Swiping at Clancy for his assault, the monster received a paw full of quills. The porcupine hit a tree trunk head-on, cutting his cranium just above his left eye. He fell in a heap, concealed in underbrush. A branch filled with green lichen landed on his unconscious body. He lay long after battle ceased. Paula's porcupine militia were used to disappearances after the fighting ended, so thought nothing of it.

Paula, with her own thoughts, wished she could meet the handsome warrior. He was selfless in combat, wise, swift, decisive. Hers was a life in limelight, his shadows. With war over, it seemed she and he were destined not to meet. She was famous, but alone. Clancy was alone, but famous. She felt lonesome for the first time since her performing days. It would be nice to have a boar who loved her for who she was, not the star on stage.

Before scavengers arrived, but while vultures began to hover above the battle scene, medical personnel came to retrieve wounded. The area was bloodied, full of bodies. Any enemy body was left. If one were faking death, the otter or squirrel who came to help might be dead. Raccoons are like that, you know. Clancy was located under the tree in a pile of sisal bushes, covered in lichen. The cut across his left eye was wide. It was not deep, but did swell creating a big, black eye. He was still unconscious.

Carried on his right side to keep his left eye elevated, otters got him to the field hospital. Doctors worked. Avoiding quills was difficult. If they touched him, muscle contractions would stick nurses or surgeons. This sleeping giant was even dangerous in his sleep. They got his wound mended and bleeding stopped. They drained the swelling, releasing force on brain and left eye. They waited. It was touch and go.

No one knew his name. No other porcupine could identify him. His family was long off, unaware to his condition. Clancy was unconscious and alone. In a ward full of other patients, he was soon forgotten. Sleep was what he needed. A screen was around his bed. He was with other 'sleeping' patients, not to be disturbed.

Paula came a few days after war ended. News from north confirmed victory in the Olympics. She wanted to give fellow combatants a bit of sweetness in such a sour place. Many knew Murray. No one wants to be in bed while the rest of the army celebrates triumph. Paula felt it right to entertain troops. A few musicians came along. She got a team of acrobats to flip around, good old chipmunks. While a small show was all she had, it was enough to brighten spirits of those unable to go elsewhere.

Clancy was out, missing her performance. As usual, she brought the house, or should we say hospital, down. It did her heart good to see those in pain, who sacrificed for the betterment of the peninsula, get some smiles and enjoy songs. She promised a huge show in a few weeks.

"You had better get better. Every veteran is welcomed for free. Come and let me entertain you. Let me make you smile. Come and see the show. We want you with us. Transit to Carnival Island is available. Please, come."

Clancy awoke a week after she departed. He was fuzzy. Memories came and went. He was unable to say his name at first. The blow gave him a massive concussion. The scar made his left eye droopy. Sullen as a rule, he lay in bed miserable. He thought only of when release would arrive, so he could go back to his tree fort and be solitary. This fighting stuff was for birds. Frankly, crows could have it. Left alone, he would not have to fight. He was too gullible.

Sometimes he could remember combat, but three battles merged to one in his mind. He saw vivid images of bears, raccoons, and blood. He saw crows and ravens fighting above him. He recalled some foolish sow exposing herself to danger. Then, hanging his head amazed at his own actions, he remembered attacks on bears. Not one bear, oh no not him, it was three bears he faced. No wonder he wanted to forget.

Transported on the back of a grateful elk, Clancy rode across the gap between land and island. He enjoyed the ride. Still weakened, he was taken by the same wapiti to a place of honor. He had a nice bed of leaves and pine needles for rest. Food and water were handy. He had a spot up front. The stage was a little more than a meter away and a handful of centimeters high. This way, performers were seen by the whole audience. He had never seen a show in his life.

Ivan Trobosky was emcee, duh, introducing acts in a quick array. It was decided to have a bunch of various acts. They did a shortened version of Fur. My Fair Sow was coming next summer. Synchronized ducks had a marvelous display, uniting five dozen ducks of five different types in a massive cluster of circles, each larger. Teals in close, Northern pintails on outer most ring.

Acrobats, a mime performed. Scheherazade and the Fisher Three played jazz. Clancy did not care for this, partially because he feared these hunters intensely. Next came barbershop quartets. This was soothing to his soul.

Clancy liked harmonies. The Willapa Bay Blues Band performed. They missed Murray, who was to come soon. Corinna was on her game. Dylan was in rare form. The whole band took off, leading the audience in a musical journey to Paradise. It was their best performance to date.

"Now, forest friends, we saved the best for last. She fought the good fight. Her prickle slew bear bad guys. She led, they fell. We love her as our favorite sow of all time. She performs for our pleasure. Yet, in a time of dire need, she answered the call. She fought bears, a much larger foe."

"So, it is with the utmost gratitude to humbly introduce the star of this or any other show, Prickly Paula, the Pugnacious Porcupine!"

Per normal, she burst from the shadows, arms above her head. She went to the far side of the crowd, waving still, grinning all the while. She paused in the middle, jumping up and down, acknowledging the crowd. She tossed garlands of edible flowers to rabid fans.

Faking beginning, she laughed, aiming to the other end of the stage. Everybody again broke into caws, squawks, and grunts. Nobody could milk a crowd like she. Paula would make a dairywoman weep. Front row was all injured veterans. The explanation for the show was to celebrate sacrifices. She identified, fighting alongside many of those infirmed.

Paula planned to dash to the end then circle back in a wild fashion to center stage. She did this regularly, so the back up band was ready with rim shots, trumpets, and kazoos. This method built the anticipation of the crowd. They were ready for her signal. She worked them like a baker kneading bread. Too hard, the loaf will not rise. Too gentle, you are left with oozing mush.

Nearing the end of the stage, she spied an injured porcupine. Not many of her kind were about. Most wounded were squirrels, mice, muskrat, and otter. She knew all the other porcupines. Three prickles went to battle, she knew them well. A few were dead, a number injured, but no longer in hospital. So, to see a battered porcupine laying in the front row perked curiosity.

Nearing the patient, she saw his face at once. It was marred by the scar over his left eye, but it was him. It was the boar who twice saved her. He rescued her a third time, distracting and injuring the final bear. He had vanished as always. They thought of him no more. Save Paula, who held him close in her heart.

Now here he was, withered, curled up on a bed of leaves and pine needles. She froze. The audience saw her face. The facade faded. Her true self emerged. This was no performance. Clancy called the true porcupine to emerge without a word. He demanded honesty. She was compelled to drop the act. This was easily done. She wanted this boar. He was real. He did not know who she was. This was him. He lay gazing, awestruck.

She leaped from the stage, the crowd gasping in amazement. Paula ran, throwing arms around him, showering with kisses. He was stunned. Memories now flooded his mind. The levee broke. She was oblivious to animals oohing and ahhing. Clancy was all she saw. Ivan came to the edge of the stage, peering over. Another wedding soon he wisely deduced.

By the time Momo and Murray arrived to play the festival, Clancy was well. His recovery combined a wedding and honeymoon. This took place in his expanded tree fort for two. Performing could wait. They had discoveries to make. They called it, 'quilling time'. Paula and Clancy were perfect, purring in harmony above the din.

She gushed unbounded love, introducing him to woodrat friends. He was not accustomed to being the center of attention, choosing to bow out that his bride might shine. As much as she adored performing and making animals laugh, he abhorred it. While on stage, he remained in his tree. He utilized time to meditate. This was more beneficial. He heard an interesting form of poetry called, *haiku*, and thought he might give it a go. They were opposite in many ways, but made the union a lively one.

While they lived, times shared buoyed with, a profound, deep love. She found one who loved her in spite of, not because of her position. Many tried to use her as a stepladder to their own wants. Clancy simply loved Paula. Her work was just that to him. Once Paula left the stage, once light no longer shone, she was a simple sow in love with a bodacious boar. That was more than enough for her, more than enough.

How sick is this? They lived happily ever after. I know, barf, but true.

Chapter 7 Miss Bluesberry

Hibernation was essential to Miss Bluesberry and all skunks. It was their way, same as marmots. After the final show at Grand Meadow, they made for their cave. It was Ralston's, but cleaned, set up for skunks. The entrance was large, narrowing quickly. It was easy to shut for the season. They were cozy the whole nap. A happier boar and sow were hard to find.

When lady skunks married, they became, Mrs., but Miss Bluesberry kept her name the same for professional reasons.

Mr. Loganberry of Grand Meadow and his trio of brothers scoured the countryside all summer to find homes for the lads. Mr. Salmonberry, the kid, stayed closest to big brother. They found him a large hole in the side of a rock ridge. It was an old tree stump, hollowed out by flickers, then home for bobcats.

It was barren, perhaps the previous inhabitant died in war. It did not matter. Mr. Salmonberry met pretty, sweet, available Miss Gooseberry. They married. He swore allegiance to his brother, initiating the Grand Meadow surfeit. It became the largest, most powerful collection of skunks in the Olympics. Brothers Mr. Elderberry and Mr. Elderberry Blue or just Blue found a large cave east and south of little brother.

This was up closer to the Valley of the Stones. The effect of having homes scattered in such a fashion made a huge territory. Their home was big enough for two families. Miss Mountain Strawberry, and Miss Cranberry were lovely sows from different surfeits. They met the brothers at the Festival at Grand Meadow. Wedding bells were on their heels with the speed of a peregrine falcon.

Their cave was at the far southern end of their territory, it's base at Grand Meadow. North, war companion, Mr. Raspberry of the Lowlands, remained behind at war's end. He wanted to be a part of Mr. Loganberry's family. To this end, he and wife, Mrs. Raspberry of the Elwha, another family altogether, pledged to move north, close to the curve in the rapids.

They found a home in a tree drenched knoll a kilometer from the main roadway. Set in darkened reaches, they had good protection and a warm interior. The way they formed chambers gave several levels. Large rocks had to be dug around.

His sister, Mrs. Huckleberry and husband, Mr. Blackberry of Mt. Queets came from home to help build this fortress in black and white. They moved to the rocky area vacated by Miss Bluesberry. They loved quiet. Our mystery critter lived there, too. He welcomed neighbors. It was lonely after the massacre.

When Ralston led forces through, skunks chased him up the road. It was at this time Mr. Blackberry of Mt. Queets, answered Lloyd Joseph's call to help, he found the abandoned site.

He claimed it, moving post-hostilities. By good fortune, his wife's brother, Mr. Boysenberry of Lost River moved close by, enlisting Mr. Blackberry of Massacre Pond (his new moniker) to help Mr. Loganberry of Grand Meadow's new army. Happy to join old fighting buddies, he and his new wife established the west edge of the kingdom. Tied to Crescent Cedar and the crow empire, this region of the Olympics was solidly in the camp of righteousness. Ravens stayed away.

The only land behind this vast property was higher up, out of what they would use. Polecats do not do snow. It was not skunk country in elevation so high above snow line. This gave them a huge swath. The river north, as it fell to the pond around to far side, was in their control.

Here they had more polecats. They had a strong alliance with crows down the Elwha a flap or two away. The result of war drove away evil, replacing it with victors. Grand Meadow, center of animal life through the region, had a strong presence in Mr. Loganberry of Grand Meadow and Miss Bluesberry.

They created a music school and performance venue for animals wanting a shot. She named it: Mr. Mizithra's Music Academy. He made it possible. It was he, a marmot who could not play a note for the life of him, who inspired it. Her husband led the school. They trained many musicians and singers in these hills.

She contacted the Covey of Quail Choir. Mrs. Queisha Q. Quail agreed to perform annually, and train others in operatic singing. These ladies, quail and skunk, were a force to be reckoned with. They founded Grand Meadow Musical Symposium which trained animals in the classics. Short tailed weasels thrived at strings and soon quartets were all over the mountains and valleys.

Bluegrass was taught by members of the Beaver Bluegrass Boys, who survived war working on the dam with Bette and Tobias. Bluegrass sound filled hills with country music, the other side of the school churned out harpists, cellists, and violinists extraordinaire. Fame spread swiftly, grew, and over time, was the preeminent musical institution in the Olympics.

Miss Bluesberry, aka the Darling Doe, waited another year to have kits. She devoted her spare time to teaching voice and helping Molly. Ever since Kelly the cantankerous Cooper's hawk had a good recovery under her care, raptors came when needing help. A lot of slamming hard into tree cases came. Cooper's hawks simply do not learn.

Most Douglas squirrel nurses would not go to this upper area, but Miss Bluesberry had no qualms. One look at her and every patient did as she said first time, no arguments. Even meat eaters would not challenge this caregiver. Off days, she sang to them, bringing joy to an otherwise unenjoyable situation.

The surfeit set, life pleasing, she and her boar beau were happy as polecats get. While Lloyd Joseph manipulated their meeting to further his war plans, neither complained. All things considered, with every good and bad twist and turn war presented, they came out stronger than when it began. No skunks died, though she feared for her life at first. The sight of destruction in the Beaver Blues Bungalow Massacre was a mental scar she bore, but did not hinder her life.

Singing to an audience, her husband, or massive crowds at the Festival, she loved to entertain with her voice. It was her way to serve fellow forest denizens like no other. This was not ego at work, it was a realization of talent and how it could help others. Patients in the hospital who heard her wailing were lifted from pain temporarily. How can you measure the value of such a gift shared with others?

Due to war and the unpreparedness had crows, marmots, skunks, otters, muskrats, woodrats, Douglas squirrels, and Keen's mice in the area, Mr. Loganberry of Grand Meadow organized a militia. Crows were enlisted as messengers in time of war. A monitor of the enemy: European starlings, raccoons, Eastern gray squirrels, long-legged bats, Western heather voles, and ravens was enacted.

Tolls were paid by any who ventured into the area. Carnivores could not be stopped and no one would try. Now should a coyote want to have a go at a skunk, it was up to the hunter how much he wanted to stink to get supper. Lady Satin did little to provoke crows as Empress Neferteri proved capable and ruthless whenever need be. Her queen regents had no sympathy for larger past antagonists.

BOOK 4: PEACE

The summer after war, an unkindness of ravens snuck into crow territory. It was off limits. Only by entreating crows could they be given permission. Violation meant harsh treatment. The goal these fools had was obvious: see if they could get away with it. If so, more forays would take place. They could loot smaller, weaker crows as before.

They harassed a murder in a low region, near a small tarn off Jorsted Creek uphill from Hamma Hamma. Ravens holed up, spreading down to Liliwaup. If they could hit and return without consequences, they could expand operations. They did the level best to intimidate crows, bullying, forcing them to desert lovely shiny items retrieved from human campers.

When the incident was reported to the young empress, her answer was swift, decisive, memorable. Two days after the attack, ravens were tracked back to a hideout outside Hamma Hamma upstream. They were toying with booty. Bright, glossy objects fascinated, attention addictively drawn to perpetual, silvery flickering. Counting on this, learning much from training, she burst a thousand crows strong, killing all in moments.

Each item taken was returned to rightful owners, dead left in a row. At one end a single raven beak stuck out of the ground. This was the sign of war. If they came out of hiding again, if they disturbed Olympic crows, it would be all out war. Patrols were increased. Any raven close to the border was killed. It was relentless for the entire summer. By fall, when they moved away from the snow, ravens were vapors.

Neferteri sent forces down to Forks, up to Neah Bay, to the far north at Port Angeles and Sequim. She sent crows to Port Townsend and down to Quinalt, Humptulips, and Matlock. Covering a 360 degree radius, she pinned her territory down tighter than her great-grandfather ever did. It made them impenetrable. Her forces ruled the mountain range totally.

In time, Lady Satin made inroads, not before humbling herself to the superior forces Neferteri possessed. The Ladies of the Lake enjoyed seeing her take over. After Lloyd Joseph brought her to them, they were active for a year, pouring all they knew in her. They brought in tutors to help in their only weak area, combat.

This was held by old war master, the only one Lloyd Joseph whose advice would heed: Sarge. Once owned by a retired Marine Sergeant, he knew more than most. The soldier talked war constantly and watched all the war movies he could. Riding to the bar on his shoulder, Sarge heard war stories. At the VFW, more information filled eager crow ears. One day, the old soldier failed to answer roll call. He was buried with full honors. Sarge took off, ending in the Elwha River delta.

A passing crow told him about the pool where Ming Ye and Ming Yen lived. He met Graedy, who sent him to Lloyd Joseph. There, he was hired as a military adviser. It was he who recommended the brilliant counterattack on reinforcements under command of Cheston, General Thackelberry, and Vasilis. A cannon pipe was genius, pure genius.

Sarge stayed on, charged by Lloyd Joseph to help the new empress. It was he who demanded retaliation of the raven raid with extreme force. Knocking it off early, nipping it in the bud, chances of further invasions evaporated. He developed a blanket effect over the whole empire. He took the charge of his master: she had no worries during his life.

It was up to Sarge to build a military so strong, no raven would dare to try war again. He had an army of murders all over. A warning system employed was the envy of all. By organizing and coordinating squads, no other animals challenged them. They achieved air superiority. Marmots, woodrats, and skunks benefited from the crows.

Empress Neferteri loved music as did her great-grandfather, but it was hard rock for her. Once the strong beat hit, up and down went her head, banging in perfect rhythm. Any group that came to play at Grand Meadow or The Lodge who specialized in this type of hard driving music and she was up front jamming.

The old Murderers Row still had it's place, though members changed. Winston, Beverly, and Lloyd Joseph retired. Neville died in war. Derrick was killed flying into a car driven by a drunk human going the wrong way on I-5 in Seattle in fog. The Ladies of the Lake retired. The task of regents and own offspring running talent bookings was a full plate.

A whole new Murderers Row was selected. Clive, Delroy, Everton, and Clifton, the only four remaining selected: Solomon out of Tacoma, Cecilia, Gabrielle, and Rapunzel from Lake Crescent, by choice from Agwe and Roje:

Tremont and Silver (for the silvery tone to his feathers) who represented Willapa Bay.

Clive chose cousin, Edmond to for his lost brother. Finally, Narcissus, daughter-in-law of Lloyd Joseph, who while old, held sway in Crescent Cedar. She and Neferteri had a close relationship. The empress had one request: more hard rock. Talent scouts scoured hill and dale in search of players whose head banged when playing. Canadian geese had their wings full, but appreciated the musical direction she chose.

While she had to hibernate, once in a while our sweet skunk sow snuck down to The Lodge to belt out songs with the band. It was an unavoidable addiction. She would bundle up all warm and fuzzy, then dart over to The Lodge. Mr. Loganberry of Grand Meadow always knew where to find her. His love allowed her get away with a few bad habits.

Going in the cold was not a good idea, but he was not about to tell her. Besides, he thought, how unhappy would she be stuck in a cave with no one to sing to but him? He smiled, the old love bucket. Happiness is found in silly places sometimes, but he knew she was his and he hers. That was all that mattered to these skunks.

Chapter 8 Spring Springs

After a year of war, rough winter with wounded dying, a late spring, it was a happy time above ground for woodrats above Grand Meadow. While not handing out cigars because animals do not smoke, Murray was, however a proud papa of six, count 'em, six wee ones. Momo delivered a girl, then a boy. After a rest, three boys in a row, finally after a bit, a little tiny girl as the finale`.

Mama rodents have a litter. This held six, a good number. Momo had enough, though as her health suffered. Delivering this brood was it for her. Molly saw it was due to her body structure, this lady woodrat was not designed for giving birth. She was down for a long time after her babies snuggled mama for milk.

Nediva helped as did Murray. Frema passed before they were born. It was sad she never got to see them. Age and illnesses caught up to her. She breathed her last, safe from the Cossacks. Young ones fed on mother, a sight which left Murray bewildered. His own mother only had two in his litter, his sister dying at an early age, before weaning was completed. Here were six healthy baby woodrats. He counted his blessings.

Molly informed him of Momo's weakened condition. He took care of her very gingerly. She was the most important animal in his life. She meant everything. Having met her on that foggy day, she owned his heart. Her health was paramount. He doted on her and the babies until such time as Molly said all was well.

Naming babies was easy: Fumi, after her mother was the firstborn, Mizzy, after Mr. Mizithra for the first boy. Shadrach, Meshach, and Abednego were the triplets. For the last daughter, Momo named her, Frema. Of course, when it was time for sleep, Murray always called them to get in the covers saying: Come on Shadrach, Meshach, to bed we go. Abednego loved this because it was so funny.

Murray never wanted to be a father. Yet, it was hard to find one who took more satisfaction. However, not a one showed the slightest talent or interest in saxophone. This was ego deflating. Nope, not good old Murray, a bunch of flute playing rats is what he got, a clump of stinking flautists!

The Gliding Ghosts flew in to bless the children. The kiddies were in awe of flying squirrels. They wanted rides, wanted to learn how to fly, and wondered aloud at the folds of loose skin on either side. Love flowed as long as Ninjas could stand it. They gave hugs all around and were gone in a flash. It was their way. They worked it out with Horace and Boris and the Canadian geese.

The Humboldt Flying Squirrel Security Air Force would handle stage security. Canucks were too apt to violence. Douglas squirrels were simply too timid. Allessandra was in charge with Edoardo and Valentino. Absent were Nicolo and Sofia, who finally tied the knot. Nicolo wore down her resistance. They had the fine training school to run. Dozens of trainees needed oversight. Intense training had hospital beds to fill. The honeymoon was held during a ten day survival camp out atop a pinnacle near Mt. Olympus. Ah, love.

"Oh dear, do you see what I see?"

"Yes, darling, we probably should not have named our number one son, Mizzy. He is already trying to get out of Grand Meadow."

"I caught him yesterday scaling the rock wall behind the stage. Do you know what YOUR son said when I asked him why he was climbing the wall?

Do you? 'Because it's there!' YOU, you are the one who taught him to say it didn't you, husband?"

"Stop smiling, it isn't funny. He could fall and break his neck. Next, you will have him running around with a stick in his paw. He could put an eye out. You don't care".

"All you will say, dear husband, is he is a chip off the old block. Do you want him to go falling off cliffs, evading coyotes and foxes? Do you want your son to die alone wandering all over the Olympics climbing mountains just because they are there? Murray, do you want me to get your mother", now it was all over for him. There was no way to fight her.

There it was, the telling mark. If she called in reinforcements, mama, he was done. He pried his wall clutching youngster loose and set him on the ground. Murray did note quietly to himself, 'little fella has a strong grip'. He will need it if he ever decides to go for a little stroll hither and thither. While he mused on this titillating thought, Mizzy was already up the wall again even further.

Yup, a chip off of the old block. Heh-heh, he would have to tell Mr. Mizithra his namesake is a mountain climbing fool, just like papa. All the boy needed was a good buddy and they could have adventures. He could hope, in total silence, he could only hope. Hey, without a sax player, he had to have something to hang his kippah on.

Thinking back to travels, he did miss them. Aches in his left leg told him, think twice. No more critters squashing his hip. Still, memories were amazing. They really did have some fun times. At this, he immediately missed his old chum. What was going on at the Valley of the Stones? Early spring was still underground. The difference in elevation was significant.

They did not thaw for a few weeks. He had time to send invitations. He wanted Mr. Mizithra, Mrs. Saffron and youngsters to meet his brood. Murray was very proud of his family and wanted his best friend and family to mingle. It excited him to think what he and his old marmot buddy had done since the days of high adventure.

Grand Meadow in spring was all life and fun. It was infiltrated with dynamic personalities. The whole scene rocked with life. Animals, birds, bugs, and fish thrived. No humans seen. Carnivores seemed to avoid the area. With so many, it became impossible to sneak up on prey. Raptors tried, the only one of success was Ollie the Osprey.

He loved to come visit Molly. She made his life so much better. Once he got his eyeballs straight, he was a fish catching fool. Osprey mate for life. Now with perfect eyesight, he could provide for a family. With a wife of extraordinary air, he made Grand Meadow home base until time to head south for winter. It is rather difficult to catch fish through thick ice.

Ollie wed an osprey from down south. They met after his medical care. Needless to say, with such good fortune, he was a real hit with lady osprey. He was the only one unmarried, so besieged with eager, eligible ladies. It was an early morning out for a breakfast catch.

It was a warm day. He had not settled in his southern digs. The nest of last year was badly damaged in a wind storm. Rebuilding was a chore, but nothing else to do. It took time, but he had plenty. Osprey gathered here for winter. It was an ideal spot to find a mate. A meal and a mate, was a good goal.

First, food. Having vastly improved fishing skill, he took his front two talons and back two (only osprey and owls use this ability by design) aiming for a lazy trout lying just below the surface. A female had the same notion, hidden by a tall elm leaning over the shoreline. As Ollie hit water and fish, so did another osprey. Shocked, both fled in opposite directions. The trout dove straight down to safety. In a flash, Ollie turned to see who attacked. The female osprey had the same thought.

She was in an elm, he a cottonwood on the opposite shore. Spotting each other, laughter ensued. Marriage followed days later once courtship ended. Ollie met and married lovely, Onyxa from Idaho, wintering in the same place. They moved home to the Olympics. She was the perfect mate. They were happy. She adored Grand Meadow. The new aerie had two eggs, young ones were on their way. Life always continues. The clock never stops. They made time count.

Ollie introduced Onyxa to Molly, his heroine, and Graedy. She met Murray, Momo, Matilda, and Boston Charlie. Ollie was proud of his friends. A new home meant fitting in, Onyxa did so easily. She was a proud osprey. Why not, she exemplified the raptor breed with aplomb. Her folks raised her beaming about her species. Ollie consummated normal bird urges to reproduce. This is nature's way.

It is age old, messing with this way of life would spell disaster. Life for osprey had bad times with the DDT issue. Once solved, the population exploded. Osprey filled waterways of the Snake, Columbia, Yakima, and Willamette abundantly. Here, man corrects his error for the betterment of nature. Chalk one up for the Thumbs. Yah.

Finding the hospital tending raptors made him happy. He was first, but not a threat like Kelly the Cooper's hawk. Onyxa was struck by the harmony between species. Once she heard music played on stage, it was clear why. Music has it's place. More music means more calm, more peace, more love. Onyxa had no issues. Her young would be born to the sound of music wrangling it's sweet way up to her brood.

Hatched early, they would fledge in eight to ten weeks, then the family would head south for winter. Fish were abundant in this lake. The atmosphere Grand Meadow offered made this home. She told Ollie it was imperative they build up the aerie. Fishing was easy. Foolish mountain trout lay on the surface, practically begging to be eaten. She obliged, her young feeding on only the best.

Daily, skunk, chipmunk, otter, muskrat, beaver, and woodrat filled Grand Meadow. Transient critters and birds were given a place to rest, a hovel, tree branch, regardless of need. Food was copious. The west side provided a solid backdrop. Nothing could invade this way. The same was true east. The river led to a natural barrier. South, the thick tree line and dense ridge, a lovely shelter from storms. Here, visitors congregated.

They were given what no other rest area could, live music. It was on stage or in The Lodge magic came to life. If you had family or were simply spent, the open air music festival was ongoing. If you liked entertainment up close and personal, The Lodge was a slice above the rest. Here party time was in high swing. Animals and birds designated this a tourist must, thanks to efforts of Boston Charlie, Matilda, Darby, and O'Gill.

The area reeked of fun. Soon, animals came simply to hear music and sample culinary classics created by Chef Ripple. His talent, it was found, lay in meal preparation. Dishes had to look as good as taste, his artistic flair at work. Shy by nature, the muskrat hid in the kitchen, whipping up gastronomical sensations. Bette and Tobias gave him full rein. As beavers, their focus was bark and blues.

He made outer casings of standard firs, pines and cedars stand out. A natural, he had a special trout recipe for Ollie and Onyxa. He took it, leaving it intact, roasted, placing herbs in the mouth, and under the skin. The taste, delicate. While osprey normally eat fresh, raw fish, this muskrat rustled up a culinary delight beyond imagination. Onyxa pecked Ripple with her hard black beak, stating he had to serve this meal every time.

Chapter 9 Dublin

The retiring wapiti army went home after saving marmots. For some, this meant following K'wati back to the Hoh River area, west of the Valley of the Stones. Grand Meadow was spoken of in very positive terms. They regaled beautiful views, the air, taste of lake water, and incredible music. For any animal or bird, it sounded like Paradise. If you played music, it was the ultimate goal. For some, it was a pilgrimage to a hallowed, holy place. Music, for a select few, is religion.

For any critter with musical ability, it sounded like the spot. Of course, talent was key. You may be the best trumpet player in town or school, but out in the big world is where the true test is found. Thus, to be the very best, you had to pass the gauntlet. This was The Lodge and the Grand Meadow Festival at the end of summer. Perform here to prove yourself able.

The Hoh River empties it's 90 kilometer (56 mile) length weaving west to the Pacific in a very cloistered area. Very few see this zone. It is home to some people who like privacy. The delta is base of the Hoh tribe. They settled the south side of the river. This valiant family fought off whites: Russians, British, and Americans.

Unlike many others, they defended home with righteous violence. When white people came calling, they were wary. A sham treaty was signed, yet they knew no English. Lies added to surrendering and leaving their sacred homelands. The old forked tongue bit whites were famous for with natives. Their answer: NO!

Over time, endurance by these coastal people was rewarded. They kept their land. The only impact on this temperate rainforest was Highway 101 cutting a curvacious path through lowlands. The original inhabitants do not cut trees, no timber harvest. Here nature rules. Hoh adapt to it rather than reverse. This honors what nature, god, provided.

Attempts by natives and whites to climb the river's course and establish human settlements met sogginess. So much rain falls, the Hoh Rain Forest park is for the most willing to combat water. Greenery is dense, slick, and tall. The pathway upstream is difficult, full of pools, waterfalls, wild twists and turns. It is a hiker's dream, settler's nightmare.

Few whites live here, most are native Hoh people. One such white man lived off the beaten path, a cabin set against a small pond. A beaver dam and lodge set to one side near a patch of firs. Beast and hermit lived in harmony. He was an odd duck, choosing to live as far from humans as he could. He worked a small job at a local restaurant to help his pension.

Injured at work, he became a recluse. He had no TV, entertainment center, or video games. He had a turntable, great speakers, and a huge selection of English blues rock music. He played all day, broadcasting across the pond for nature to enjoy.

The mated beaver pair over in the lodge, Silas and Norma, had a few pups, but one stood out. Silas fell in love with one musician, a guitar god from Ireland. He was a performer who poured every ounce of energy in every song. Rory Gallagher was a talented Irishman unmatched. His music filled Silas' heart, naming his son, Dublin honoring his hero, since Ballyshannon would be ridiculous.

They had other whelps, but Dublin was special. From birth, he wanted to hear music. He snuck close to the cabin window to hear tunes better. The wail of guitar, the strength it provided, filled his young beaver heart. He knew what he wanted to do for life. He formed a local band, playing until good enough for gigs. They practiced after dam chores. Even local wapiti approved.

When elk returned from war, stories were vivid and frightening. They told of a place to play over the top of the Hoh, Ice River, White, and Blue Glaciers down around north. A huge lake lay in this sweet valley. A river headed downhill, eventually ending in the Elwha River. The Hoh has an east/west flow. A dozen smaller brooks and rivulets feed in the Hoh.

An idea hit Dublin. No other band member wanted to go. They thought him a fool. His parents did not want him to leave, but saw rigid determination. On a fine, warm, dry Hoh River morning, he began a trek up river to his prize: playing at the Grand Meadow Festival. If he trod day and night, he could make it in time. This was his insane plan. Off he went, guitar strapped to furry back, heading into the unknown alone.

The journey east took him in elevation over two thousand meters. His family lived above Highway 101, so he had a climb. The best thing for him, it was all water. Pools and beaver lay upstream. He had no trouble the first days. Soon animals were few, birds his only company. Crows provided directions for a song. Dublin pressed on, the goal his only thought. To play with the best, he wanted this more than life itself.

He suffered, walking and swimming for hours each day. Soon, fur was scraped, paws worn raw. He ate when he had to, sleeping in any safe hovel. Coyotes soon thinned, replaced with scents of cougar and bobcat. Dublin kept aware of environs, pushing his body onward and upward. He would not be stopped. The weathered guitar on his back his lone friend.

Once up high, strumming, he was alone. Snow remained, clinging to cliffs and rocks. Loneliness was one thing, cold another. Reaching the summit, Dublin wisely rested for several days to convalesce. Health weakened by a lack of normal life. While rest seemed foolish, he took time to let his system recover. He swam in a nice pool eating all the bark he could to put weight on his bones. His tail drooped. Necessary rest was a good idea for days after which he began downhill. He picked a bit.

A late spring storm whipped up, blanketing with snow. He had been on the river for weeks, becoming disoriented. A passing Cooper's hawk named Kelly steered him in the right direction. Soon, he saw black rocks, brown tree trunks, and green ferns. He sped, driven. Within two days he was on dry ground, a warm air current wafting his way. He rested a full day, then began his final trek to Grand Meadow.

Along the way, he met other animals coming from his Shangri-la. Each had a tale of wonder and amazement. Dublin could not wait to arrive. The journey took a toll. His paws were worn raw by ice and rocks. He lost a lot of weight pushing so hard. During the trip off the summit, his guitar fell off, crushed as he clung to a ledge after sliding on some lose stones. Losing his ax, he wept.

The ax was his only way to show the judges at Grand Meadow. Undaunted, he pressed on to the goal. It took longer than thought. In a short time, he was worn to a frazzle. Each day met frustration. Where was this elusive Eden? He could not go back, so pushed his fragile self along. Were he not to arrive soon, he would not play on this most famed stage.

"If you've come for tryouts, you are too late. They ended a few days ago. Best go back home, stranger, and try again next year. Acts have been chosen and are practicing for the big show. They do not make exceptions. Pardon me, you don't look like a musician and have no instrument".

"Why don't you get food, a night's rest, then head home. You are too late", came painful words from otter guards greeting him along the path.

Bedraggled and beaten, Dublin finally reached the southern outer limits of Grand Meadow as dusk concluded it's daily vivid display. He came to the water's edge, not a beast in sight. He slid in the warm wet and rested his weary body. Paws ached. He was hungry, which made him irritable. Unable to see the stage or The Lodge in the dark, he asked a slowly swimming muskrat where he could sleep for the night.

The response was low. Ripple was beat after a full day of preparing culinary delights. Dublin, exhausted, launched into the muskrat. Defiant, he shoved the whacked out beaver, swimming away in silence. He had no time for tourists who lost their way.

Weak, unable to find a place to sleep, Dublin meandered across the lake to a setting in some reeds and cattails. Here, he reasoned, he would find a nice spot to sleep. In the morning, he would find those in charge and tryout. He could not accept they were not willing to hear him. Defeat did not register.

He had come all this way to play. He had a sound, a talent. Dublin was unwilling to simply give up and go home. Frankly, the prospect of returning never entered his mind. He was going to make it at Grand Meadow or die trying. This was one determined beaver.

The others did not know how he slid down Ice Glacier. They did not know how he hung by strong beaver claws to a rock to keep from falling to his death alone high in the Olympics. They did not know the suffering, lack of food, of shelter, hiding he did from hunters seeking fresh beaver for dinner. No one saw his lonely nights, snow and rain pelting his determined rodent soul on to the goal. Setting his mind, he would not be swayed or deterred.

Finding the lake and reeds, he swam ashore. Pain endured would have reward. He could feel the end in sight. No, was an answer he would not accept. Even if full with talent, all he wanted was a chance to play. If he stunk, that was fair. If not, they needed to find a spot for him to wail away. They just had to, he had come so far.

The pathway at the end of the reeds was not natural. It was animal made. It wound it's way to the south end of a straight rise. Since it was too dark to see, he fondled about until locating a warm hovel. It was quiet, no one seemed about. Every critter in the area was fast asleep, Ripple the last animal on the lake. Only fish were about. They did not talk, ever.

Stars twinkled above, dancing a spectacular nightly display. Finding a soft spot, Dublin fell to the ground in a heap and slept all night without moving. In the morning, he arose at first light. He was rested, but hungry. Food was the first order of day. After he ate, he would find the authorities and demand a listen, if they could but lend him a guitar.

Without realizing, Dublin crept on stage in the dark, coming to rest in backstage rooms for stars. Here, they prepared for shows. This was the haunt of Cloris, Delores, Horace, and Boris. They slept not far away in a tree fort built during war. It was deep and secure. All they had to do was shinny the trunk and land on stage. Dublin laid down in exhaustion.

Now awake, he scanned the scene. He was on a large, flat area. Water shimmered ahead, out half a dozen meters. Reeds swayed in soft morning breezes. A few bugs ventured forth. A robin red-breast came, yanking an early bird special worm from bejeweled soil. Good old Rayen.

He walked out of the small alcove. Out of the corner of his left eye, an object caught his view. It was…a guitar. It stood on a stand made of branches and twigs. He looked further, a few more axes, a bass, then he made out a drum kit. He was on stage. He managed to swim the lake, find backstage, precisely where he wanted, only no audience.

Haggard and wasted from the road, he had little strength. He had not eaten the day before, wasting no time. Food could freaking wait. He grabbed the guitar most like his, lost journeying east. A few picks and it was in tune. Dublin took a deep breath, stepping forward in dawn's early light, guitar, paws and fingers ready. Alone he began.

Bette was dead. Spent from the previous day's playing and running of The Lodge, sleep was instantaneous. In deep recesses of her private chamber, noise never got through. This was by design. Sleep was essential to one so driven.

Toss and turn. Sound. A guitar kicked her awake. It was like a riff going through her mind and she could not get rid of it. It was like a tick. It would not go away. Waking so early was not common. Bette and Tobias had rooms next to each other, with a habit of rising late in morning. Musicians are built that way they would say. So, to be irked alert was not cool. She was not happy.

The noise would not cease. She clearly heard it. It was undefinable. She heard a riff, a rhythm…a guitar wailing away in the far distance. It did not originate inside the club. She headed to Buford's Billiards room. Silence. Nothing moved. Stunned, still hearing music, she made for the front door. Opening, strumming intensified. Bette made out notes clearly.

Someone was on stage, illegally, playing HER guitar. Whomever they were dead. Not waiting for guards, she flew across the plank walkway, dashing for the stage. Morning's light shed a few rays through trees, but she could see a figure wailing on her ax. Anger swelled her head. She ran on stage. The quartet of slumbering Douglas squirrels stuck heads out of holes in unison, darting on stage to examine the commotion.

Bette stopped. Sound. Notes. Finger play. The music was incredible. It was guitar playing like none she ever heard. She heard it all a hundred times before. Each note opened a wound. Each note wept. The guitar sang. Whomever was playing her guitar made it sound like nothing she ever did with it.

Music rippled across water, soaring over the river, cascading to the valley. Other animals came out to listen to the moving morning opera. The dark figure stood alone. Bette saw no other animal, just this fellow wailing away on her ax. Horace and Boris looked at each other and saw dollar signs. Wow.

Without a word, she grabbed her back up instrument and moved beside the mystery man. He was a beaver, beat up, done in, but made the guitar literally emote. Bette slid up to his left eyesight, but his eyes were shut. The guy was wailing. He did not look at the neck, he played from heart, his soul attached to every pluck.

She started playing, evoking a jolt from the fellow to her right. He did not stop, but increased the flow. She jumped in and began to duel away with this dark stranger. Twigs swam up and sat at his kit. In a moment, after shaking off water, he hit a solid beat. Tobias, yawning, climbed on his big bass and began to thump, not really sure what was happening. Soon, his tail was slapping wood like never before; who the heck was this guy?

The visitor kept pace, hitting note after note with beauty. Others, seasoned players, were in awe. This guy, whomever he was, was freaking awesome! Tobias looked at Bette. He did a double take. His sister wore a look he had never seen. He knew her all his life; from the same litter.

It was a look, dare he say it, of love. Bette was not a flirt. She avoided males. She got proposals daily from ardent fans, not all beavers, either. Ick. Her only love had a long neck, funny shaped body, and six strings. Brother witnessed this look, turning to the dark beaver on the end.

He was okay looking. Bedraggled. Looks like an errant gunslinger come to prove a point. Yet, when he felt the music this bum produced, Tobias was struck as well. He judged book by cover, a foolish act, shades of his late father. Sound erupting from his sister's instrument was nothing like she played. Bette was not in love with the beaver, but the music he played. Only a soul of beauty could produce a sound like this.

The impromptu jam session awoke the entire Grand Meadow crowd from peaceful slumber to a musical sojourn in love. Without stopping to breathe, they kept playing. Soon shore and water filled with eager ears, drinking the melody coming from some stranger playing Bette's guitar.

She never, never shared instruments with anybody, but here was a new beaver jamming on stage. He was better than anyone else, even Bette. While the whole final festival had it's musical acts selected, here was an exception. And we know how exceptions work out, do we not?

He had not spoken a word, but Bette fell in love. She looked at his face, focused only on playing. Eyes shut, he kept playing, wearing a loving look of satisfaction. She saw it in every pore. He only wanted to play music, to make sounds to soothe the savage breast. He and she shared the same passion to play the six string with fervor and love.

This amazing creature felt every note, an emoting weathered face. He was a guitar genius. Whipping hands back and forth along the neck, he paused on frets only long enough to hop to the next. It was foreign, but held such appeal others wanted to simply listen, rejoicing in playing along side.

After what seemed like spellbound heavenly hours, the sun burst on another Grand Meadow day. A soft, cordial zephyr wafted lazily from lowlands to bless dawn with a coat of warmth. It looked full of the same promise, the same joy found every morning in this peaceful nook. The sky reigned azure blue, a dash of anemic clouds passing without purpose. Music engulfed to the brim with emotions of love, passion, and joy.

Early risers were fed a steady flow of notes filling ears with sounds so sweet, it excited the dawn. Woodrat Murray heard and bolted for the stage. Water's edge packed animals and birds of every species. Not waiting for Momo, he ran to the music like a zombie. He heard a new sound, sax and harmonica awaiting his lungs. Murray had to find his sax and play. This was going to be a blast.

Murray neared the stage. The stranger to Grand Meadow collapsed, fingers still finding notes as he fell. Dead before we even got to jam, he thought? Who was this beautiful hunk of beaver?

Chapter 10 Bette's Beau

Every animal instantly stopped playing, rushing to the mysterious heap. Bette grabbed her ax, blood drops drenching it's body and neck. Horace and Boris, Cloris and Delores, were at his side in a flash. Used to players fainting or needing help, they were prepared. Trained by Molly, they began administering first aid. The good doctor was at hand. Molly was drawn to the music as well.

Graedy grabbed one limp rear leg, Twigs, Tobias, and Murray had the others. Bette at his head, held it gently in paws. Molly tended wounds. They found a table. Cloris and Delores prepared the room. Horace and Boris cleared the surgical theater as they would a crowd on stage.

Front paws dripped blood from his long ordeal and playing in a weakened state. Back paws had embedded gravel. Scrapes, fur ripped in places along all limbs. Scratches showed tail abuse, face weathered. By the smell, this beaver was not from the area. He bore an odor familiar to Graedy and Molly: like K'wati's elk gang. They were from Hoh River.

He wore this identical scent in his deep fur. Evidence told a tale of a long trek over Hoh Summit, down the pass to a turn north. He followed the long path to Grand Meadow. That he made it, unbelievable. To play with such fervor, miraculous.

"I never knew a guitar could sound like that.", muttered Momo, the precise thing everybody was thinking.

Ripple attested seeing him late last night, but the beaver was rude. The muskrat simply headed to bed. He had no time for tourists. He did not notice injuries to the stranger in the dark. Otters coming off guard backed up his story. Dublin just wandered into Grand Meadow. Due to the lateness of the hour, he met no one, no place to sleep for the night. He did not know The Lodge or plethora of holes in the ground to utilize. He slept backstage, then arose and began to play rather than eat.

"Well, well, well, he's one of ye, ah Molly. Ye ere workin' on an Irish musician. Be careful me loverly wife, by the look on her mug, Bette is in love wid 'im", laughed Graedy.

Bette blushed. She never blushed, ever.

Any animal who would rather play than eat was destined to join the band. Spots were filled, no room at the inn. Well, how about the manger? Every musically inclined animal in Grand Meadow devised a plan to get this fellow in their group. Of course, first he had to live. Bette had not moved from his side, cradling a limp head in paws of love. Molly gave her a wet cloth, gently wiping face and muzzle.

Though worn and wasted, Dublin was handsome and strong. For any beaver to travel so far upstream, cross glaciers, and down the valley just to play was incredible. He needed two things now, rest and food. Rest came first. Exhaustion overwhelmed. He slept all day. Bette remained in a chair next to the bed.

Cloris and Delores were on hand, tending to his unconscious body. He was wounded in several places. All four paws were wrapped, a cut or two patched over, tail soaking in saline to seal cuts. He showed evidence of strife. The trek east took him high in elevation, higher than any beaver would normally.

The only conclusion was this fellow had an extreme desire to play for the Grand Meadow crowd. He had undeniable, burning passion. The Lodge was one thing, Festival another. This beaver traveled a long time. Molly reckoned his long journey was several months.

Graedy, with Horace and Boris adjusted his corpse. Bones were out of whack. He needed attention. Graedy worked several times before Dublin was back to normal. His physical health proved slower to mend. Pushing himself to a frazzle it took time to recover. Molly took extra care. Bette visited daily, Horace and Boris as well.

Another who came often was Murray. He took an instant liking to any animal who would make such a pilgrimage. The beaver had purpose, he ventured far from home to the unknown on a righteous quest. Murray liked this young beaver quite a bit. Mr. Mizithra had to meet this one. The three of them had a lot in common.

After he came out of the semi-coma, Dublin was showered with a barrage of questions. Molly kicked all but Graedy and Bette out. Horace and Boris watched guard. He was weak and did not need every animal in Grand Meadow make the poor fellow sick again. Out, out, out, the cry.

No animal dared cross an angry doctor Molly O'Shea O'Grady. The room emptied in a flash. Once the chamber was still, doctor gave water, sat him up, then asked a few basic questions. He was a complete mystery, save he was a beaver from far away.

"Yes, I'm from Hoh River. Returning wapiti from war told us about the music. I just had to come see if I was good enough."

"Name, Dublin. My folks have a small dam, lodge, and pond up the river a piece, east of Highway 101."

"No, dragonflies? No, we don't send messages by dragonflies. That sounds weird. Word to my folks? I guess. Silas and Norma, just behind the hermit's music cabin in the small lodge off the Hoh. Easy to find. Sure, let them know I am safe and never coming home."

"I got lost in a spring snow storm. A Cooper's hawk, Kelly, told me how to get here."

"My guitar fell off my back, down the glacier when I slipped. It was guitar or me. Had to let Diamond, my guitar, slide away. I wept."

"What do I want? Just a chance to tryout. The otter guards told me they were over. All I want is for you to listen to me play and tell me if I am good enough. If I can but play, you will hear my talent, my wares."

"No. How can that be? No. I do not recall playing. On stage? There is nothing, no memory. An hour? Wow, sorry, a complete blank. Listen, since I don't have any way of recalling, would it be okay if after I get bandages off my front paws, a few days eating and sleeping I could get a tryout? I lost Diamond, so I need to borrow an instrument if possible."

His humility broke poor ol' Bette's heart wide open. She was utterly smitten with a massive case of first love. He could not remember playing? Well, how good was he when awake? If he could rock the whole lake first thing in the morning, what would he do healthy with the band? Bette's heart went pitter-patter.

Without being aware, everything he said melted her heart like ice by a blazing fire. She called her guitar, Ruby, so knew the attachment. She went to her vast collection of prized axes, selecting a beauty. Presenting it to Dublin, he blushed, nodded, then strummed.

"Perfect, when do we jam?"

Yup, Bette was in love.

Chapter 11 Dublin Fits In

After a week, Dublin was ready for discharge. Horace and Boris had a nice spot for him, but Bette insisted on the best The Lodge had to offer. Cloris and Delores were given extra pine nuts to watch him and tend to his every need until fully recovered. Bette doted on the fellow without him realizing. He was a country bumpkin if ever there was, good heavens.

For his part, Dublin could hardly imagine a sophisticated guitar goddess with a huge reputation like Bette of Grand Meadow wanting to play with him, much less find him interesting. He could not imagine a greater honor than to play as her backup. Clueless is what the poor fellow was, absolutely clueless.

As soon as he was able, they got him on stage to practice with the band. Murray was there, Ronaldo and Miss Bluesberry as well. These three did not have the pleasure of playing with the unconscious beaver from Hoh. Twigs sat at his kit, Tobias at his bass, while Bette got Dublin set. It was sickening to her brother, but he glistened at her happiness.

Her famous gruff exterior faded every time she caught sight of her 'beaver'. Finally happy with his guitar, Bette nodded to Twigs. He laid down a beat heard round the world. What happened next became legend many claimed to witness first hand. The actual crowd was rather small.

Twigs and Tobias set down a nice twelve bar blues rhythm. Dublin was told to jump in any time. Ronaldo laid a quiet backdrop piano to soften the mood. Bette and Murray waited, hoping what he did in a coma was no fluke. It was not. Oh me, oh my, what he did curled fur. Seizing the beat, Dublin set forth a flourish which froze Miss Bluesberry like stone.

No one played like he. It was unlike any other player she had heard. For a few moments, everyone let Twigs and Tobias guide him along. It was ethereal, mystical, magical, hypnotizing. Other musicians were in a state of awe. Notes were crisper, tighter, sweeter. The sound caused the air to glow. Again the guy's eyes were shut, what a trip.

Murray awoke, blurting into his reed. Something came out, but he was still trying to wrap his head around the guitar Dublin laid down. Having never heard saxophone, Dublin did not know a good saxophone note from bad. For once, sour. Murray recovered in an instant and went to work showing his talents to the kid. Ronaldo took a spin, shaking the black and white keys for all they were worth. Finally, bravado back with gusto, Bette tore Ruby to shreds.

The guitar responded to the familiar strums, letting loose chords and notes aplenty. The sound swelled to a crescendo. Miss Bluesberry, feeling left out, began to improvise. She sang so sexily, her husband, by now part of the lilliputian audience, maybe thirty animals, had a redness to his black and white face. Wow, so sexy, dearest, he thought, wait until we get home. Skunks in love, who can explain it?

While they played, Horace and Boris were busy with dragonflies. Messages went to Darby and O'Gill. They appeared and heard the reason for the summons. This beaver had natural ability not easily defined, sweet and vicious simultaneously. You could see he loved music, instrument an extension of soul, visible part of the iceberg.

Inside the player was a complex, untaught talent aching for release. He had to play, simply had to, identical to breathing. He played with morning's first breath to last exhale at night. The ax lay at the edge of his bed, so he could play at first light. This is why Bette loved him, who he was as a beaver sustained them for life.

After the extemporaneous jam session concluded, Dublin was taken to The Lodge. Here, they played again, known songs. They did not know, Million Miles Away, Bullfrog Blues, or Bad Penny. He did not know most of their music, but all knew, Party Up Them Hills, Boys. Off they went. Next came an insane version of Crossroads. Practice produced a list of ten songs, he taught them: Messin' With The Kid, Bullfrog Blues, and Nadine.

Once they got Dublin's sound, laced with early Fleetwood Mac, Peter Green (his favorite), Wishbone Ash for progressive rock and blues, Long John Baldry, John Mayall, Eric Clapton, and Jeff Beck, it was easy to adapt. Miss Bluesberry LOVED Bullfrog Blues. It took time for her to get past, 'Did you ever wake up with those bullfrogs on your mind'.

They would jam and jam, approaching the first line, but she milked it like a dairywoman on a binge. By the time they wrapped the song, a full ten minutes elapsed. Miss Bluesberry insisted, nay, demanded he be in the show. Whatever, he was going to play for her. Bette was all smiles. No problems sharing dual billing with this budding genius. He threw himself into the song; you would think he wrote it. All he did was imitate what he heard on the hermit's turntable back in the swamp off Highway 101.

Upon hearing his story, Murray and Momo wrote the song which defined Dublin for all time: Hwy 101 Swamptown Blues. It told the tale of his life, saga over the Olympics, across glaciers, to Grand Meadow, and jamming at dawn, guitar drenched in blood. It was a powerful, driven song, whose bridge drifted into a dual guitar lead. Here, Bette shined with her slide work. Funny thing, at the end of the guitar lead was a bit for harmonica. Hmm, who wrote that portion, do you think?

Murray had no problem inserting himself in a song for this guy. He and Dublin had adventure and music in common. Modest beaver meet humble woodrat. Aside from Mr. Mizithra, his very best friend, Dublin would occupy the closest male relationship he developed, due to music.

BOOK 4: PEACE

Momo began playing more percussion with the group. The children were doing well, Nediva more than happy to watch the kiddies while mom and dad played. Mizzy was the handful. A solution for the dilemma of the shows was solved by Momo. Working the show in Carnival Island taught her a thing or two. Start playing earlier, shorten breaks, and work the show until dark.

The dramatic effect of late performances heightened music for the audience. Miss Bluesberry loved the idea. She did not need to see to sing. The crowd would be adult, young and old in slumber land. This way no one was excluded. Dublin was in the big dance. He was happy. Boston Charlie asked Momo to join future plans due to her past knowledge. She obliged. Dublin named the ax Bette gave him, Topaz, after her eyes.

Shows began, taking off without a hitch. The whole Grand Meadow was alive with buzzing. K'wati and his wapiti gang came up, several knew Dublin from back home. He became their champion. For a beaver to ascend the Hoh, cross glaciers, then make it to Grand Meadow was a feat beyond compare. He was from their neck of the woods, so adored him.

Neferteri and her entourage loved Dublin, the dual guitar sound. It was perfectly vicious. It produced a harder edge on the sound. Darkly lovely. Kelly the Cooper's hawk was granted permission to attend. He and Sergio had a fine time with the music. Sergio stayed after his friend, the Whiskey Bend Kid, left. Grand Meadow was more to his liking.

Lloyd Joseph, aged and nearing death, came to watch a few of the performers. He was honored in a lengthy ceremony, then he retired. Not long after leaving, he flew alone to a solitary ledge. There, at the end of a brilliant life, he passed. Mountain goats, Gottfried and Gerta tended to him with as much love and respect as he earned in life.

Fame Dublin created at this show went back to the swamps. Silas and Norma beamed at their son's success. His former band mates were stunned he made it and became so famous. His drive paid. He was a true guitar god. They thanked the hermit by chopping up a whole tree. The entire family stacked it outside the shack. His love of blues led their son on a dramatic quest. It succeeded, so they were grateful. The hermit was bewildered, but thankful.

Animals from all over the Olympics came to listen, play, have a really good time. Other than fear of the occasional raptor dropping in to snag somebody, or coyote venturing in, the whole affair was one big fun fest. Love blossomed. Friendships began. Animal business was conducted. Food consumed. The Lodge rocked. Molly and Graedy kept busy fixing minor injuries, especially due to Humboldt flying squirrel training mishaps. Music reigned supreme.

Toss in the humor of Slick Sammy Seagull and you have a winner. He and Murray had a lot of catching up to do. Seagulls do not fight either, so he stayed away from everything. No one faulted them, it was the way it was. He was happy to see Murray and sent love from Paula, Clancy, and the band south on Willapa Bay. He sent the same in return.

Her name, Tamika. To male beaver, she was hot. Coming from far away, her family and friends stayed near the stage. She and her girlfriends loved music. Her parents were pillars of their dam and pond. Her mother, Florida, was a hero from Bette's dam. The recent war disrupted lives. A few adults answered the call. For devotion: lifetime accommodations, all the bark they could eat was the reward.

As veterans, they did not fight for later benefits, but survival of the species. What they were offered after seemed almost embarrassing. Those who fight honorably deserve praise. They defend our way of life, our nation. An extremely small percentage of soldiers go in for benefits. Most do so to ensure the survival of a country, never seeking recognition. If you do not know, you cannot understand. It is a very personal thing for soldiers. Pride of a nation, this is why they join and face untold horrors.

Tamika's father, Norman Sylvester, was an old blues guitarslinger. He wrote classics: "Damn the Dam Blues" and "Bad Beaver Blues". It was a hoot. Blues riffs are beautiful to musicians. He had a slick style, tight in fact. His band mates were ready for the show. The problem: bass player, Ox, got injured falling off a rock playing with other young beavers.

He showed off, slipped, hit a stone, ending in an arm sling thanks to Molly and Cloris. Delores was busy with other duties. Since he could not do the show, Delores whispered in Tobias' ear. With a smile, he offered to play. Norman Sylvester was happy. Tobias filled in for Ox.

The young, eligible daughter of this legend saw Tobias where buddy Ox was supposed to be. Her friends saw. Gasping, she caught sight of her future husband. Ox was forgotten in that same breath. Her eyes saw only this young, dashing bass player. He played tail with bass, a double threat. Ox could not do it. She was starstruck. He was the most.

She was in love. Again, it does not take animals as much time as people. They have no time for a long courtship. They were married at the end of the festival. Norman Sylvester and Florida were pleased. She would live at The Lodge. Bette and Dublin would have an entirely new section. Tobias and Tamika took over and expanded the original apartment.

Tobias saw her in the audience and was instantly entranced. Then, after the show, Norman Sylvester introduced his daughter. They met. They held paws a long time rather than shake. Since the attraction was mutual, it was a foregone conclusion they would marry.

Norman Sylvester did not even need a shotgun. Love was obvious. Miss Bluesberry sang at the ceremony. It was a special song just for her best friend and his new love. "T&T" (Tobias and Tamika) was the tune. A love song so precious, it was her gift to them.

Mr. Loganberry of Grand Meadow, silent but deadly mayor of the area, performed the service. Molly and Graedy loved them, promising free health care for life and any little ones added. She was a blessing, making the band more stable; Tobias would not look at ladies any more, just four strings on bass and Tamika.

As for Dublin, carried up in the rapid heat life injected, he did not notice the love Bette bore. They played on stage, but the hayseed had no clue when it came to ladies. He was from a small pond. She was a big lake girl. Thoughts of romance lay distant. All he knew, when they played guitars it felt special.

She was very talented. He loved her solos, eyes closing, lost in music. It was the same for him, though he did not think himself in her league. When he overheard Twigs commenting on her love beaver, Dublin finally got a freaking clue, about time. In a flash, he proposed, she accepted. Mr. Loganberry of Grand Meadow wed them like Tobias and Tamika.

Beaver's marital futures decided, Twigs saw no such horizon. He and Ripple never married, happy bachelors. Ripple gave up on matrimony when he lost a violent altercation years before. Twigs loved skunk and beaver, so could not marry. No muskrat ever tickled his fancy.

He remained in the group forever, in love with a set of drums. Music was mistress, who could never leave or be unfaithful. Sitting behind the band on his kit, he saw it all, having the time of his life. Drumming kept him young, fit, happy. He saw no need to confuse it with a lady muskrat.

When the festival wrapped, newlyweds went on honeymoon to a landlocked resort up from the rock backdrop behind the stage. It was a two day journey, then alone in Paradise. Dublin and Bette had one cozy lodge, Tobias and Tamika the other, pond between. After a week of rest and getting to know each other, the couples headed to Grand Meadow to catch up on gossip and tend The Lodge. It had been a most wonderful year, past fading quickly from memory. What war?

Chapter 12 The Valley News

While Grand Meadow enjoyed salubrious summer days, marmots in the Valley of the Stones were living life quite nicely, thank you very much. Post Sleepy Time activities were booming. Mr. Milk Chocolate and his colony broke ground on new tunnels, chambers, and escape passageways.

Mrs. Lip Gloss had those free and able help repay the sacrifice made for the good of all. The previous winter's free time was spent improving digs. Spring sprang. They sprang from darkness to the light of the Valley of the Stones. The view was ever breathtaking.

Work finalized the colony's interior with aid offered Mr. Milk Chocolate and family. The rest of the settlement repaired bulwarks. It was staffed with guards all 100 days above ground. Work done this season would serve both colonies and Keen's mice flourishing in the valley.

As a permanent home, they did not want half measures. Each room was dug to perfection. Defensive walls and tunnels were reinforced or completely redone in the case of the most devastated part. The bear tore it to smithereens. Tunnels on the south side were filled, so no one would fall to death. After a while, tough as it was, Fort Ballerina was restored.

As if the action of the supernatural, after completion, a raccoon, not knowing about war as she was from far away, sniffed the ramparts. She detected fresh, tasty marmot and began to ascend the wall. From four separate slits in the wood, sharpened pine spears jabbed soft belly meat. The raccoon fell to hard ground, writhing in extreme pain, bleeding from four deep wounds.

Shocked, bewildered, she lay a while. Gathering herself, she hobbled a sore path home. The story echoed the Olympics. Never again would raccoons venture near. Others, bobcat and raven, did the same, never coming to this valley. These were seriously twisted marmots.

Homes for Keen's mice were laid out with an eye to the view. Zor, Zum and the rest created a real paradise, mouse style that is. They had a path up and around the tallest fir in the valley, base of their central complex. A few choice nests on the way up made the journey safer, more homey. Tunnels underneath, embraced roots and wove through the area to fantastic home chambers filled with happy rodents.

They thrived, very welcomed by marmots who had only praise for them and efforts in war. The tallest tree served two purposes: pleasure in seeing forever over the east edge of the ledge, and highest guard tower in the valley. A lofted nest overlooked the vale. Certain times, dawn and dusk, it was spectacular.

They saw west over ramparts to land on the other side. While they could see nothing immediately in front, no flying object would evade their purvey. Guards were on 24 hours a day the whole 100 days period of summer fun. Bats were never trusted. They took this duty to heart and maintained their vital importance. Life with the marmot population was a mutual relationship.

The other cool thing for them and squirrels: climbing this tallest tree in a wind storm and hanging on for dear life. You can have a roller-coaster ride all day long, trying to hang on to a swaying conifer in a 100-150 kilometer per hour wind blast. Wind hits the Olympics, up from the Strait of Juan De Fuca, increasing speed. By the time they reach high elevations, deadly. Firs sway, nature providing rides for the kiddies. The Humboldt flying squirrels thought it a good launch pad were they to fly down to the Exception one thousand meters below. Whee.

Marmots saw it the same to benefit both species of rodents. Big or small, getting along with others depends much more on you than them. Only you control your attitude. You determine your response to anything different. Fear generally drives bigotry. You cannot blame others for your reactions, that is silly and not realistic. You form your beliefs. They may be influenced by other factors, but ultimately you are responsible for you.

Mr. Murray and Miss Paprika were in their second year, full of vim and vinegar. They were a handful, eager to do everything all at once. Their spirit was a tonic to the old man. Mr. Mizithra loved them and reveled in the days of summer. He basked in sunshine, watched children grow, loved his wife, and told stories in omnipresent warmth. He lent a hand to defenses, helping lead repairs to wall and floor.

The bashing Orlaith and her angry otters performed split most of the stones to the dismay and death of a whole lot of voles. She moved to the pond formerly occupied by Beulah and Buford. It was her reward. Here, she built a strong otter presence. Her battle prowess was a major factor in defense of Fort Ballerina. She raised a family, filling the tarn with happy otters. She worked perfectly with Silas and Clementine.

Mrs. Saffron wanted to have one more litter. Her beaming husband agreed. This Sleepy Time would be lovey dovey time. He glowed. Marriage and fatherhood suited this scarred old marmot to a tee. They made plans, knowing the children would marry next summer. A new set of pups would give a finale in splendid fashion. In spite of war, Mrs. Saffron was in fine health. She was in her prime, a vital force in the colony's populous.

Heroics in war, when bobcats neared the lair, were regaled through the marmot world. At the celebration at Grand Meadow, she was singled out for bravery exhibited when bobcats came within a scoop full of dirt of uncovering the future, the youngest. Becoming mother again, though, was her greatest desire. Her husband heartily echoed these thoughts.

Watching the wee ones whenever Mr. Mizithra had chores on the wall, she planned winter's activities. She would not neglect her older children. They were like teenagers in the marmot age chart. Any interest in a pregnant mother would be minor. They would eat, bulge, go to sleep, get up, listen to dad tell his stories, and let her blow up with a new batch of kiddies. Next spring both would seek mates, nature taking it's course.

Anxious for siblings, but acting vacant as teenagers do, Mr. Murray and Miss Paprika had a marvelous summer. They worked on walls, on the pathway with friends their age, and had fun splashing in the pond. This was a full 100 days. They were happy in this peace-filled wonderland.

Mr. Milk Chocolate's colony took shape quite well. The passageway between colonies at back was reinforced. Walls were weak in some spots. Specialized groups led by Mr. Beef Stew on one side, Mr. Hop Sing on the other, both excellent engineers, had the task of making this most secret escape route predator proof.

Once in perfect shape, two tunnels shot at a steep angle with only one big twist cut back toward the surface. These were totally hidden from view. The only way to find either would be to stumble across one. This was not likely. Mrs. Lip Gloss and Mr. Root Beer pronounced them sound. A round of applause went up for the two who led the work.

Female marmots were excellent at certain areas. Here, smaller size paid big dividends for the colony. Example, Mrs. Pebbles, Mr. Little Joe her hubby, created and dug passageways to the surface. Mrs. Wilma, Mr. Hoss' wife, surveyed and dug tunnels to meet those digging upward. Male marmots came behind to increase the size of burrows. Working as one, the difference in genders was an advantage, not reason to quarrel.

A few small chambers were dug by ladies, so ones traveling up might conceal themselves below to ensure safety before going above ground. Such wise measures ensured ultimate safety. Great pains were made to rid the area of scent. Openings were covered up top, plugged at the start so no odor of marmot might reveal their most protected area. In a time of trouble, all they had to do was dig a bit at the bottom, to open the top. Then they could dash to safety.

Mr. Mizithra had a fine time. New arrivals from both colonies collected about the sage, Flett's violet in it's rightful place. Baby marmots had to learn basics. Two-year olds wanting to bone up on lessons were welcome, too. He continued to teach the same tales. War stories faded, lessons on defense remained. They needed what was crucial to survive.

No marmot desired glory for combat. To them, killing was not a pleasure. Talking about it in a glorious light was untoward. Instead, the four whistles, stories of the Thumbs, Kai, were bread and butter. It was embellishment he specialized in, for which audiences gathered.

Out of all marmots to help, one Marauder stood out from the rest. He was a most valiant leader, rising to lieutenant by war's end. He began as a common foot soldier. He was known for assistance given in the trek from the old colony. He was a marmot with a gold star for everything.

Standing out early, shuffling marmots to freedom, is no reason to think of a great leader. Many others did their job in this same capacity fading to obscurity. However, when war was declared, the bravest made the most impact. When walls were breached, many died, but others fought. Of all marmots on the wall, Mr. Beef Stew rose above.

Chapter 13 The Saga of Mr. Beef Stew

War gave Mr. Beef Stew opportunity to shine, which he did not seek. This fight was thrust upon him. He fought along the pathway, personally killing two raccoons on the journey down. At the bottom, he was severely injured. Only by immediate surgery did he survive. Keen's mice got him in an empty, hidden chamber, Douglas squirrels tending his wounds.

This secreted tomb was the sort of place young marmots would use to sneak off and do whatever it is teenage marmots do. Mr. Beef Stew was sliced open, unconscious, and bore a severe gash across his nose. The final cut never fully repaired.

Battling in Fort Ballerina proved bravery in the heat of war. For him to plunge down the pathway, into the fray, certain death, was why he was regarded with honor. His valor was obvious. Others drew strength from his guiding stand. War was won here, along with the wapiti cavalry.

Were it not for Mr. Beef Stew's efforts, many more would have died. Voles and squirrels fell to his spear, then two raccoons in succession. At the bottom of the footpath, he was cut down by another raccoon who carved his stomach with one pass, muzzle on the return slice. Mr. Beef Stew fell, left for dead.

Keen's mice saw him go down, quickly grabbing his limp body. They took him down to the small chamber as Douglas squirrels worked to save his life. The slice across his muzzle left a scar on top. It was a clean cut. It looked like a part in his fur. It was a telling mark on such a big fellow. Mr. Beef Stew was the largest male Olympic marmot anyone could recall.

BOOK 4: PEACE

The cut across his belly was quite another injury entirely. If they could not stop the bleeding, death. Red blood flowed. Three different teams of Douglas squirrel physicians, graduates of Molly's, finally got him stable. Surgery took hours. Battle over, he rested comatose on a blood riddled bed of leaves.

He slept three days, coming out groggy. By this time, celebrations were winding down. Doctors told him about his muzzle, it was covered in a bandage. He pawed it, but Keen's mice nurses stopped him from fussing the dressings. He felt his stomach. The large bandage and extreme pain meant he was battle-scarred severely.

In the animal world this meant death. He would be fed, no need to move. He could stay and be served by ready volunteers. Any injured too badly to feed themselves was cared for by a bevy of mice, marmots, and squirrels. Mr. Beef Stew wanted for nothing. He had a lovely view of the valley when the sun shone. He gained weight at a steady rate requisite for dormancy. Doctors eyed the trauma to his stomach. The one across his nose was tended by nurses. Mice fussed over him like a rock star.

Once the bandage came off his hooter, the scar was covered with a salve provided by Molly and Graedy to Douglas squirrels. He had a really cool battle blemish, but his schnozzola would sniff again, not a worry. It made him look a whole lot tougher, too.

The stomach, however, presented a problem. The raccoon's slice did major harm. Recovery occurred slowly, with one rough aftermath: he was impotent. The cut made it impossible to produce children. Marriage was out of the question. The rule of nature demands mating for the purpose of making offspring.

If it was not possible, there was no reason to marry. No female would want to, not because he was a bad marmot, quite the contrary. It is the way things are. You cannot change what is the way of nature to suit that of man. It is not the same, nor can it be. Reality is reality.

After struck by the raccoon, wounds healed, damage remained. Mr. Beef Stew went through a long period of sad reflection about life. He wondered at his function in the colony. It was Mrs. Lip Gloss who spent the time working with his broken spirit. A need presented itself, someone capable stepped forward.

Mr. Beef Stew found a renewed purpose in existence. She was an elder for a reason. This was a most complex rodent. From the beginning of our story it was noted she possessed compassion in buckets full. Here was a strong, brave, proud male marmot. His very size intimidated other males and impressed the ladies.

The reason he did not marry the first year was he was busy moving the masses. By the time they arrived, most eligible ladies were on some other marmot's arm. He felt it best to wait until next year. A few lady marmots were ready to wed in Sun Days. War disrupted his intentions.

Injuries sustained in defending his people completely changed his life. To fight was one thing, living with wounds left behind, quite another. He fought and was injured for his colony. No one shunned him, but knew he could not marry. This was awkward for all concerned.

Mrs. Lip Gloss took him under her wing. She gave good council. Not condescending to her patient, but treating him as an adult, she was able to give him real reasons to go on. He had a real value to the colony. She told him procreation was not the only job in the settlement. He had a myriad of talents to exploit.

She had not had a litter in years, yet was still vital. Would he disagree with her? No, not if he was smart. Slowly, he saw truth in her words. Wounds force us to look at another facet or two in the diamond of life. If you can no longer do what you used to, find something else. This is not easy, no sir, but the alternative is to surrender to change. No thanks, folks. If you can no longer hike, encourage others. Teach what you learned, so they may enjoy what you miss doing.

He had two major talents which had nothing to do with fighting. He had an eye for engineering second to none in their colony. His folks were good diggers and taught him well. Once he moved with the colony to the Valley of the Stones, it was he who began the initial tunnels. So, when they needed to shore up the rear tunnel, he was first choice.

This stalwart, young marmot healed, recovering physically, more important, psychologically. No bad thought came to him, for he knew his reason to be. Mr. Beef Stew had a laser focus on the task at hand. The passageway he dug with his charges was the strongest any could recall. As time passed, he became the one called on any tunneling job.

The other endowment this fellow held in copious sums, the ability to tell a tale. From an early age, he was considered the next Mr. Mizithra. When laid up in hospital Mr. Mizithra came by. He listened more intently than as a kid. No marmot could hear the crowned head of story tellers and not learn.

As a young rodent, he heard lessons, the four whistles, the rest of the fables. They tickled his fancy. So did wrestling. He was a courageous fighter. No male marmot wanted to wrestle this big boy. War proved his bravery and ability to fight. He almost gave his existence for the cause.

With a need to devote life to something else, Mrs. Lip Gloss had a challenge: what can you do other than wrestle and dig? He did not ponder this long. The answer was soon found. It happened just this way.

Mr. Mizithra was on the ramparts slaving away, telling everybody else what to do. Needing to teach the kiddies, he sent word to Mr. Beef Stew to take a load off his 'brain' and teach the story of Ollie the Obtuse Osprey. He did so. Young ones raved. A star was born, a reason to live found, a necessary place in the colony discovered.

Mr. Mizithra took him into his confidence, sharing every story he knew, encouraging Mr. Beef Stew to develop his own style, his own flair. The secret, the mentor imparted to his young, large charge, is hold them spellbound, in the palm of your paw. To be a great yarn spinner, one had to add a splash of color in an otherwise blase` world and to know when.

"Timing, my young charge, is everything", he said at least a million times until Mr. Beef Stew was shoulder to shoulder with his mentor. He could tell a juicy whopper about as well as the old pro. Mr. Murray and Miss Paprika adored "Uncle" Beef. Mrs. Saffron helped with motherly advice about life.

Not everybody has the same role, she said. In her life, departing her old colony was something raccoons forced. She did not wish to leave. In the distinct minority, she held her mouth, waiting to see what happened. She wanted to remain in a wrecked homeland and rebuild. Mr. Milk Chocolate would have none of it.

Once she met marmots from the other colony, she fell head over paws in love with old, beat up Mr. Mizithra. Both got a big laugh out of watching him when she said this to her chum. Her husband, love him, was playing in the water with the kids. He had droplets dripping all over while he shouted, four yellow teeth jutting out of his gaping gob. 'If I can love that', both laughing, 'you are doing fine'.

Encouragement from fellow colonists was real. It was heartfelt. They did not say one thing and do another. He could trust what they said. Real results blossomed. He got healthy inside his head. It is never easy to come back from extreme trauma. Some cannot. They who do, do because those attending truly cared. It was not a chore. It was not a facade. Fakery fools no one. It is as lucid as clear water. He had real help from real animals.

Given these positives, he still had down times. It is only natural. To always be happy is not normal. We all have highs and lows. If someone you know dies, being cheerful is not a realistic way to respond. Death is tragic, cry a bit, let it go. You will always have a scar to remind you. So, Mr. Beef Stew had struggles. He saw couples with young ones and wished he had that opportunity. Coming to grips with truth took time, ups and downs, but he saw the light.

In time, his place was next to Mr. Mizithra as number two narrator in the colony. He was valuable again. The best thing was he saw it, too. No need to fight wars. It would never rise again. He had a purpose: teaching the ignorant of the world. He had seen war, battles, good times and bad. Taking real life experiences and using them to teach made him as whole as was able. He had some cool scars to go along with his own stories.

Mr. Beef Stew could look at life with a true joy in his heart. The tunnels he engineered were a fine legacy, as were the fables. He owed it to his own spirit, Mrs. Lip Gloss, Mr. Mizithra, Mrs. Saffron, and the colony. As a family, they helped their brother heal. Everybody won.

Chapter 14 Mrs Lip Gloss's Charge

Spring handed off to an early summer. Everything down in the Valley of the Stones was going well, save one problem. With the death of Mr. Savoy Brown during war, they needed another elder. Mr. Milk Chocolate had Mr. Gilligan's Island, Mr. Marathon Bar, and Mrs. Ellie Mae installed as overseers.

Mr. Milk Chocolate was not yet prepared for the obligations a supervisor bore. He had more to learn.

The same was true with Mrs. Lip Gloss's colony. Many like Mr. Egg Flower Soup and Mr. Macaroon, ones expected to take the position, were slain in war. The depletion of those qualified was whittled down swiftly. The breach of the wall and savagery down the pathway took many. They did not make any provision for Ralston's invasion. Who would?

Mr. Savoy Brown's courage aside, his death left a huge gap in the leadership of the colony. If they did not have a good marmot in this most important post, the family would suffer. Elders were essential to their way of life. This was their most ancient family structure, their foundation.

Numbers born versus those who passed during hibernation due to old age, too weak at birth, or not fat enough for winter, was a goodly amount. The next Sleepy Time would generate many babies. More of the living would survive. They would regrow, as K'wati predicted.

They had to eat, eat, eat, drink, nap, then eat more. Peace meant better chances. The new, improved safety found here compared to the old colony was gargantuan. Everybody was better off over time. This truly was ideal. Mr. Mizithra and Murray found it thanks to Mr. Jon Deer.

A great deal of consideration went to Mr. Mizithra. She knew him since birth, knew his family, his entire life. He married her daughter, fathering two lovely granddaughters. He was a leader. Others in and out of the colony respected him. In time of war, he was most vocal, active, and successful. His leadership at the Siege of the Great Wall of China was a familiar tale. Others told it before he, proving humility.

After the murder of his family, her daughter and grandkids, it is true he took off for a long time and became a very different marmot. He indulged in a rather extended period of mourning. Once returned, other colonists forgave misgivings and embraced their brother back from the edge. It took longer than others, perhaps because of the brutality of seeing bears devour his family. Knowing pain, he was at once loved and helped by them as best they were able.

Soon, relaxed and settled in mind, Mr. Mizithra had a revelation: I shall tell stories. He did. Everybody flipped. He was an instant sensation in the marmot nation. The sadness forcing his travels paid dividends for his colony. With time, stories gained more color as the palette grew.

Coyote was still coyote, no bigger or meaner, it was his delivery, his cadence which made stories so memorable. The dude knew how to slap on color better than Vincent Van Gogh. The one real similarity was the whole ear deal. Animals who heard his fables left in awe. Just like he did with herons, those who had the good fortune to listen to this sagacious marmot felt blessed.

Well, folks, the same became true of ample, old Mr. Beef Stew. His abilities soared with nothing else to do. There were times, Mr. Mizithra would bow out, allowing his protege` to take the spotlight. The sooner he was able to take over, the sooner Mr. Mizithra could get to the job of becoming a papa one more time. This was a chore he would have a great deal of fun accomplishing with Mrs. Saffron.

One night, there came a tapping, gently rapping at their chamber door. Two shadows stole the light. Mr. Mizithra called whomever it was to enter. It was Mrs. Lip Gloss and Mr. Root Beer. They were invited to sit at the table. Mrs. Saffron went to fetch some clover honey tea, the favorite of Mrs. Lip Gloss. She shooed the kids to go play with other two-year olds. The elder's faces were extremely serious, no distractions now.

"The thing is, Mr. Mizithra, Mr. Root Beer and I have spoken often of your qualifications. With the death of so many ahead of you in the list of those potential elders, it became inherent we select someone this year. If we are to go underground without three, it is not wise."

"We surveyed the entire adult population of our colony. Mr. Root Beer did his own, I did mine. It was unanimous. Every single adult in our family, whether they came with us from the old settlement or joined after we arrived in the Valley of the Stones, is in favor of your ascension to be overseer of the colony."

"This is a most serious responsibility, one we feel you are worthy of fulfilling admirably. You were my son-in-law. Since then, you proved a most estimable fellow. I knew your parents, their hearts would fill with pride at you and your life. Other marmots hear tales you share and love your warmth and incredible story telling style. Some know you from war. You never speak of it, but you did well. No shame there in any way."

"This is a sizable choice. Please discuss it with your wife and family, then let us know. We want you to accept the position, but something this large isn't simply about you. Mrs. Saffron's voice is just as important as yours. Let us know, we await your word."

Any promotion, at work, in church, the military, or sports have that good old yin and yang stuff working. You have more power, more money, and a brighter future. You are a step further up the long ladder of life you heard tell about from ma and pa and other old folks. Getting a promotion is fantastic. Your self-worth soars to heights never realized. You can see a bit more, plan the next step, see the world with more color, fun stuff.

The bad, well, it's not fun. With a higher place on the pile, your load increases. People are now 'under' you. More responsibility means you are the jerk who tells people what to do. The joy of this is found in how you treat those you direct. Bad bosses do not get good results. If you have an atrocious supervisor, looking for a new job becomes an occupation of it's own. A lousy foreman quickly has new employees, more, then more, until management gets a clue and gets rid of the real problem.

Extra fun with a raise: meetings. Meetings for this, meetings for that, it matters little. You may even have meetings to prepare for upcoming meetings. It is a lot of fun going over the same thing again and again. Yippee. Another benefit is by sticking your hand up and saying, 'Yes, I will be a leader', you are an open target. Raised hand, means the head goes along for the ride. You stand out above the rest. Good-bye privacy.

You are sought for answers. If unable, you are not *their* leader. You are ignored, demeaned. Stewardship has many pitfalls. Leaders have a huge load to bear. Only the very closest ally can be leaned on in crisis. Think High Noon. This is true whatever post taken. The price is always the same. It is very lonely in the rarified air up there in charge.

He and his wife had to figure out what to do. Did he want to? Yup. Did she want him to? Yup. Did this mean she could not have this next litter? Nope. Did he still want pups if he took the huge task of elder? Yup. Was he going off on one of his journeys to think on it? Yup. Would she watch the kids until his return? Yup. Was he going to stop and discuss this with Murray? Yup.

Mr. Mizithra loved his wife. He hugged and kissed this lovely lady long and hard, then set off to see the elders. He told them of his plan to head for Grand Meadow to speak with Murray. They approved heartily, adding a single request. From behind the leaders came a large form. It was the kid, Mr. Beef Stew, tete` hung low. He did not so much as raise up when Mr. Mizithra approached. This was odd.

"We understand the need to go think on this grave task. It is lifelong and often thankless. I know you will take your time. We ask you to go see Murray the woodrat. He has a message for you. Your desire to see him was anticipated. We prepared accordingly."

"The elders have but one request," this Mrs. Lip Gloss said pulling Mr. Beef Stew in front, "take this big galoot, you even bigger galoot. I want him to summer in Grand Meadow. Murray has a little word to share with him as well."

"Well, Mrs. Lip Gloss, thank you for giving me time to go and think about this offer. I am humbled. As you know, life threw me a curve ball completely unexpected. Healing taught the value of reflection. I find it interesting you know me so well that Murray was contacted. You are wise, very wise, indeed."

"If I have to take this big lug, I will," Mr. Mizithra said with a grin.

He grabbed the shy giant, hugged the elders long and hard, then climbed the pathway. They spent the first night in the ramparts. He had to make certain all was good before departing. The sense of pride he took in his job on bulwarks was one reason many felt him more than qualified to become an elder. He had to see it for himself, the sign of a true leader.

In day's first hours, as earliest cracks of burnt orange split the black, Mr. Mizithra and Mr. Beef Stew were on their way to Grand Meadow. No hurry, they took a few days, taking in sights, lollygagging for all they were worth. Mr. Beef Stew did not know any better. He was in a coma on the way to hospital. After war, he returned by speedy elk. He had no real sense of the beauty of the walk down to this fabulous Paradise.

They arrived at the rim of the valley. It spewed out, a splay of colors and life end to end. Animals, birds, and bees changed the panorama with every breath. It was a living mural of nature. Mr. Mizithra sighed. Each time here moved him. Mr. Beef Stew stood in silence, mouth agape, four curry colored teeth jutting out in awe.

In a few paces, they were at water's edge. A long sip of sweet lake water quenched parched throats. Each took a ride on a beaver ferry to save time, then climbed the little brook to Murray's place. He was so glad to see them, welcoming them to see family.

This was Mr. Mizithra's first time seeing the kids. He hugged Nediva a long time, weeping with her at Frema's passing. He had to admit, Mizzy was his favorite. The little fellow, full grown as woodrats mature in a moment, was all over his namesake, asking questions. They soon formed a bond lasting all their days. To Mizzy, he was Mr. Mizithra, godfather. Momo hugged the big galoot, then gave a big welcoming hug to the bigger galoot, Mr. Beef Stew.

After niceties, Mr. Mizithra motioned his old chum outside. They stood leaning on railing of the lanai outside his home. The view Murray had of the lake, to the trees, and beyond to blue sky was breathtaking. In a normal setting, he would grab Mrs. Saffron, tell old buddy Murray to split, and begin working on their ongoing romance.

This time, such was not the case. Murray knew the offer made by the elders. He was not surprised knowing marmots as he did. The reason for letting the woodrat know, they felt Mr. Mizithra would need to discuss it with someone close. No one was closer than Murray. All the adventures they had, love shared, wives adoring one another, Murray knew this best friend's heart better than anyone else.

"We gotta talk, Murray."

"Yeah, we do."

Chapter 15 Old Friends Talk

For some time, they stood outside, discussing the ups and downs, ins and outs of a marmot elder. Woodrats had no such structure. They had a nest, a small family unit. The culture, way of life for Olympic marmots, four whistles, tunnel systems, demanded strong leadership to carry it through. The family demanded it. For a colony of forty to endure, they had to have different voices carry out chores. Training, maintenance of chambers and passageways, security, had cooperation between various members of this extended family.

"You know, Mizzy, vat my opinion is. You are ze best marmot I know. Ze times ve had is proof to me zat you are qualified. You hesitate, vy if I might ask? You cannot possibly believe yourself to not measure up to vat zey vant?"

After all these years, Mr. Mizithra released his most inner pain.

"Murray, old chum, you know me better than any creature alive. My wife knows me in many ways only she can, but our shared history is long. We have been friends since early days. Remember the old colony when we met? That was long ago, but holds key to worry about becoming an elder."

"Do you recall how we met? The time was my returning after the mourning period for the passing of my first wife, Mrs. Christmas Special. You know the fate of she and my daughters. Visions of the bears' slaughter still haunts. The evil scene never leaves my mind. It takes no vacation. You know the inner turmoil the aftermath caused."

"The reason I sought adventures after returning was my lack of desire to be around my colonists. Seeing my parents was heart wrenching. To have others look at me knowing my wife and young were eaten, yet I lived, was shame I found hard to bear. No one spoke, this is not the way of our species. It was the guilt I bore for living."

"The first hibernation alone was torture. I had to get away. Since it became apparent life was going to be the road, I turned my new way of existence into something positive. The colony could take care of itself, I would get out and see the world."

"Telling stories was the feeble attempt to pay back, but my dishonor could not tolerate being an intimate cog in the wheel. I was like a visiting minstrel. After a journey ended, a new batch of fables to weave presented itself. I kept busy, not dealing with festering pain inside. The ignominy, Murray, the shame of living drove hard. Turning it to something positive was the only way to endure. Fighting deep seated disgrace, loss of face, meant driving myself to extremes."

"That circumstance went on and on until you fell off the ice on our way to White Mountain and my selfish quest. There was no real reason to do it, other than finding of our new home, more an afterthought for me. You saw the display of my intense anger when beaver invaded Beulah and Burford's place.

The vicious action exhibited surprised even me. Violence is not our custom. This is not a real leader for peace loving marmots."

"I sat on the frozen cliff edge all night when you fell, thinking only what the selfishness of my inability to stay home cost you. Did this intense drive, coming from my guilt, the life of my best friend was dispensable? How egotistic I felt. What conceit I possessed."

"The sole thought was the White Mountain. That we found the Valley of the Stones was secondary. Once the White Mountain was climbed and we settled in our new land, my plan was to find another traveler. You were happy at the Beaver Blues Bungalow with your family. For me, it was still the open road, the next horizon. What was around the following corner held my interest, not my colony."

"I did them a favor by finding the place. Again, it appealed to my sense of adventure, my need to push myself further and further, to find my limit. I have to admit it is a most spectacular spot. That aside, my goal was to get everything set up for the new settlement, then take off again. We have Mr. Beef Stew. He would have been my choice. Finding a good woodrat is not so easy, my friend."

"When Mrs. Saffron came in my life, thoughts changed within. She made me whole. This marvelous, loving lady gave me the freedom to heal. Her love was real. She did not want only to be beloved. She wanted to share her love to someone seeking to be showered in her essence. This made our needs and wants identical. This blessing showed me to cease my travels. Happiness felt inside came from accepting her love, knowing she wants mine. Our children are an extension of love."

"So now, old buddy, old pal I come to the worry. Do I deserve to be an elder, when I really never intended to be part of it? The colony was a rest stop for Sleepy Time, a place to tell stories. The 100 days of fun and sun was my escape from pains of the past. Near death events were my tempting fate. This is not healthy. We both had a death wish of sorts. You for your reasons, me for mine."

"Ralston, Olaf, all of that…We did well, you and I. You, wow, you are one bad dude. The fight in you is admirable. The strength I found was in my wife. The love she gave, the fact she allowed me to be myself and open up about the pain of the past, is why I fought so hard. No longer did I seek escape. I faced ancient days. I wanted a life with her, my children, and the colony. And you could not be left behind, my friend."

"So post war, Mrs. Saffron approached me to inquire if we wanted another litter. My reaction surprised even me. The idea of more pups did not enter my mind. Rebuilding, strengthening walls and tunnels was the agenda, not more kids. She asked me and joy overwhelmed. Suddenly, I did not wish to go away. It became important to stay and live with her and my young."

"It was simple to sit, eat, drink, and tell stories. Watching wee ones grow was a boon. Being a dear old dad was the only thought in my little marmot brain. Young are the biggest blessing of all. They are both good. Your namesake will be a fine addition to the colony. His sister is a bit of a wild one. We shall see with her, but her mother is trying to tame her. Ha-ha-ha. It was good. I was happy. My only ambition was to be a good husband and dad."

"The death of other eligible candidates, like Mr. Egg Flower Soup and Mr. Macaroon, left a gap in qualified choices. Certainly, I was no first choice. Call it fate if you like, but fading into history vanished. Still, I felt they could find a better marmot to fill in after Mr. Savoy Brown's death."

"When Mrs. Lip Gloss and Mr. Root Beer came offering this position as overseer, it was not expected. After so much craziness, the intent was: fade to the background, tell stories, and love my wife. I wanted to be a forgettable marmot in the crowd. My days of standing out were over. The spotlight could turn someone else's way. The goal was forget war, past, and forge a new, fresh future. When they came to me, it opened the scar, tore off the scab, and left me with a realization: my pain is not over."

"They sent Mr. Beef Stew. You have a message in regards to him, too? Okay. So, here I am. Please, Mrs. Lip Gloss told me to seek you out for answers. She specifically said to see you. She sent a message ahead. You are awaiting my arrival. You know I do not like mystery. Besides, I am here for help, not talking behind my back. What is up, Murray?", he was on edge. This was clear, so the woodrat took a deep breath to speak.

Just then, Momo came silently, a tray with tea and snacks for them. Mute, she smiled and nodded, not wishing to break the spell. They had been out, talking low some time. She had no idea what was happening, but saw a break in the action. Murray was about to speak. Mr. Mizithra spoke a long time, very seriously. Only wishing to help, she prepared a tray and took it with a pause in the conversation. Her mute interruption gave Murray a moment longer to gather his thoughts.

Pressures Mr. Mizithra felt hung heavy, a perpetually raining, black cloud. It was a pallor over him. Even in his most joyous of times, a sense of sadness clung to his shoulders like hungry vultures. The truth, even his wife knew privately absorbing the pain, is he never fully recovered from the death, murder of wife and daughters. He saw them torn asunder playing in snow. Meandering over the Olympics, welcoming adventures, tight situations, and near death experiences was madness, utter insanity. Both belonged in some form of mental therapy, but lived.

Now, he sought peace. He wanted to grow old with Mrs. Saffron, raise kids, and spin yarns, spun fatter with repeated telling. He honestly felt his service over. He discovered the Valley of the Stones. He fought for the colony. He built bulwarks and ramparts. His wife was equally brave. Lessons shared with young ones of the extended family were of vital importance to the whole body. Having done his bit, Mr. Mizithra simply wanted to take it easy forever and enjoy life.

Nope, now they wanted to engage him in a highly responsible post. Settling disputes, organizing defenses, assigning duties, lovely. He would be under constant pressure, in the limelight. Belief in him weighed heavy, humbling further. This was a serious position demanding full attention. While their litter could still come along, he would not be able to devote all his energies to them.

This would increase the burden on Mrs. Saffron. She was willing. She approached him to have another batch. He felt the position of father as important as mother. Children need two parents. The burden of this yoke would not be equally borne. This is simple common sense borne over the span of history.

"You know, Mizzy, zat my opinion isn't vorth analyzing to you. Ve are just too close. Zis is vy ze elders sent ze dragonfly to see me. Zey are in earnest. Zey vant you to become an overseer, but know you are schtill trapped in ze past. Mrs. Lip Gloss, truly ze most compassionate of all marmots known, felt your pain und gave me ze answer."

"I am to send you to see Erasmus. He is ze vone who organized all ze behind ze scenes operations in ze var. He vas a very close friend mit old Lloyd Joseph, K'wati, und Buford und Beulah. He lives a schmall valk from ze old site of ze Beaver Blues Bungalow. He vas zere und got out just before ze raccoons arrived. It is zere you must now go. I vill send off a dragonfly to Erasmus. He vill be avaiting your arrival. He is ze visest animal of all time. He planned ze entire var. K'wati's rescue of us vas not by happenstance."

"Tomorrow, head off to ze old pond, near vere ze club vas. Zere, you vill be taken to Erasmus. He does not move any longer. His traveling days, like ours, are over. I am happy here, mit Momo und ze kinder. Mama Nediva loves zem, too. Ze past is over. Zey had aches und pains. My left leg tells me ven ze veather is about to change. But I have a pond full of vater bulbs, a loving vife, six kids, mama, ze band, you as my best friend, und a place to live vich is pure Eden."

"Ve found a hot schprings not far, und are going to build a nice hot tub. Ve vill line it mit schmall branches stripped of bark. Jugs of cool vasser to pour over fur. It feels great. Momo is so happy. She loves ze hot vasser. Mama loves it, too. Good for vat ails you, she says."

"Ze other message sent concerns me, Momo, you, und Mr. Beef Stew. Let us go und essen. Zis schnack is not enough. Ve vill go eat, have a hot tub, zen discuss ze second part. Ve need to take a break. Zis is all very heavy schtuff, Mizzy."

He agreed. They went inside, ate a huge meal, then the hot springs. The path had handrails, lined in colored stones. Insightful. Heavy rains these hills offer need drainage. The depth of stones and slight angle to the side kept the walkway drained. Nediva could easily climb the pathway. They went, soaking in warm, soothing water for some time. Mr. Mizithra loved the release of tension. Looking over at Murray, he nodded, Murray smiled. Right again.

Chapter 16 Murray and Momo's Sacrifice

After the lovely, needed respite from the previous serious interlude, the old amigos ventured back to the porch. They could watch the sun dip over the mountains behind the hospital, showering the meadow in a panoply of color. After work, Graedy and Molly always came down, but were informed tonight was not a time for a visit. Nediva fell to sleep immediately after these warm water escapades. She was given a large sip of water to rehydrate, then tucked in by her loved ones. Their respect for her knew no bounds.

"In a moment, Momo will be joining us. Ze kids need to be put to bed. First, zough, I vill tell you von zing, my friend. My fazer vas killed, you know, ven I vas a kid. I never told you how he died, did I? He vas killed by a coyote by protecting me. I schtepped out of our old home mitout looking. Ze hunter saw an inschtant meal und lunged. My fazer did vat he could for his son, he jumped at ze coyote. He died und I lived."

"You know, Mizzy, I never vorry about it. For me, it vas vat life vas. I cannot change it. It is impossible to go back und make a different choice. I survived. I vas a fool, my fazer paid mit his life. So, to thank him for his sacrifice, I live each und every day am happy to be alive. He died that I might live. It vould be a pitiful vay to repay his forfeiture mit misery. Zat vould be poor currency for a life so valuable to me."

"I live knowing tomorrow may never come. I know vat death is, I have seen it up close und personal. I vould gladly jump in a coyote's mouth to save my kinder. Until zat time, I enjoy each meal und hug from Momo. Und here she is."

"Mizzy, ve heard from Mrs. Lip Gloss und she has a problem. She has a marmot damaged from a traumatic situation. Mr. Beef Stew is here for a reason. Ve vere asked to allow him to take our son, Mizzy, as his traveling companion."

"Stories of him climbing to the top of the rock wall behind the stage at four months had others notice", shared Momo, taking over the conversation. She was about to burst with information and emotion. "He was off with any animal going anywhere at any time of day. He took off with a pack of woodrat teenagers to climb the peak south. It is the same one Dublin barely made it off of alive. That was at six months."

"It is clear to every animal in Grand Meadow, he is *EXACTLY* like his father. He has ants in his pants. The boy cannot sit still. He wants to go explore the world. We can no longer control his desires. He can best serve all concerned by going to the wild blue yonder, as the *father* before him."

"We will have long meetings with Mr. Beef Stew before they are off on their first adventure. They are going to go back to the same spot he tried with the other pack. It is felt Mr. Beef Stew to be big enough to make the difference. They will map out the area and increase what we know about the great beyond. Their work will benefit everybody, but will worry Nediva and me. She is already getting the moldy potato skins ready to eat in the corner."

"You know, Mizzy, my Murray is your best friend. Mrs. Saffron is mine. We chat via dragonfly all the time. My opinion: you can be a great father and still help the colony. I look in your eyes and see reflections of my pain. The difference is I cast evil inside my heart to the winds. You keep hurt bottled up inside. This is why you are not free."

"Go to Erasmus. He can help. Please pay attention to what I say. You must pay strict heed to each word, every syllable he utters. He does not speak as does anyone else. He is ancient. Erasmus is older than any animal known. He is a philosopher. He is omniscient. His form of speech is very ancient. It is, therefore, not what you will expect."

"Erasmus learned an ancient art of communications from his elders. His species is very solitary. Fully trained, he settled in a small hovel near beaver, living as a hermit for the most part. He came to the bungalow, to eat and divulge wisdom. What he speaks is called: *haiku.*"

"It is a poetic way of sharing thoughts. It always takes him a moment or two to gather his ideas. He is unlike us in what he eats, his manner, and lack of manners. Erasmus speaks, you listen. He does not say anything without a reason. Furthermore, you must appreciate one truth, he has no favorites. He does not care if you are the lowest bug or the most exalted king. K'wati seeks Erasmus often"

"When you hear the *haiku,* take the words and syllables in, then go digest them in your heart and mind. He gives peace of mind. He knows you, saw you at the break in by Jimmy Joe John Brown Jr. and how you responded. Your eyes. He said showed light into a broken heart."

"Go to him Mizzy, he is vaiting," said Murray, "But let us go now und rest. You vill say good-bye to my son und Mr. Beef Stew in ze morning. Ve are both frightened und proud. Ve hope he vill come back safely und zat Mr. Beef Stew vill recover more from his demons. Each of us is on zere own path, my friend. I love you, Mizzy. Go und get some shut eye. Send a dragonfly off to Mrs. Saffron in ze morning. Zen go und listen to Erasmus. You vill be happy you did."

He agreed and made for bed. Knowing where he was going did a great deal to relax his mind. Mr. Mizithra slept soundly. He would go to the pond. Everybody who loved him, even his wife, wanted him to go and meet this Erasmus character (we have been waiting long enough as well, don't you think?).

Come morning, he arose and ate a massive meal. The jaunt down the road along rapids would take a while. He had to go downhill alongside that nasty human lane and Gilligan's Island. Once there he had to locate this mysterious Erasmus.

Mr. Mizithra met Dublin and hit it off well. He had heard so many stories of this amazing marmot and he of this gifted guitarist, it was inevitable they be friends for life. He was happy Dublin married Bette.

He first went over and had a long talk with the boy. Mr. Beef Stew had a bit of time to train before taking off on his first adventure. He and Mizzy had an instant bond. Marmots and woodrats seem to hit it off well.

Mr. Mizithra charged him to ever be cautious: first, next, and always. He told the big guy to watch out for Mizzy. With a serious look he learned how the rest of the colony respected the young member. They loved his story telling abilities. The last thing Mr. Mizithra advised, never underestimate the woodrat because of size. Often the smallest member of the team is best for a given task.

He took young Mizzy aside, slapped a kippah on his adolescent noggin, and gave him the best godfather advice he could muster. Watch every direction. Remember, you can swim, Mr. Beef Stew cannot. Use the buddy system at all times. You can climb a tree, Mr. Beef Stew cannot. You cannot fight, Mr. Beef Stew can. Keep an eye on each other. Always sleep in a safe place.

Finally, have fun. It is a real blast on the road. They would see new stuff, other animals, waterfalls, mountains, all kinds of neat and groovy things. He wished Mizzy the very best, then gave him a big squeeze.

Mr. Mizithra turned to his oldest friend, hugged he and Momo, then set on the path to the pond. A beaver ferried him back across. He turned left towards the path following river and rapids. Nabbing a parting snack of purple lupines, he sent a dragonfly to tell his wife where he was going. He paused, turning to see their faces once more. A tear blurred his vision. This journey was his own. Mr. Mizithra felt alone, very much alone.

Chapter 17 Mizzy & Mr. Beef Stew

After Mr. Mizithra left Grand Meadow, those able, shared wisdom with the travelers-to-be. Mizzy was under a year old. Mr. Beef Stew would lead due to age. It was only common sense. Once Mizzy got to it, they would share duties. If it was best of Murray to take the front, okay. If they needed a big galoot, the marmot was up. This worked for Murray and Mr. Mizithra. It was logical to do the same for the new guys.

Major time was spent in three areas: Safety, food and water, and shelter. Safety is paramount. They knew this as kids, but venturing to the wild unknown, looking out for enemy was number one priority. Murray and Mr. Mizithra nearly died a number of times on these journeys. Take the example of the red-tailed hawk who hit marmot butt, aiming for a quivering woodrat. This proved the need to think on your paws, and improvise to stay alive.

Tearing out roots beneath and dirt above as a team, they dispatched Borya. They taught young travelers the need to think outside norms. If they were to survive wilderness alone, no back up, they had to rely on each other. Murray dove on the log first, Mr. Mizithra just behind. They avoided death by coyote. When on these hearty adventures, they learned safety first, last, always. One misstep meant death.

Food was second on the list. While Mizzy had to eat, for Mr. Beef Stew, it was essential to eat constantly with an eye to hibernation. He was a big guy anyway, but the need for the marmot half of this traveling duo to gain weight was incumbent. Mr. Mizithra already shared his ideas on the matter with his young charge. For Mizzy, his dad had what they needed down pat. They needed to eat and find water close to wherever they ended each night.

Third, shelter. This was drilled in their heads: you must stop each day in time to dig a hole complete with at least two escape tunnels. It is to be one simple chamber with an entrance. Under rocks is best, tree roots second, a steep hill last. Whatever hole you establish must be on ground easily dug, hidden from predators, away from danger zones. These would be aeries, caves, and the like. They were home to those who eat rodents. Mizzy had it worst because raptors like little woodrats. Yummy.

Other teachings included nuances with other species, telltale signs of traps set by hunters, and quick medical tricks to help when injured. The last item got a rise out of the guys, but was a reality. Eventually, one or both would face pain inflicted by predators, man, or a fall. They needed to learn to take time. This was key to making the entire journey. Pacing one's self was prime.

After thorough preparation, the parents hugged Mizzy good-bye, gave a long, stern look at Mr. Beef Stew, then sent them the first time together. They grabbed a beaver ferry across the lake, then turned south past the stand of trees which leads to Grand Meadow. A last glance and wave to the teary eyed family, then gone, full of joy. Smiles these two bore were furry ear to furry ear.

Adrenalin surged; they flew up the path the first two hours. A break, Mr. Beef Stew quickly learning the value of eating on the run, then back at it. They soon cleared five kilometers, a limit for a day uphill per Murray's instructions. They found a nice spot, dug for an hour, were soon full, safe, and fast asleep. Their initial foray into freedom was a success. The discussion between them before slumber was an agreed to need to follow instructions given. They were right. It was important to remember everything and implement them.

They rose early, ate, got bearings, then headed for their first goal, a large, snow covered outcropping a few kilometers up. This was not far, but a steep climb, nosebleed stuff. While early summer, bold winter snow happily clung to this large obelisk popping out the side of the mountain. White stuff owned the top half of solid rock. A crack or two, the odd, small pine, and a few small ledges were what kept the pillar from looking like an upside down ice cream cone.

Elevation was the problem. For the first time in his life, Mizzy had to go uphill at a steady rate. To scale this steepness, you hunch forward to keep from falling backwards. It was windy, cooler than down below. They had the first experience with suffering for no good reason. This is how one learns if camping and hiking are merriment or forgotten. They saw this as pure fun. Mizzy was grinning.

Mr. Beef Stew hit the road moving the entire clan from one spot to the next. He recalled the thrill he got seeing rapids the first time. He helped move the caravan, but was escorting a family of young and old marmots. Speed was not a focus, keeping everybody united and safe was. It was a guided, known route. This journey with Mizzy was something completely different.

It was nearing time to stop when they reached the pinnacle of the escarpment. A handy little hollow and an assurance no animal would be crazy enough to climb this high provided assurances of safety. Scents do not travel in the high wind. Mr. Beef Stew carried a large wad of lupines, lilies, and greens. They had food, melted snow for water, and a safe shelter. They did not have to dig either. Something had used this cave to sleep in before, so a bit of dried leaves made a nice bed. Back to back, they were warm all night.

In the morning, solar rays of fervent light shot in the entrance of the motel room. Up in a flash, they shook off the night, grabbed a really good look at the spectacular view, and scurried down the obelisk. They got to the bottom, took a look up the cliff from whence they came, gave a high four, and headed to the second destination.

The next stage was to find the waterfall Dublin and other animals described. This meant uphill again. They turned west southwest. What was unclear was how far it was. When Dublin traveled to Grand Meadow, he was virtually out of his mind. HIs only goal was to get to tryouts. He had no desire to return home. Mapping the course was of no concern.

They left the broad path, turning up a narrow chasm. They located the stream Dublin mentioned. The brook babbled, flip flopping down the gulley. Trees clung thick to one side the entire length of the vale. The other side had a few stubby firs, the sun not hitting in spots. This was an easy jaunt, food and water plentiful, chances of enemies slim.

No sign of the waterfall was found the first or second day up this trail. Shelter was easy, so they had a good time. They did not, however, relent security measures. One mistake and you were somebody's supper. Marmot or woodrat, carnivores were not picky, just hungry. The second day up, they dug in some roots and had a nice place. They chose to stay an extra day to rest and reconnoiter. The temperature cooled in this higher clime. They used wisdom offered by Murray.

Mr. Beef Stew had map making materials. They took time to draw path lines, landmarks, signs of hunters (coyotes were most prevalent), and sleeping areas. The waterway was charted from Grand Meadow up to the split they followed. The stream traveled the general path.

BOOK 4: PEACE

They took the divide Dublin spoke of leading to the waterfall. It was important to map so they could use the information for any travelers in the future. Their work was fun, but for the good of all animals. It is like George Washington, who surveyed the wilderness in early times, before the United States was it's own nation. He had a blast and helping the community.

The marmot shoved food in his mouth on a continual basis. He soon learned the value of this from Mr. Mizithra. The only way to gain weight for Sleepy Time was to constantly eat. Traveling burned calories, so he had to consume a whole bunch. Food was plentiful. Mizzy had diving in pools to retrieve a water bulb down to a fine science. It was dive, bite, turn, and hit the surface. Once on shore, a meal fit for a woodrat was at paw.

Languishing along the shore of the stream, sunlight began to go behind trees from camp. One last check of the area, they got in the third night in their temporary home, and were soon fast asleep. Each had an ear towards the front entrance. The second day, they improved their digs. They tunneled a third escape hatch and padded the chamber for a softer bed. Noise from the babbling brook kept sound and scents down.

Sun set fifteen minutes prior, when a rustling of leaves stirred the hikers to full alert. Something was moving outside their abode. By the sound of the movement, it was not a large animal. They sniffed the air, Mizzy possessing better olfactory senses. The smell was unfamiliar, same for Mr. Beef Stew. The one thing was, it did not smell like a carnivore. They who hunt have a distinct odor of blood and death. Since these were both primarily vegetarians, meat eaters were easily detected.

Still, they did not wish to tempt fate. Choosing to stay quiet, not to stir a hair, patience had them glaring at the entrance for a shadow. The closest escape hatch was a few steps from their position. They had a plan to incorporate for this situation. Were this unknown critter to enter, each would take a different escape route. Fear of separation gripped both. This was the instruction from Mr. Mizithra and Murray. It was tough, but for one to survive an attack was better than both becoming a meal.

Sounds quieted. Whomever was outside was listening intently to the surroundings. The noise of sniffing came in the entrance. After a bit, it disappeared. The animal who passed by went on it's merry way. They were once again alone. However, they knew to check to ensure the visitor was gone. More than once, a marmot or woodrat did not double check and wound up history for lax security.

They waited a while, a long time. A nose and cut up muzzle jutted slowly from darkness of their home. A smaller proboscis did the same below. They hovered for a bit. Moving as one, they emerged from the black tunnel into soft, summer moonlight. It was again with caution they peered to see what they could see in a full circle slowly turning.

The brook babbled, a light wind tossed fronds and leaves about in a sweet evening salutation. Mr. Beef Stew's outer fur wavered about. A long period of waiting finally ended. They relaxed. Whomever it was left. The scent was not a predator. It was not skunk, rat, or mouse. The odor was not marmot, either. A mystery animal passed by and kept on it's way.

Weary, they turned to go back to bed. In the morning, the plan was to head upstream in hopes the waterfall would soon appear. A noise unfamiliar exploded from above. Looking up, something passed by their faces to the ground in front of them. In an instant, Mr. Beef Stew, large as he was, lay on the shore, Mizzy held by a strong arm against the wall outside home.

The assailant relaxed the grip on Mizzy's throat. Hopping off of Mr. Beef Stew's stomach, they saw a sight which froze them in place. It was a Humboldt flying squirrel. They are dark creatures who venture at night. Daylight sightings are rare. They were allies in the war. Mr. Beef Stew had no interaction with them. Mizzy was born after hostilities ceased.

Stunned, they stood together against the wall. Their attacker stood a meter away in a menacing mode. The squirrel was much smaller than the marmot, especially this big boy. It was two against one. They noticed it was a female flier. Thoughts turned to retaliation. Mr. Beef Stew squinted, focusing to attack. Mizzy caught the vibe, recalling Mr. Mizithra telling him to let the marmot fight first. He could help, not start any tussle.

"You two. What are you doing in my territory? This is Humboldt country. We leave signs all over the freaking place. Are you too stupid to read them? Hey, hey, hey look up at the face, not the folds."

"The penalty for trespassing is death. You did not ask to pass up our stream. Last year, some stupid beaver slipped by me. I got in trouble. This time

I am going to hold you for trial. We cannot let any old marmot and woodrat pass through our nation without paying homage."

"We saw no signs and have been here three days. We are from Grand Meadow and are on our way up to the waterfall and beyond. Our mission is one of discovery. We want only to help our animal populace and have fun. If we trespassed your land, we are sorry. As to a sign, we saw nothing which bespoke ownership.", stated a diplomatic Mr. Beef Stew.

"Very well, ha-ha-ha. I have followed you two since you first took the split up this way. Your intentions are clear. You do need to work on looking around, though, you are sloppy. My name is Lollobrigida. My family runs the training school north of here. I left to get out on my own and see the world. The Olympics are huge. We are a small family. The few scurries of our species on this range are it for the whole world. They want us to marry and make babies."

"That is fine for others, but for now I want to get out and find new, interesting things. I can have babies next year. This year, I am out to have a good time gallivanting around. Do you mind if I tag along with you fellas?", a Humboldt flying squirrel has large, round ears, a sweet fur, and huge, circular, soft, inviting, dreamy eyes. The boys were entranced by her all-embracing orbs. Could they say, no to her? They could not.

In the morning, the trio set off in the direction of the waterfall. Lollobrigida was not a walker. She glided from tree to tree in the virtual darkness though it was day. Flying ahead, she soon returned, stating the waterfall was a short distance. They rested by water's edge, had a bite, a rest, finishing the trek to the falls.

Splendorous, was the word. The distance top to bottom was a mere five meters, but it was broad. Water shimmied as it fell. Colors formed a constant rainbow, glistening as mist fluttered away. Landing on stones below, spray splattered in a vast array. A pool quivered with unceasing waves. Here the stream gurgled, heading a few kilometers below.

Lollobrigida disappeared in trees over top of the waterfall. The boys trudged to the top on the south. Here the stream was again a pool. It was slow climbing the hillside. The waterfall was it's lone site of excitement. Here they remained, a handy old hovel used by former travelers making a fine room for the night. They dug it out, had a huge meal, talked a bit with Lollobrigida, then turned in, pooped.

As a nocturnal animal, she had to work hard to sleep as well. They were able to eventually reach a compromise. Thankfully, it was the time of year when days were long and warm, so they could travel and make adjustments. The key was these three amigos got along well. They quickly adapted to each other's strengths and weaknesses. After this evening, they moved to the stream's source, mapping it entirely.

They returned, bringing Lollobrigida to Grand Meadow. She had a short time before going back home to find a mate. This was her time to establish memories for the future. They escorted Mr. Beef Stew back to the Valley of the Stones. She got to meet Mr. Mizithra, Mrs. Saffron, their kids, and the whole colony who loved her species.

Climbing the Keen's mice tree, she got a good ride on a truly windy day. She had the time of her life. It was she who formulated the idea to fly off this tree down to the Exception. What a flight that would be? It is true she was a bit nuts. It would be a long way down, wind on the way playing havoc with these light gliders. She had a bit of the devil may care.

Lollobrigida made it back, married a fellow named, Geno, creating a family of really tough Night Ninjas. For her whole life, days with the marmot and woodrat on the trail up the lazy river were her favorite of all memories. For Mr. Beef Stew and Mizzy, this first effort was the most fun either ever had. Mizzy had stories for his family; mom and grandma could put down the potato skins.

For his part, Mr. Beef Stew began to heal his heart and soul. Time away gave him opportunity to think apart from familiar haunts. He was with a friend, relying on him for help. He saw unknown parts of the world, met a new species, made friends, and had a really good time.

Plans were already under way for next year. They were carried out, as was so for years to come, these two becoming as close, having as many close calls, as many hair raising adventures as their predecessors. Mr. Beef Stew's yarn spinning rivaled Mr. Mizithra's. He had stories the old man did not. He continued the legend established by Mr. Mountain so long ago. Marmots are madcaps, yes, they are.

BOOK 4: PEACE

Chapter 18 Erasmus

Heading downhill took a few days to get past Gilligan's Island and rapids to the pond. As he waddled along, a rack of antlers appeared on th horizon. This was followed by a grinning black-tailed deer. Jon Deer. It is glorious watching a stag jaunting powerfully through the woods.

Directly behind him came My Deer. His lovely bride had news. Two fawns were with her, hiding shyly against mother to view this curious, new animal. Jon was a beaming father. Unlike Mr. Mizithra, his duties as father were more limited. Their culture was for males to help make the babies, then let the doe do the bulk of the raising. It is their way and not ours to question.

"Mr. Mizithra, I would like you to meet my young fawns. We have told them the stories of your fame, in battle and the yarns you spin. If you have the time, perhaps...?"

"This is my daughter, Yes Deer, a strong young doe, sure to make any stag sit up and take notice. Yes Deer is as powerful as her loving mother. Next, well, Mr. Mizithra, this one is right up your alley. My son is much like you, a wanderer, a madcap. His name, Gadzooks Deer. It is because he is such a shocking buck".

"Children, this is the famous Mr. Mizithra, the hero of the Great Wall of China, and the best danged storyteller you ever heard".

Giggles abounded from the fawns. They were awestruck, fawning all over him and sniffing his fur. The scars were frightening. Introductions ended, he obliged his friend, telling the young ones the tales of Borya and the close call. The youngsters were so frightened, they ran to hide under My Deer's stomach.

The time with Jon Deer allowed Mr. Mizithra the opportunity, at long last, to thank him properly for sending he and Murray to the Valley of the Stones, his forever home.

"One thing about you, Mr. Mizithra, you sure don't look around much. It was Lloyd Joseph's dictum to me directing you there. He got his advice from Erasmus, whom you are going to see. Mr. Mizithra, you are a most honest, sincere fellow, but you have very little on the ball at life more than a breath from your face. That crow loved you to the utmost, bless his memory. You need to realize it".

"Mr. Mizithra, you are much more important to the entire Olympic animal community than your understand. Those as valiant as you do not exist in large quantities. I, for one, did not engage in war. It is not our way. You, the stories abound. A marmot of distinction. That is you, sir. Do not discount your value. It is not just your yarns, it is you".

Mr. Mizithra bade them well, again on his way down to meet this mysterious Erasmus character. The interaction with Jon Deer was truly disturbing. Perchance Jon was correct, he did not see the whole picture. Lloyd Joseph really did a great deal for he and Murray. It was a lot for him to take into his mind and comprehend.

Approaching the shore, he spotted Orlaith and her family frolicking in the warm sunlight. She had a huge smile and massive hug for her old commander. The shillelagh used in war hung on the wall of their lodge.

Silas and Clementine and four pups were swimming around and gave hellos. These two worked for Bette and Tobias to build the dam. Now, the family lived in a quiet, peaceful place with no problems. They had a good life. Dandy bark was plentiful, the way it ought to be.

Nearing the far corner, he spotted his old friend, Boston Charlie perched on a limb, enjoying the sunshine. He waved Mr. Mizithra over to speak. He was not certain, but this could be the same limb the Belted Kingfisher from when he and Murray first met. Mr. Mizithra waddled over, shoving some handy lilies in his mouth.

"You were sent to meet Erasmus. I was sent to do the introductions by Neferteri. Lloyd Joseph knew this day would come. He made provision should Erasmus live post war. It was touch and go. He barely escaped the Beaver Blues Bungalow Massacre. It was unbelievable he got away. He is old and slow moving. The speed of the attack caught all by surprise."

"Erasmus wants to speak with you. His *haiku* is for your benefit. The life he lived parallels yours a great deal. Ralston, not Igor, killed his entire family. He alone lived, as you did. Erasmus studied the art of philosophy to recover from grief. Those who teach in the high caves, in snowy places, imparted knowledge. The wisdom of crows and wapiti came his way, too. He councils them."

"Take his words, meditate, pray on them. The reason for each of the four *haiku* you hear are for healing. The first will deal with *Sympathy* for your loss.

The second, *Forgiveness,* for yourself because you did not die with your family. The third, *Love,* that you may learn you are loved and need to remove the barrier you placed to prevent love from reaching you. The fourth, *Leadership,* the truth is you are the very best choice. You must accept this truth to go on a happy marmot."

"This Erasmus wants to do for you. He knows your pain. He knows your history. You are someone he cares for though you never met. When he speaks, you listen. You may not speak or ask questions. These are the conditions he extends. The wisdom is powerful and offered out of love."

"He is a Sewellel, a Mountain Beaver as the Thumbs call them which proves their stupidity. He has a mere stub of a tail, a few centimeters, does no dam building ever, and lives in a small tunnel, not a lodge. As a Sewellel, he is the most ancient of rodents, in fact, nothing like you or Murray. The Thumbs consider them a living fossil. His is a most difficult species to understand."

"They eat rhododendrons, things you cannot. If you did, you would die. They swim, climb trees, and burrow. Sewellels are brown, have a small patch of white under their ears, and live very solitary lives. It is a fact Erasmus has been alone since his family died, but in an extraordinary move, especially for a Sewellel, he invited others to heed his teachings."

"He has infirmities, is old. His assistants are Sewellels. His injuries, an arm is missing. We will go in a bit. The way it operates is a dragonfly will come. He only uses orange ones. Private, very private is he. To begin, you must now go and rest. Eat, drink, take a nap. Please. It will be when he is able to see you it will happen. We are on his schedule. Until then, I am going to stab a fish and enjoy the sunshine."

Mr. Mizithra nodded, found a nice spot close by, ate a bite, then took his friend's advice and slept. While understandably worried about the unknown, he felt safe. Those who loved him from top to bottom sent him to see Erasmus. His wife, colony, best friend, felt this boon essential.

He was private, pain well hidden. No, it was not. Reminders of his past were visible to all who met him. There was a hollowness, a sadness which permeated his aura. The horrors seen were etched inside his eyes, spirit rent. He felt these facts secreted from the world. His kids wondered why daddy seemed so far away sometimes.

The problems were his and needed to be worked on by him. Were he to have the distraction of his wife and children, success was impossible. This sacrifice Mrs. Saffron gladly made, hoping for a whole husband in return. Given the opportunity, he was determined to make it count. This stumbling block was getting old and wearing thin.

Chapter 19 It's Time

"Mr. Mizithra. It's time.", came Boston Charlie's soft voice.

Together they wended the pathway to the southern portion of the pond, where Graedy and Molly's waterfall spills in the tarn. Here, a small stand of trees set on a broad knoll. The top held greens and bushes. A large lip, glistening emerald in sweet sunshine, curled over, exposing a series of modest sized tunnels. Positioned perfectly to garner absolute sunlight, a gnarly, old animal lay basking in sun's welcomed warmth.

It was here Mr. Mizithra caught his first glimpse of Erasmus. He was nothing to look at, but a Sewellel is not what anyone would normally deem a cute critter. The aged one was propped up on a ledge, a padding of leaves aiding necessary comforts. His front left arm was gone. He had a scar in front of his eyes, a slice designed personally by Ralston. When injured, the invaders left him for dead.

Recovery came from Molly. She held him as Graedy seared his arm shut to stop bleeding. They utilized fire in a helpful way. The wound was bandaged, more bloody than harmful. Now, he lay on his side, clearly of great age. He was white about the muzzle. His eyes were a bit hazy. War and pain painted his body for all to see. After this bad experience, he focused completely on getting rid of Ralston.

Was it a bit of revenge? Yes. However, the broader problem was the union with ravens and others. The McGuffin was the bear. No one in planning: Sarge, Lloyd Joseph, Mr. Mizithra, Murray, especially Erasmus factored this weapon in the mix. Were it not for this bear, the system Erasmus and Sarge devised would win the war. It was this unexpected turn which created the problem. Sarge remained with his rodent chum; he and Erasmus had a strong bond.

The initial fight by Mr. Mizithra due to his recognition of Igor and the final assault by Mr. Guar Gum tilted the scales. The bear died, war was won as it should have been. Mr. Guar Gum's sacrifice loomed high in history. The wapiti were fortuitous in their arrival. The Exception was regarded with esteem by the mountain animals.

Now, with time passed and peace reigning, Erasmus felt it the right moment to repay Mr. Mizithra with the benefit of his wisdom. After conferring with the powerful long before war began, he informed of his desire to help this stalwart marmot. They had very similar pasts. The clear difference, Erasmus moved forward. Ahead lay reflection, acceptance, and rebirth. We are talking the whole Phoenix bird deal.

Four young Sewellels attended Erasmus. His comfort was their sole concern. A murder of crows kept guard nearby with a handful, not quite a surfeit, of skunks under command of Mr. Blackberry of Massacre Pond and his wife, Mrs. Huckleberry. Private, yes, but protected. Those in the know on the dark side would love to kill these important figures. To slay Mr. Mizithra, prince of the Battle at the Valley of the Stones and deadly leader at the Siege of the Great Wall of China would be a huge retaliation. To add Erasmus, secret genius behind the opposition's planning would make the successful one a hero for all time. Sarge was on top of this, too.

No raven, vole, squirrel, or bat was seen. Raccoons avoided this area like the plague. The true narrative of what happened to Ralston was the reason. No way were they facing the wrath of the mad, united animals at the pond. It was worth it to go around, avoiding them.

This made the time for Mr. Mizithra most beneficial. He had no worry of being attacked the whole time they were together. His only concern during his journey was from above. Raptors had no concern about war between animals. They just wanted dinner.

> **"She lived, I died**
> **Warm memories**
> **Lined with ice"**

The words came slowly, painfully. He had a halting way of speech. An animal he was not introduced to simply blurt out these words. They tore his soul, each syllable more powerful. Uttered, they lay upon Mr. Mizithra's emotions like a galactic, dense slab of slate. He was completely crushed. In nine words Erasmus pierced his facade. He burst the balloon.

Mr. Mizithra nodded, bowed, then left on his own. He heeded the charge to remain silent. Boston Charlie did not follow his dear friend. Erasmus, relishing in a few more rays was soon taken gently back to his room inside the burrow. While the sun was welcomed, it was not wise to be out too long. It was necessary for rest. He did not have much time. He was the longest lived Sewellel known.

Finding a quiet spot west of the meeting place, Mr. Mizithra sat and thought long and hard on the words. What Erasmus spilled from his mouth came not merely from wisdom, but his own grievous experience. When his entire family was killed, he lay, shocked and wounded, unable to give any support. Instead, in a daze, he witnessed terrible death at the paws of Ralston and a gaze of raccoons killing for sport.

"She lived", Erasmus began with true words. Erasmus did not know Mr. Mizithra' family, but Mrs. Christmas Special did precisely what was stated. From the moment they met wrestling in fun and dust, until the moment she was murdered, this was one lively marmot. She got the most out of every day.

Mrs. Christmas Special drained the grape each day represented until it's skin was dry. Happy beyond belief with her marriage to a big galoot in Mr. Mizithra, two sweet daughters were the blessing she treasured. Not one day did she exist with dread or sadness. She lived a full life. Yes, it was short, but until the three bears came ending her magical story, she lived.

He had not. Truth be known, a door shut, a wall erected, and moat placed around his heart to never again be hurt so deeply. Mrs. Saffron alone was able to build a drawbridge. The pain of murder in front of his face, unable to defend his loved ones, was mortar which held defenses in place. Replacing this was traveling. He escaped death to feel closer to lost ones. Unable to be with those who were happy, he stayed in the black of hibernation and told stories to help. He was surrounded, yet alone.

"I died", yet he breathed air, drank water, and ate. Selfishly, he took the precious life wife and daughters embraced, and hid in his own living death. This was passing time, waiting for the end. It was a way to hide, so pain could not penetrate. Sadly, he did not realize this is impossible. The Exception had animals visit him. It was people he could not stand.

For Erasmus, it was a mirror. Injured, but alive. Why did he live and the family die? After reasoning this out he realized the stupidity. He was still breathing. Any other conclusion was self-defeating. In other words, live or die, but do not linger in limbo. For man, this is a familiar theme. Facing problems, realize you are still here, so do something good rather than destructive or merely transfer the pain by heaping it on others.

The sadness was Mr. Mizithra did not see beyond pain. Glimpses of fun with Murray or getting married, climbing the White Mountain, were things he participated in, yet did not fully appreciate. He only lived a portion of pleasure in a moment because fear gripped him. This did not permit enjoyment. Any moment tragedy could strike. The truth was, and he realized it, he was one uptight marmot. Safety is vital, no question, but he never relaxed, not really. Even laying on a rock in the summer sun, watching a red-tail circle below, he was conscious of bad things.

By existing in a constant state of fear, he did not truly live. Guarded, he ate, breathed, drank, slept, but did not dive headlong in living a full life with all he had. It kept him bound to the past, unable to fully live in the here and now. The realization reached him. He was dead. She lived and he died. Humbled, he saw the poem for the value it provided. Buoyed by this pealing back of one layer of his pain, he now thought more deeply on the second line.

"Warm memories", the mountain beaver hit pay dirt. Mr. Mizithra, in spite of bad stuff, had a most wonderful life full of excellent episodes. The memories were very good for the most part. If you live for a while, it is impossible to avoid people passing away. You cannot escape seeing bad events happen. The opposite is the same. You get to see the most glorious things in life. You can see the sun rise and set. Trees, forest, prairies, seas, the world, you can see the freaking world. Yes, you.

Travel and realize how much there is to enjoy. The good fortune in this is while bad acts happen around this old globe, positive outweighs evil. Focus in on good stuff, not bad stuff. Easily stated, difficultly lived, I know, but possible.

By taking time to meditate, reflect on his experiences, Mr. Mizithra began to see the good in his life.

A corner was turned.

"Lined with ice", here's the rub. While memories held promises of good results, the reality of past pain could not be sidestepped. Here is where a barrier lay. He placed it in love's path long ago. It was an old, well built barricade. *"I will have fun, but not fully enjoy it because of the bad stuff I experienced"*. This was his innermost mantra. He selfishly took pleasure out of existence.

This way made him miss the true pleasure of life. This impediment buffered from exploiting the soft underbelly of his heart. *"Hurt me once, okay, but it will never happen again. It is too painful"*. He crawled inside the pain and closed the door tight. Their slaughter was enough, thank you.

This is common. No one likes to be hurt. When loved ones die, the agony is inconsolable. You never get past the death of someone loved. It is key, here is a good thought, to lock love up, not imprison yourself. Take love received from someone and lock it in your heart. If that person dies, show their love for you to the world. Do not hide. They gave you love so you could enjoy it. Their passion did not perish when they did, it remains very much alive in your soul. That is good stuff right there.

Mr. Mizithra felt ice protected him. It kept pain alive, so he could not entirely enjoy life. Now he saw it forestalled him from enjoying any love others offered. Never would images leave his mind, but the freeze was melting. Ice on warm memories faded, leaving glistening, soft beads of clear liquid, allowing him to see the love he already had. He could see more clearly and was thankful.

Chapter 20 The Lesson Continues

The next day, he returned to Erasmus. The mental turmoil of the prior day had not fully alleviated the mystery of his complex life. It was something in the cadence of Erasmus' words. Knowing more was to come, he came happily to receive hard medicine. It was a bitter tonic, but was there to cure what ailed him, sort of like castor oil. It was a cloudy day, ominous clouds roiling above, passing wickedly by in great haste. No sign of precipitation, but it was a dismal day.

"Hurt her beloved
Freezing in time
But I remain warmed"

The lack of fanfare was evident. This was no ceremony. Erasmus had powerful, painful truth to deliver. He had no time in his old life to waste. This is not someone you chatted lightly with about the weather. He was again taken back inside, today not a good day in weather conditions. The Olympics have wacky atmospheric phenomenon. They knew best not to mess staying outside. The *haiku* delivered, news received, he retired.

Mr. Mizithra again measured these few words. Here was his poem, his message on forgiveness. At first he felt this a mere repetition of what he received yesterday. The sympathy was there. Wait, it was not the same. It was quite the opposite. The Sewellel was ordering, not sympathizing. This was a lesson narrowing on the individual.

After a long period of personal reflection and self-pity, Erasmus got his head turned around. By not forgiving himself, he hurt the memory of his beloved family. This was a slap in the face. Our dear marmot reached the identical conclusion. He felt shame. His own self-pity denied his past resting, allowing full appreciation of what his personal history meant. It blocked any true love from ever reaching his hidden heart.

"Hurt her beloved", was a saying encompassing both women. His first and second wife were each still in love with him. One could no longer give physical hugs, yet embraced her sweet memories of a marvelous past. A moment's evil lined an otherwise delicious painting in a loving frame. His wife, Mrs. Saffron, was in love with him and had a ton of the stuff to unload were he to forgive himself.

Were he to remove blinders placed on his vision personally, what a beautiful life he had: a loving wife, kids who adored big papa, a colony full of students enlightened by his fables, and friends like Murray who considered this big galoot something special. Erasmus came to the same conclusion as the marmot: both were selfish. It was pride, ego, which fostered these notions. How dare Mr. Mizithra assume to be so wonderful the whole world had to stop for his emotions?

Sure, pain needs sorting out, but clinging to it like a hoarder was wrong. It is a balance, he realized, betwixt pleasure and pain. He felt guilt enjoying life, so allowed but a sliver to enter. While the rest of his family and friends were in full party mode, he would be detached, aloof while not realizing it. His children asked mom why daddy seemed so sad. How do you tell a happy child why their dad is unable to be so?

"Freezing in time", yes, yes he was. Unwilling to grow, he caused his health to suffer by chilling in a blizzard from the past. Shivering, you do not move, yet live. He could not see beauty in front because he would not turn from pain in the past. Have you ever been outside in a snowstorm, unprotected against cold? The only thought in your freezing brain: find a way out of the nasty stuff to some place warm and dry. Any other consideration can wait. You first must get warm.

Here was his life, not frozen but freezing in time. He was actively engaged in keeping himself from being warm and dry. Why? So he could suffer alone with the dead, who feel nothing? A fool, he knew how he was a sap. By his actions, others did not have a ghost of a chance of getting to truly know him. Only Murray and Mrs. Saffron had a real clue. Now he felt shame and anger with himself.

"Only I remain", now Erasmus lowered the boom on his poor soul. By this statement Mr. Mizithra realized the only one who could forgive him was himself. He alone held the answer. He placed obstacles in the way to his heart, to protect from further pain and show the world a fool's pride. Only he could remove them. No one else is you. You are you. No one else can be you. You alone answer for what you do. This weird sentence is truth. If you accept praise, you must be willing to receive criticism due.

There is an old military saying, "Face the music". It came from when a disgraced soldier was 'drummed' out of the service for conduct which brought shame to himself and the military. A musical procession led the dishonored service member out of the post, hence he faced the music. It was a way of saying, you cannot hide bad deeds. This is the same Erasmus spoke. Only Mr. Mizithra could answer for his feats. They were not bad. He alone could forgive himself. He did nothing wrong, except live. He was merely a victim of circumstances.

A wash came over him, absolving the dirt of selfishness, pain, shame, misery, and other nasty stuff accumulated in his brain. The power of the surge cleansed his spirit. Mr. Mizithra felt a wave of joy, finally clean. It was the first time since he was a pup he felt this way. Realizing he was alive made the difference. Forgiving the past did not mean forgetting. He could hold his past precious, but live a life with those still in existence.

He did remain. He was the survivor. Erasmus mirrored him. This Sewellel took endurance, using it to better others. While Mr. Mizithra did the same, he was still reserved. Erasmus was not, crippled and aged, he was very approachable. Other than obligatory hugs for friends and kisses for his family, the truth was this marmot was rather remote and rarified. He watched, he did not partake.

No longer would this be his character. Forgiveness is very powerful. Now, he felt it's wonder. The long toted burden was lifted. The yoke to the past shattered, burned on a funeral pyre. He saw his wife and children in his mind's eye. They loved him. His past family loved him when he was who he was then, not today. How could they? Hindrances in his mind rested. It was okay to tie past and present. They did not interfere with each other. It was not hurting his first wife to love the second. He could afford both in his heart. There was room at the inn.

Light shone in his heart. It opened. He was actually happy in a new way: unguarded and innocent. It was a weird sensation, but he liked it. This was virgin territory for his old, knotted up heart. A fresh feeling of love filled his emotions and burst forth. Alone, he shouted for joy at the revelation. Forgiveness was possible. He knew it for certain.

Warned by Boston Charlie, even though he was on cloud nine, Mr. Mizithra did not send off a dragonfly to let his wife know he loved her. It was best to wait until the entire session was over. They were only half way through. He had a lot more to consider once the topic turned to the third and most arduous subject: love.

Chapter 21 Tell Me About Love

Rain pelted the following day. Their meeting was canceled. Mr. Mizithra took time to stroll about the tarn and enjoy a big meal all day. Feeling very good, he had a long talk with Orlaith and her family. The otters were a fun loving bunch. This pond gave them a real home. They loved beavers tending the dam. It was odd, but she was one of the very few he would talk about war with because she was such a powerhouse.

Her shillelagh wielding minions were a stroke of genius. She had no qualms about smashing vole brains in with grandpappy's stick. She had a mean streak perfect for the matriarch of her bevy. Their lives were full of fish and fun. They loved their marmot pal and his leadership during war. He loved her battle fury. Her troops fought and died with honor.

The next day, he went to the tunnels with Boston Charlie. The Belted Kingfisher was most solemn. Love was a biggie. In his mind, Mr. Mizithra wondered, not being demeaning, what an unmarried mountain beaver could tell him about love. Love is personal, private. There are universal things, like the emotional portion. Love is far more than feelings ever-changing as water flowing downstream. Emotions can often move up or down on a whim, thus unreliable. True love takes much more effort.

> **"Misty clear blue**
> **Hurt no more**
> **Wait, light shines"**

This was love? What? Blue, light? What was this? Confused, he left, shaking his head in disbelief. What Erasmus said did not register what he thought of as love. Passion is the ultimate emotion and achievement. To be both humble and bold enough to give love is an honor, a pleasure. To receive love, real love, is the greatest gift one acquires. It is treasured, it's flame kept fervently alive. Armed with love, nothing can stop you.

He had it, but felt what he sought could not be found in these confusing words. Boston Charlie saw the body language. Erasmus did not, he was very tired. He rested an extra day. Health was not good. Moving about philosophizing was exhausting. As a good friend, Boston Charlie waited reverently for the old rodent to retire, then beat wings over to a branch near his marmot chum.

"You know, Mr. Mizithra, it is obvious you do not understand the words Erasmus graciously offered. Please, do not consider love as you always have. Open your mind wide as you do your mouth to feed it full of lupines. Take time. You have two days. Think on his words. Take blinders off your mind. Open your brain to thoughts he imparted. I am certain you will soon discover their incomparable beauty."

Mr. Mizithra considered his friend's advice and sought his home for the duration. He improved his little spot. A nap was wise, so he got comfy and some shuteye. He had to wrestle with words. The bird said to open his mind, take off blinders. How to do this was his concern. He was who he was. He could not change, but might do this: take a detour. He allowed it to seep in his mind listening as each word came.

"Misty clear blue", Erasmus' initiation into madness. Alright, what does this mean? He pondered three words. They say What? Misty, something moist, wet, hidden. He sorted through his mind all day. He took another nap and stuffed his mug with lilies. Arising, he hit upon what Erasmus meant. One definition for misty is inconspicuous. Love is not what shines, it begins unclear, shadowy, indistinct. It takes form when applied. The marmot suddenly realized: what he meant is love is not haphazard or in plain sight. It is discovered, exposed, labored over, but never alone.

What formerly was out of focus and indistinguishable, now could be clear, whole. Love was made between two, hence always new. It cannot repeat over and over again like a mold on a conveyor system, stamping out the same, consistent item. Each time love alights, it is unique.

So, for him to understand the word: love, he had to know was not what he expected. It would take form over time with two, not one shaping it. Love is not a one way street. It cannot exist if unequally yoked. Love is not political, it is pure.

"Clear", okay, easy enough he began. The first word dealt with what is hidden. It did not mean misty in the traditional sense of the word. What did this mean in reference to the word love? He slept on it. After his nap and a good yawn, he ventured out of the cave, spotting a familiar sight: a mother red-breasted nuthatch pushing her last young one out of the nest. She and her husband had a fine clutch of eggs, five, but this last little fella wanted to stay warm, safe in the nest. After a bit of effort, she finally got him to fly off, a tear in her eye, smile on her beak.

He saw the meaning. For love to take shape from a cloudless blob, both involved had to give to receive. It was, free. Mrs. Christmas Special fell in love with the big galoot in spite of himself. She freely offered herself. Some humans do not have this capacity, wallowing in misery, unable to understand love. It is because they are selfish. Love is free, you cannot buy it. He saw truth. If he wanted love, he had to offer it to Mrs. Saffron, Murray, anyone, for free. You cannot sell love.

Love can only take shape this way. It cannot be ordered. You are not able to force someone into this emotion. It can only come from an open heart, received by the same. There, form begins to take it's initial foray to a distinguishable mass. Sculptors take a big, old hunk of marble, carving the Thinker or the like. Passion is identical for those sculpting. Both have to wield tools to chip away unwanted parts. Selfishness cannot enter in or it will fail. Four eyes, four ears, and two hearts succeed versus a cold, unfeeling cyclops.

"Blue", why blue? The initial thought was of sadness, music sung by Miss Bluesberry, poured out over sad notes of Murray's saxophone. Given time, it was obvious, blue refers to the heavens, purity found in space. Nothing is evil there. In the sweet azure above, love dwelt joyfully, no reservations. Here, he could discover love in it's most distilled form.

Blue meant what is untouchable by stain of earth, man, or beast. Two who found love, helped it take form first with freedom, next with purity. You must bring no agenda to love. This emotion, if carried out carefully, will remain clean. Marmots, many other animals and birds mate for life. They work in the clear light of day, not in shadows to scheme against their mate. Humans alone kill spouses with words. We have a lot to learn from "dumb" animals.

Love means working with what you and the other individual are. You can change yourself easier than someone else. Parenting is the best way to learn this lesson. Acceptance of weaknesses and strengths in your partner is crucial in mountain climbing, paramount in love. After all, they must accept the same out of you.

This demands discovery, which means you must willingly open your heart to another. If you harbor resentment towards 'love', is not really love. If you have such feelings about those you claim to love, it is time for a long talk with yourself. Try as people might, they cannot alter who someone is. Change yourself first, then try your level best to help others.

"Hurt no more", now became simple to understand. Since he no longer hurt, Erasmus would no longer injury those around him. In the aftermath of his family's destruction, he wounded others by hiding away. He was not a friendly sort. Sewellel are not overly sociable anyway. He made life miserable for those who tried to help. He eventually chased Molly and Graedy away to his deep, inner shame. The bruised look on sweet Molly's face turned him around, breaking him to the core.

Erasmus accepted life and fate. He no longer hurt. Oh, a twinge will always remain, death does that. Acceptance is key. Armed with comfort of being pain free, he no longer afflicted others. If he hurt, then so should everybody was his old way. After enlightenment, he sought Molly and Graedy. Sending an orange dragonfly, he secretly begged them to come. Showering with tears and pleas for forgiveness, otters hugged him, gladly accepting his confession.

Mr. Mizithra realized stinginess halted his wife, children, and others of his true self. The pain of freezing subsides in warmth of fire or stove. Love compels us to reach out. He wanted desperately to get home, to show his spouse how much he loved her and happy she bore with him in spite of his obstinate old brain. Pain eased in his heart. He clearly saw the bygone time. Love improved him for the future. Mrs. Saffron was going to receive a whole new husband.

Pain is a horrible thing. It is overwhelming. One can only focus on severe, biting sensations, nothing else. The worse the trauma, the more difficult it is to deal with effectively alone. You must have help. Fall off of a cliff and break an ankle, then try to get out of hell alone. Many who go in woods alone perish for lack of respect for wilderness. It is unforgiving. Pain is internal as well. You look fine, but inside a cauldron of broiling tumult. This was so with Erasmus and Mr. Mizithra.

Each bore obvious wounds. The heart was where the worst damage occurred. An invisible, unhealed scar is still there, demanding we hesitate judging others from ignorance. With this deterrent removed, another brick in the wall keeping him from love fell by the wayside. Had he left before now, efforts made on his behalf would be for naught. Now he had a clearer understanding of love. More to come.

"Wait, light shines", final nail in the coffin on love. Love is one elusive doohickey. You cannot just go out and find it. Love is not tangible. It is rare, precious. If you are blessed to find real love, not the light version popular among the phony, you are most fortunate. If you do find love, do everything honest you can to keep it alive and happy. Not everybody feels this way, pity. Real love is not waiting on every corner for you to come by and snag when you haphazardly happen to drift into the mood.

Passion is incredible. For someone to love you, accept your love, is what it's all about. To be loved, you must be lovable. Makes sense, but so few work on being whole people it is difficult. Love comes to those who want, who need, who work on themselves to find it. Being the best you possible makes for a really good target for Cupid's arrow. Erasmus and Mr. Mizithra had that in common. Both sadly lacked wooing abilities.

For love to come twice to Mr. Mizithra was fantastic. This barely registered in his mind. Blind, he took it for granted. Fool. He met Mrs. Saffron whilst on the trek to the Valley of the Stones. He was a key figure in this effort. He was one busy marmot. She saw him and was stirred. He, the same, but without her commitment. Now he saw this. Her patience, not his was now Erasmus' focus.

She waited and waited for him to get around to getting better, but something was missing. The poem now hit the heart, he even kept his wife at arm's length to stop pain from reoccurring. By extension, his children suffered the same fate.

Light did indeed shine. What a fool he had been. He saw she loved, adored him. Pealing off this final, stubborn, protective layer, hurt. The light of her love shone on him. He allowed this passion to glow, absorbing darkness, freeing his past. In the next moment, he was weeping, paws to eyes in humility. What an incredible woman he married. She endured bearing young, fighting war, and his constant leaving. Add to it most of all, his inability to totally love her. Mr. Mizithra felt very small. Selfish ego denied her, most of all, of the passion he possessed within.

Emotional outpouring freed his soul. It brimmed with emotion. He spewed the final cerebral pile loose from his mind. Liberation feels so amazingly awesome. He was free. Pain subsided. Light shone. He was alive again. After too long, he was back amongst the living. He actually smiled.

This would be the final trip. Of this, he was certain. No wanderlust accompanied his spirit after this day. He now saw what Murray did first, love was pretty neat stuff. Momo and family were his whole world. Music came in a distant second. Yes, he saw and was prayerfully thankful.

Mr. Mizithra was healing.

Chapter 22 The Unavoidable Truth

No one likes a know-it-all. Erasmus was one, in a most positive sense. He saw Mr. Mizithra's heart like an amoeba under a microscope. The evening of the break-in, he held Jimmy Joe John up against the bar, stick in hand. Erasmus saw a thing few possess: real courage. Not braggadocios swagger of someone telling a big whopper. He was real. Stories were true. He added a flutter of flowery language to enhance the saga, not lie.

Erasmus saw him explode from the unknown, the shadows of the club, from nowhere to defend those he barely knew. Even Buford bowed to obvious leadership qualities. During war, he led without question. He assumed the role effortlessly. The siege, defense on the ramparts, assault on Igor, were qualities one seeks in those others willingly follow. There was no question of owning this portion of leadership. No, something else.

> *"Same row after row*
> *Turn a corner*
> *Who but I"*

Ah, clever Sewellel. Not only did he know the very public life of this particular marmot, but the private. On top, he knew the way of Olympic marmots. Each species has it's own culture. An observant rodent, Erasmus watched other animals with a wise eye, cocked in the direction of truth. Each different variety of rodent, they are most plentiful, is unique. He was no more like a squirrel than mouse. The same is true with all critters.

His opinions were formed by watching actions and interactions of various types, birds included. The main thing about marmots, they were governed by very strict colony rules. For example, four whistles remained unchanged over centuries because they worked. There was no reason for flair. The odd exception was during war. This temporary measure was dropped immediately after battle ceased. Interior structure, passageways, chambers, tunnels, entrances, had a certain uniformity to them.

"Same row after row", rules and discipline have their proper place. To remove these and give them a toss is unwise. It leads to demise. They have a purpose we cannot deny, though laws are not always fun. Marmots had very good elders in Mrs. Lip Gloss and Mr. Root Beer. Mrs. Lip Gloss, as noted, was most wise and compassionate of marmots.

Mr. Root Beer was a marvel. He knew every rule, bit of history, and reason for each law. He harkened back to days of Mr. Mountain and genesis of their species. Leaving lowlands, moving to hills, he led with an iron paw. Mr. Root Beer was not unfeeling, but held a tight rein for the betterment of all.

The past was essential to present and future. Like a good history teacher, he embraced the past knowing it can help today. Sad but true, we repeat the bygone far too often to ever learn it's lessons. To Mr. Root Beer, varying from ancient ways was wrong. He fought change vigorously.

Holding the past, he forged a promising future. It was he who put forth the notion of moving, having Mr. Mizithra find a new home. He did this to cling to ideals. Intrusion by Thumbs was not good. Man destroyed their way of life with cameras, numbered tags, and radios.

"Turn a corner", these, like late Mr. Savoy Brown, ruled with love and discipline. The result was success. War took everybody aback. It was not easy to recover from this torture. The death toll was immense. It effected marmot colonies, woodrat nests, and Keen's mice homes. The Valley of the Stones completed work to rebuild home and defense. The future of the colony was secure. Both overseers were getting elderly, though.

One of the worst results of war was lack of a future in what we call middle management. Too many in the age group like Mr. Egg Flower Soup and Mr. Macaroon died in the struggle. Finding any qualified to lead was difficult. It would be a couple more years. They already had two in mind for replacements. As the colony settled and grew, more would be essential. It was clamant they continue to prepare the next generation to take over once old ones passed. This is the fashion of the colony. The way it had always been and had to be for this method insured survival.

The only lifestyle they knew was set in stone. They were unable to change. This was impossible. Progress of the whole demanded looking from a different angle, another facet of the diamond. Think of a great leader. This person did not act like everybody else. By standing out, they became obvious above the din. They were mocked, shunned, and hated being able to see vistas we cannot. When 18, Alexander the Great led his father's army to victory. Others were stunned. He conquered much of the then known world. His little force made it all the way to west India.

Mr. Mizithra had something Mr. Baklava and Mrs. Marsha Marsha Marsha also had, but in smaller quantities,: he had been outside the settlement. No others could say the same. As a traveler, he was a survivor. Killing the red-tail at Gilligan's Island, dodging the coyote at the rapids, proved his ability to think and act his way out of any problem. Bravery was not a problem with this tough guy.

"Who but I", yeah, exactly. No one else could turn that corner. He operated in fear on his journeys, but was brave when called upon. At no time did he become a coward. He hid from Kai to defend himself, not to cower. He killed the predator, not something a coward would do. In the teaching he did for the colony, no one could deny how effective he was with stories. While Mr. Beef Stew had a great style, he was no Mr. Mizithra. Ha-ha. This he saw quite easily.

A leader had to see beyond the pale. The one selected had to think of the good for the entire settlement, not selfish interests. This made love stronger. This type was universal, not personal as with his wife. Love was for his species, the northern community in the Valley of the Stones to be precise. Mr. Milk Chocolate had his own colony to care for. Having been in war for the viability of the nation, he had a vested interest in it's continuance. This was his family. He wanted to serve.

Okay, I am the only choice, he realized. Locating three dragonflies, he sent orange ones. These were secrets. The first went to Mrs. Saffron, to let her know he had one final task to perform. He would be home to love her for the rest of his days, never more to roam, honest. He also told her their love was going to change and was truly sorry for the past. Eyes open, his passion exploded for her.

The second dragonfly went to the elders, informing his intention to become overseer once he arrived home. The last went to Murray. He had to share the good news with his best friend. He told Murray what was up. He, of all, had to know.

He made his way back to the tunnels. Thanking Erasmus, he was humbled. The Sewellel nodded, waving his stubby arm and smiling. Clearly worn, he asked to lay in sunlight for a bit before going inside. There was nothing better than nice solar rays for aged bones. Boston Charlie was given the task of informing Neferteri and K'wati of his decision. They would be pleased. He as a truly honest and brave rodent, a most rare quality. He took a crow escort. This was still serious business.

This was a gratifying season. Mr. Mizithra was free of loads carried for so long. The first instinct was to head home and family. However, he had just one last personal mission to perform, an indulgence. It was a bit risky He wanted answers to questions begging his mind. He headed along the shoreline, west, up the pathway he and Murray came down so very long ago.

Chapter 23 Fade To Black

It took three fat, sunny summer days to waddle up the path, past the area known as Bear Town. Stopping frequently to eat and burst into joyous laughter, Mr. Mizithra moseyed along, love in heart. Being free is a righteous thing.

He dove in paws first, reveling in cerebral vibes. Ancient stomping grounds evoked forgotten memories of his former home. The colony was clearly vacant. No other marmots moved into their old digs.

Cameras installed by the Thumbs were long gone. From evidence he saw, the area remained empty since they left. It was late. He dug his way into the old place, finding his past bedchamber. He held a large wad of lupines at hand. He ate and slept in his old bed. While odd, he knew where he was and that was a comfort.

While he slept, the last few days went through his mind. Upon long reflection, something truly miraculous transpired: all the bricks in his old wall, obstructions to love, vanished. Mrs. Saffron had a clear path to his most inner marmot. His bare heart lay naked and exposed. He was happy. The ancient way of hiding within his own carefully constructed torture chamber felt foolish. He would be home, he would be soon enough. Once with her, love would blossom as never before.

The biggest beneficiary was them. She got a whole marmot and he a lady marmot who loved him for who he was. They had the rest of their days to embrace this truth. They would be the best danged couple in the colony. Another litter would be most welcomed. Soon Mr. Murray and Miss Paprika would be eligible to marry.

Then, grandchildren. It would be time for Mr. Mizithra to torment them, bringing only the noisiest toys to their chambers to drive his own children crazy. This is a grandparents' revenge. He would take his progeny on his ancient knee and tell all the tales. True to this, he aged in a fine fashion, at length earning the colony's moniker of Pompa. No higher honor of love and affection exists in the Olympic marmot world.

Yes, forever he would spin yarns. He filled his remaining years with love. The help of others proved not only their love of him, but his worth. They would not waste such a precious emotion on someone undeserving. Truth emancipated, gave him clear purpose. It included service to the body as a whole. They helped. With immense love and wisdom, he repaid tenfold. This he did with boundless joy. Freedom needed to be shared. He determined this his life's mission.

Mr. Mizithra contemplated his heroic friend, Mr. Guar Gum and the selfless act which cost his life. A character in the background, at brass tacks time, he arose. Every marmot for the rest of time knew his story. Mr. Mizithra told of his sacrifice. It was, in fact, the only tale about the old war he told. Glorifying Mr. Guar Gum, his childhood friend was honored.

When they did have another litter, the first male was named Mr. Guar Gum to honor him. The entire colony approved heartily. The edge of the cliff falling off the valley floor was, 'Mr. Guar Gum's Leap'. This shows the value of people or animals may not be discovered solely in leadership positions. Often, it is the lowest who rise in times of crises to leave the true mark on history. Mr. Mizithra never forgot the willing sacrifice his lifelong friend offered.

The other pups were: Miss Strawberry Jam and Mr. Hawaii Five-O. These three came to be regarded as the strongest family in the colony. The rest of history saw them as key to the survival of their marmot village.

He arose the next day, meandering to Murray's old shack. No sign of the ranger, wood pile looking pretty much the same. A wry smile crossed his mug thinking of Nediva and Frema worrying over their high dramas. Those days of wild times were done. He went home to love his family. This was a most noble facet to his life.

The colony survived when he found a new home. Their hamlet benefited due to his oversight. He had no lordly thoughts, only of service. He did his best for those who made it possible in his past and future. It was for those who now needed the best leaders to step forward and serve.

Coming over the familiar hill took him past his old home to the left. he saw something move in the spot to his right. Summer sun glistened. It was barely visible. The length of grass did not help. Greenery wavered tall in soft, summertime breezes. Utilizing caution, he crept forward to the unknown stranger. At once, the figure was identifiable.

"Mr. Mizithra", slobbered good old Pacific Northwest banana slug Ernie, "long time no see. Anything new happen with you of late?"

Mr. Mizithra humbly bowed his head and smiled.